Microclimate, vegetation and fauna

Ph. Stoutjesdijk and J. J. Barkman (†)

Contents

Originally published as 'Mikroklimaat, vegetatie en fauna'
[J.J. Barkman & Ph. Stoutjesdijk]
by Pudoc, Wageningen, the Netherlands, 1987.

ISBN 978 90 5011 545 2
NUR 412
www.knnvpublishing.nl

Published by KNNV Publishing 2014

Cover design after figure 72 by Emma Diemont (Kokotopia)

KNNV Publishing

Preface

This book is the second and extended version of Microclimate, Vegetation and Fauna (Opulus Press AB, 1992), after a Dutch text "Mikroklimaat, Vegetatie en Fauna" (Pudoc,Wageningen,1987) written by the late Jan.J.Barkman and myself.

From several sides it was suggested that an English translation of the book would be worthwhile. Jan and I had just agreed with our friends Marijke and Eddy van der Maarel, that they would take care of the translation and have the English edition published by Opulus Press, when we received the message of Jan's sudden death, September 1989. In his last letter he asked me to make sure that the sequence of our names, which by mistake was reversed on the cover of the Dutch edition, should be right this time.

Our aim in writing this book was to consider microclimate from an ecological point of view. Herein we stress the great physical diversity of natural and semi-natural habtats, a diversity which to a large extent is created by biological forces and at the same time is a condition which maintains biological diversity. Of necessity, we also consider the physical and biological interaction of plants and animals with their immediate atmospheric environment.

Several other texts, especially physical ones, have dealt with aspects of microclimate. Never-theless, we thought that there was room for a more ecological approach, which did not overlap too much with these more general texts. We have deliberately reduced the treatment of those subjects which received a great deal of attention in other textbooks, while, on the other hand, we have been more extensive where we could offer our own new findings, or where our treatment could make a topic more understandable.

The reader who would like to read more on the subject will find much of interest in books cited in the References, particularly those marked with an asterisk.

The book is written for ecologically interested vegetation scientists, plant and animal ecologists, biogeographers, foresters and agriculturists. Biologically interested climatologists, geographers, environmental scientists, landscape planners and amateur naturalists with some physical background may possibly find something useful in it as well.

Since the present book is based on the Dutch edition mentioned above, I acknowledge first of all the Agricultural University of Wageningen, the Netherlands, for financial help with the publication of the Dutch book and I thank H. J. During, Utrecht and J. C. H. van der Hage, Utrecht, who reviewed the Dutch manuscript.

With regard to the present book, I acknowledge the permission of Pudoc Publishers, Wageningen to reproduce the figures from the Dutch edition. Marijke and Eddy van der Maarel took care of the English translation; Eddy van der Maarel and Martin Sykes, Uppsala University, provided valuable comments and literature. My wife, Martien, helped in many respects to make this publication possible. To all those I extend my sincere gratitude.

After over a half century I found back Dr. Christa Deeleman-Reinhold. I thank her very much for all her support in preparing this new edition, Herbert Diemont for communication with KNNV publishers and prof. Henk Siepel (Wageningen University) for reading chapter 6.

Ossendrecht, October 2014
Flip Stoutjesdijk

1. Introduction

1.1 Vegetation and macroclimate

Weather and climate are very important for the distribution and the functioning of plants and animals, both their populations and their communities. Ample indication of this significance is found in the reflection of the earth's major climatic zones by vegetation zones and (faunal) biomes.

Vegetation, soil and fauna are interrelated in a complex manner. Climate has a twofold impact on plant growth. First of all, climate governs the distribution of plants directly; the soil conditions, so important for plant growth, are dependent on the climate as well, and so are the animals. Moreover, plants and animals are again interrelated. Finally, vegetation influences, in its turn, both the soil and small-scale climatic conditions (the microclimate).

Soil development, leaching and podzolisation, salt accumulation, erosion by rain and wind, solifluction, all these phenomena are determined by climatic factors. They may vary along with the variation in macroclimate. For instance, from S to N Europe the mean annual temperature decreases; as a result the evaporation decreases as well, leading to an increase in precipitation excess, which in its turn causes an increase in leaching and a general impoverishment of the soil. In high latitudes frost and thaw phenomena play a part as well, leading to solifluction and cryoturbation. In relation to the general climatic zonation from S to N, i.e. from semi-arid via semi-humid and humid to frozen in the north, the following general zonation in soils can be observed: from terra rossa in the south to brown earth in temperate zones and podzol soil in the north and finally to permafrost polygon soils in the arctic. The corresponding vegetation zonation starts with evergreen lauriphyllous wood in the south via summer green hardwood and evergreen coniferous wood (taiga) to the treeless tundra.

Not only are large-scale climatic differences reflected in vegetation and fauna. Even within the boundaries of a small country like the Netherlands climatic differentiation is large enough to have a clear effect on vegetation and fauna. Many European subcontinental and submediterranean plant species are found only in some suitable areas in the southeastern and/or the southwestern part of the country. The subcontinental *Helianthemum nummularium* ssp. *nummularium* is restricted to dry limestone hills with a warm microclimate in Limburg, SE Netherlands. Much further to the north it occurs in similar dry and warm conditions, in abundance, for example, on limestone grasslands on the island of Öland (SE Sweden). Within the Netherlands the submediterranean *Cirsium eriophorum* and *Trifolium subterraneum* are restricted to the southwest, where the winters are relatively mild and the precipitation has a mediterranean-like pattern with December as the wettest and May as the driest month. On the other hand, *Polygonatum verticillatum*, occurring in Limburg, and also at one locality in the NE, is restricted to the only areas in the Netherlands with an annual precipitation of over 800 mm!

In the very windy macroclimate of the Wadden Sea islands in the north (with an average wind speed over the year of 7 m/s) salt spray limits the growth of trees. Some epiphytic lichens occur on the sandy substrate of the dunes of these islands.

Boreal-European species such as *Cornus suecica* occur only in the northern and northeastern Netherlands, where the nights are cold all year round, and the winters much

Fig. 1. Biogeographical zones in Europe [adapted from Walter & Straka (1970) and Walter (1977)] and some well-kno localities with a deviating meso- or microclimate.
1. Basalt rocks near Agde with a North-African flora; 2. Doline Creux du Van with extremely low temperatures; 3. Doli near Trieste; 4. Doline Gstettner alm; 5. W. Allgau with many thermophilic plants; 6. Kaiserstuhl, thermophilic flora fauna, e.g. praying mantis (*Mantis religiosa*); 7. 'Rainforest' with rich epiphytic flora in the extreme wet and frost climate of SW Ireland; 8. Kullaberg peninsula with isolated populations of several arthropods; 9. Öland, with a dry sunny climate and limestone rock creating steppe-like conditions. Many south and east European plants and animals; Valley with very low rainfall and xerophytic flora near Vågåmo.

colder than elsewhere. This area also has a relatively low saturation deficit and this may explain the occurrence of several Atlantic plant communities of wet heathlands and bogs of the class *Oxycocco-Sphagnetea* (see Westhoff & den Held 1969 for a survey of Dutch plant communities to be mentioned in this book, and Ellenberg 1988 for a thorough treatment of Central European communities) with species such as *Drosera anglica, Erica tetralix* and *Gentiana pneumonanthe* (Barkman & Westhoff 1969). Another 'Atlantic' aspect of this area is the occurrence in the open of plants known as woodland species in continental Europe, such as the fern *Blechnum spicant*.

Still more striking is the behaviour of epiphytic mosses and lichens, which are more or less independent of the type of soil and almost entirely conditioned by the climate. The atlantic moss *Tortula papillosa* used to occur in most of the Netherlands, but was only dominant within 5 km from the coastline (we have to write in the past tense here, since this and many other epiphytes have become rare or extinct because of ever-increasing air pollution).

Likewise many kinds of animals show clear distribution patterns within the Netherlands which may be attributed to climatic variation. The grasshopper *Ephippiger ephippiger*, a southern-European species, is common on heathlands of the central and southern Netherlands but is not to be found on the Drenthian heath. Southern-European beetles of the genus *Harpalus* are restricted to Limburg in the southeastern Netherlands, while the boreo-alpine beetle *Agonum ericeti* is almost entirely restricted to Drenthe in the northeast (Turin, Haeck & Hengeveld 1977).

On the map of Europe presented in Fig. 1 we have also shown several well-known spots where the local climate has special features which deviate from the large-scale climate and which are also reflected in the local fauna and flora. Thus weather and climate determine to a large degree the distribution of plants and animals, and of whole biocoenoses, not only on a continental scale. A more detailed analysis on a smaller scale shows many correlations as well. The question is: How does this correlation occur?

Obviously, one needs, for this type of correlative study, an insight into the biological side of the problem, and the right climatic data must be chosen and be available. For instance, not the relative humidity but rather the saturation deficit is important for the evaporation of soil or of plants. For homoiothermic animals the vapour pressure rather than the relative humidity is important. For green plants, the photosynthetically active radiation (PAR) is more important than the total radiation, etc. Some examples will follow. First let us look at the relation between fog and lichen flora in the Netherlands.

Fog is especially important as a source of water for organisms which cannot take up water from the substrate, such as lichens on trees and rocks. It is important for lichens to be able to photosynthesise continuously in the day time, which requires sufficient light and moist (but not wet) conditions. Fog is more favourable to lichens than rain, especially fog in the day time. In climates with mainly night fog and drought during the day lichens do not thrive, because respiration at night continues and is not compensated for sufficiently by photosynthesis (a so-called negative photosynthetic balance). So, the combination of moist nights and dry days is worse than a continuous drought! In the Netherlands, for example, the number of days with fog is generally the same over the country, but the number of days with fog at 14 h decreases strongly from the NW (29 days near Den Helder) to the SE (14 days near Maastricht). In the same direction the epiphytic lichen flora is considerably impoverished. Thus, maps of fog frequency are of little use if one does not know whether the fog habitually occurs at night or in the day time.

A good climate description does not only include data on average values for the important factors but also probabilities of extreme deviations from average values, for instance extended periods of drought or snow cover. Rosén (1985) has shown how the heather *Calluna vulgaris* occurring on the dry limestone grasslands of Öland, may die off locally after an extremely dry summer and then slowly regenerate again in 5 to 10 yr.

Extremes, especially their frequency and duration, are often, though not always, important factors in determining survival and hence micro-distribution of plants and animals. In southwestern Ireland, where palms can be kept in the open because of the mild winters, the wheat does not ripen owing to the cool and rainy summers. Pigott & Huntley (1978a,b) have analyzed why *Tilia cordata* cannot reproduce in northwest England. Viable seed is usually not formed because the temperatures at the time of flowering are frequently too low to permit the pollen tubes to reach the ovary. The rate of extension of the pollen tube increases from the minimum temperature for germination, 15 °C, up to 25 °C. Above 19 °C the extension rate increases rapidly. Pigott & Huntley found that when the number of degrees (air temperature) above 19 °C were summed up hour by hour, an 'hour-degree' sum of 30 was needed, within three days, to give a successful fertilization. When fertilization has succeeded, the seed crop can still fail because the weather is unsuccessful for afterripening. In Finland fertilization will usually be possible but there the absence of afterripening will be the main cause of sterility.

Neilson & Wullstein (1983) showed through experiments with seedlings, that the northern distribution limits of *Quercus gambelii* and *Q. turbinella* in the southwestern USA are primarily caused by a combination of spring frost and summer moisture stress. In a study of temperature impact on the distribution of northern-European plants, Woodward (1988) found that the northern distribution limits of southern plants are directly controlled by climate; for annuals the heat sum, taken as day-degrees above a certain base temperature, of the growing season is decisive, for perennials both the heat sum of the growing season and the annual absolute minimum temperature are critical (for heat sum, see also Ch. 4.3).

Extremes are not only important because of their absolute values but also because of their duration. The Mediterranean oak species *Quercus ilex*, for instance, tolerates a short period of frost with temperatures as low as –20 °C, but not a long period of –1 °C (J. J. Barkman, unpubl.).

On the one hand we can see efforts to refine the use of standard climatological parameters for biological studies on a continental or regional scale. A recent example is the statistical study by Moreno, Pineda & Rivas-Martínez (1990) on the boundary between the Mediterranean and Eurosiberian vegetation in the northern part of the Iberian Peninsula. Many climatic parameters were derived from 205 meteorological stations in this area and checked for their discriminating significance. The two bioclimatic-phytogeographic regions could best be characterized by dryness during the months of July and August and by the level of minimum temperatures during the coldest months.

Another recent example of the use of numerical methods and models to indicate the decisive climatic factors for plant distribution is given by Booth (1990), who used data on mean annual rainfall, rainfall regime, length of the dry season, mean maximum temperature of the hottest month, mean minimum temperature of the coldest month and mean annual temperature to describe the climatic requirements of three important plantation tree species, including the pine *Pinus radiata*.

On the other hand we should try to bring the meteorological measurements closer to the plant or animal. It is clear that this approach is necessary where more than a mere correlation is sought, but where, rather, an ecophysiological approach to animal behaviour or plant performance is aimed at. Woodward (1987) presents a clear account and many examples of this approach.

1.2 Macro, meso and microclimate

"Ostriches may live in the meteorologist's climate but few other animals do." N.N.

The examples in the preceding section are concerned with the macroclimate, which we may define as the weather situation over a long period (at least 30 yr) occurring independently of local topography, soil type and vegetation. The weather factors are measured as in standard meteorological practice, i.e. temperature and air humidity in an instrument shelter at 1.50 m (earlier 2.20 m) and wind at 10 m. At these heights the local influence of substrate and soil can generally be neglected. Influence of the vegetation is excluded by measuring over a standardized surface, viz. a lawn. Such standardized measurements are carried out at thousands of stations throughout the whole world. These measurements show clear regional differences. They form a good basis for biogeographical maps of countries and (sub-) continents.

The *mesoclimate*, or *topoclimate* is a local variant of the macroclimate as caused by the topography, or in some cases by the vegetation and by human action. Ravines and mountain passes may have their own mesoclimate, but also large lakes, extensive forests and big cities. Topoclimate describes effects due to the relief of the landscape (van Eimern 1969).

The city has its own environment, with its own mesoclimate, and this has become an important object of ecological studies (e.g. Laurie 1979; Bornkamm, Lee & Seaward 1982). According to Miess (1979) the mean daily temperature in June in Frankfurt can be as much as 3.5 °C higher than in the area surrounding the city ('urban heat island'). The warmer mesoclimate of big cities is also to some extent expressed in the spectrum of plant species occurring in cities as compared with the flora of the environs. Wittig & Durwen (1982), using Ellenberg's (1974) system of indicator values for temperature preference and continentality of distribution of Central-European plants, found that the flora of some West-German cities, including Cologne, showed a shift towards more thermophilous and continental species as compared with the environs.

The local characteristics of the mesoclimate can be measured at the normal observation height of 1.5 m for normal climatological observations; air temperature and humidity may also be measured at 50 cm and wind at 1 - 2 m. In all cases the instruments should be located free from the influence of vegetation and from the shade of trees, hedges etc. In mesoclimate research the focus is on horizontal and not on vertical variation, just as with macroclimate (Wallén 1969).

The condition of the soil surface has a profound influence on the heat and moisture budget of the atmosphere and thus on various climatic factors. At the surface most of the sunlight is absorbed and changed into heat, which is transferred both to the near-the-ground layers of the atmosphere and to the upper layers of the soil. Through radiation loss at night a considerable local drop in temperature may occur. There is a strong reduction in wind speed; over a rough surface turbulence may increase as well. There can also be a considerable evaporation from the surface which moistens the lower air layer. So-called negative evaporation may also occur in the form of dew and hoar-frost.

All these influences are strongest in the lower 2 m of the atmosphere and the upper 0.5 to 1 m of the soil. The climate in this zone is called *microclimate*. Naturally, vegetation has a considerable influence on the microclimate. The climate inside the vegetation is also called *ecoclimate*. The difference between micro- and ecoclimate can be explained as

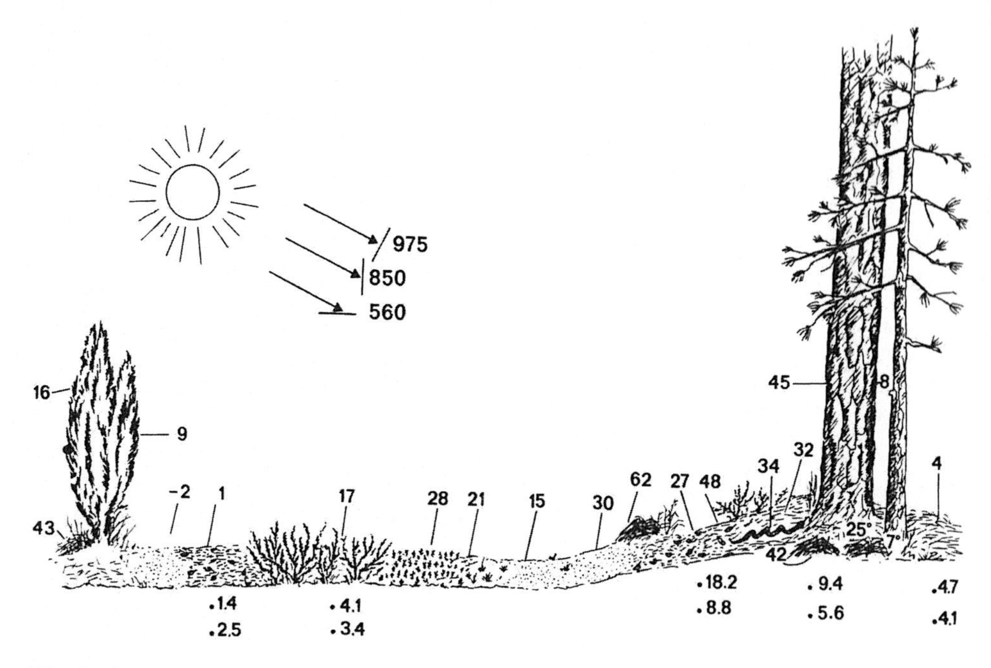

Fig. 2. Surface temperatures along a transect perpendicular to a woodland edge with adjoining heathland and juniper scrub in the Netherlands. Soil temperature at 4 cm and 9 cm depth. Air temperature at 1m height is 11.8 °C. Figures indicate the intensity of solar radiation on three planes with different inclinations in W/m² (see text Ch. 1.2). A clear sky on 3 March 1976 at midday. (After Stoutjesdijk 1977.)

follows: in a tall forest the ecoclimate can extend far beyond the 2m limit of the microclimate, whereas in an environment free from vegetation, for example a desert, city street or snow field, there is still a microclimate but no ecoclimate. For both micro- and ecoclimates it is essential that not only horizontal gradients are studied, but also vertical ones (Wallén 1969).

The climate on the smallest scale is sometimes called *nanoclimate* or *epiclimate* (Monteith 1981). This is bound directly to surfaces such as on a leaf, air cavities in litter, an ant-heap, fissures in rocks. A thin layer of air 'sticks' to these surfaces, even in strong winds. The nanoclimate of the boundary layer is characterized by its own temperature and air moisture. The height above the surface where the nanoclimate prevails is measured in mm, whereas the microclimate is measured in cm or dm. As regards the horizontal scale: the macroclimate may extend over hundreds of km, the mesoclimate over km or hm, the microclimate over m up to hm, and the nanoclimate over cm only. Of course, there are no sharp boundaries, and different authors use different definitions and scales. Terms such as ecoclimate and nanoclimate are only used when the need arises.

The macroclimate is an important object of study for physical geography, biogeography; mesoclimatology is studied in relation to physical planning, urban ecology, bioclimatology (medical climatology and environmental health). The microclimate is directly relevant to vegetation ecology and the distribution and population dynamics of plants and animals. The nanoclimate is linked to ecophysiology and the population ecology of very small organisms. Not only the climate (as an 'average' situation over at least 30 yr) but also the weather, the state of the atmosphere at a certain moment, can be studied on various scales. Hence we may speak of macro- and microweather and of macro- and micrometeorology.

The climatic scales are superimposed upon each other, or rather, they are variations on the same theme. Sometimes the variations deviate from the theme to the extent that the theme can hardly be recognized any more. The microclimate at the south side of a dense juniper scrub, for instance *Juniperus communis* (ssp. *communis*) in temperate Europe or *J. virginiana* in temperate America, is as different from the microclimate of the north side as the difference in macroclimate between a desert and a boreal forest! So, the microclimate may show the same magnitude of difference over a few m as the macroclimate does over 5000 km.

A first impression of the nature and dimensions of these effects can be seen in a snapshot of the climatic situation at the edge of a pine woodland on a sunny day in March in temperate Europe (Fig. 2). The woodland edge faces south. The woodland changes into heathland with juniper scrub still protecting from wind. Wind velocity at 1m height is only 0.2 - 0.4 m/s whereas at a nearby meteorological station 5 m/s is measured at a height of 10 m. At the same time the air temperature at a height of 1m is 11.8 °C, which is 3 °C higher than at the nearby station. Most spectacular are the differences in surface temperature. They change from dm to dm and range from –2 °C in the shade of the juniper, on frozen soil with hoar-frost formed during the very clear, cold night previously, to +62 °C on dead grass, on a steep edge perpendicular to the sun's rays. A range of intermediate temperatures are found as well: on juniper needles on the sunny side: 16 °C (against only 9 °C in the shade), 45 °C on the bole of the pine tree, on the south side, and only 8 °C on the north side of the same bole. An adder taking a sun-bath has a surface temperature of 34 °C. Soil temperatures show a similar variability.

This all demonstrates the staggering differences which may occur from point to point and which must be of great biological significance. The range of these local differences is still more impressive if one compares the noon instrument shelter temperature of +1 °C at Stockholm (ca. 60 °N latitude) and of 17 °C in N Africa (ca. 36 °N). Such differences in microclimate are of course well-known to a human visitor looking for a place to sit and rest in the landscape of Fig. 2, and "mothers with babies have a sure instinct to find the best spots in early spring" (Geiger 1961).

If we consider the impact of the microclimate on a plant or an animal, we realize that a complex of factors is involved. For instance, the temperature at the surface of a plant is dependent on air temperature, air humidity, radiation and wind, but also on the transpiration of the plant. Clearly, plants and animals are in interaction with their environment.

Each plant and animal has its own microclimate and is heavily influenced by that. If we know that the microclimate is so important and moreover deviates so often from the macroclimate, why study relations between the distribution of organisms and biotic communities and the macroclimate? In fact, even if we look at large-scale variation, it is the microclimate which is decisive. The boundaries of distribution areas of plants and animals are often linked with climatic isolines, such as the 10 °C July isotherm, the 0 °C January isotherm, or the 1000 mm precipitation 'isohyet'. As a matter of fact, this description is incomplete. What really matters is how these climatic factors are realized in the microclimate and that, moreover, extreme values rather than averages are decisive. Factors such as rainfall may not be important as such, but, rather, in combination with other factors, for instance evaporation and air humidity, leading to more complex parameters such as evaporation/transpiration balance.

As early as 1926 E. J. Salisbury wrote: "It is therefore not remarkable that the limits of species susceptible to temperature changes often show but little relation to isoclimatic lines: rather it is surprising that the correspondence is sometimes so close."

Why do ecologists study the macroclimate? For two reasons: First, the microclimate is still little known. We do not have enough microclimatic data to find correlations with distribution boundaries. Second, the macroclimate is a factor complex independent of vegetation, while the microclimate is, to a certain extent, formed by the vegetation itself. In order to find out the extent of this we need data on the macroclimate.

Even when using mainly macroclimatic data it is of course important always to keep the microclimate in mind as was done by Woodward (1988) and others when studying correlations between air temperature and distribution of plant species. The link with the microclimate is also obvious in the species-energy theory (Wright 1983; Ryrholm 1989; Root 1988) which concentrates on the importance of sunshine duration to explain the distribution of plants and animals.

In the case of micro-mapping of species or community distribution, ecophysiological, distribution-ecological and population ecological studies, as well as synecological research of phytocoenoses (plant communities) and synusiae (layer communities) it is necessary to study the microclimate. In practice, however, we can work with macroclimatic data in chorological research, or with vegetation mapping on scales of 1: 25 000 or smaller, albeit that the correlations found can only be of an indirect nature. Lamb (1977) commented on the use of beetles as palaeoclimatic indicators as follows: "It has been found strange that even beetles which sometimes live in microhabitats with their own microclimate can be used as indicators of the broad scale distribution of world climates but these microclimates presumably only occur at all widely within certain macroclimates."

How the climate affects the situation in the microclimate can be demonstrated by referring to the example of the woodland edge. The enormous range of temperatures is only found on a sunny day after a bright cold night. With overcast weather the range of temperatures is only a few degrees C.

It is clear that the frequency and duration of sunny and cloudy periods are of great importance. After a period of heavy rains the soaked surface is only moderately warmed up by the sun. A heavy snow cover at the same time of the year could make microclimatic conditions completely different at the same time of the day with the same sunshine, wind and air temperature.

There is another indirect effect of climate. A forest edge - artificial or natural - is only found where there is forest. With the same weather (air temperature and sunshine) the microclimate conditions that are realized may vary greatly with the type of ecosystem (and this with the macroclimate) of the desert, the steppe, the deciduous forest, the taiga, and the tundra.

Even for an ant or a seedling there is a correlation between its own weather and that of the meteorologist, but this correlation is complicated, not only for the ant and the seedling, but also for the meteorologist and the biologist.

There are basically two ways of studying the microclimate. First we can perform direct measurements. In view of the considerable variation in space and time this will be a time-consuming undertaking. However, some simplification can be achieved by measuring for each organism or biotic community at representative localities, only on days of the year and at times of the day which are significant, and preferably to concentrate on extreme values.

The second approach is to search for quantitative relations between macro- and microweather. The temperature distribution presented in Fig. 2 is quite predictable with this type of weather. It is now possible to build simulation models to derive microweather parameters from the standard macroweather data provided by meteorological stations, and this has actually been done in relatively simple cases, e.g. by Mitchell et al. (1975) for the desert environment and Unkaševic (1989) for the influence of vegetation on ground temperature.

In the search for correlations between macroweather and microweather it is important to concentrate on the processes at the surface of the earth such as absorption of solar radiation and transformation into heat. Here the ecologist meets the meteorologist: the processes at the surface not only make the microclimate but also affect the state of the higher layers of the atmosphere.

When studying energy transformations, the meteorologist is first of all interested in global figures, valid for a larger area. The ecologist, on the other hand, will look more at the local differences causing environmental differentiation. This implies that relevant microweather parameters must be obtained for each single habitat. This is still a weak point in simulation models.

The present book is about the biological aspects of microclimate and its variation in horizontal and vertical directions. It tries to stress the great diversity found in natural and semi-natural habitats, also as far as microclimate is concerned. First the main microclimatic principles and processes will be described. This treatment is fairly concise. For further reading we recommend the textbooks by Geiger (1961, 1965), Berenyi (1967), Mattson (1979), Rosenberg, Blad & Verma (1983) and Oke (1987). The physical interaction between organisms and microclimate is described in the books by Monteith (1973), Campbell (1977), Grace (1977), Gates (1980) and Etherington (1982). Further recommended reading on climatology and meteorology include McIntosh & Thom (1969), Lamb (1977) and Barry & Chorley (1982).

For an explanation of the many biological terms used in the book, the reader is referred to Holms (1985) or similar dictionaries.

2. Microclimate: principles and processes

2.1 Energy balance of a dry surface

"How do you do know all that", he said. "It's obvious".
"Well then why didn't I see it?" "You have to have some familiarity" "Then it's not obvious, is it?"
R. M. Pirsig: Zen and the Art of Motorcycle Maintenance

We all have some notion of the physical processes and factors creating a microclimate: solar radiation warms the earth's surface, evaporation increases air humidity, wind cools, a tree casts a shadow, etc. This chapter is about a systematic and quantitative survey of physical processes and factors operating at the earth's surface. We start from a simple situation without clouds, rain, or evaporation. Dry sand is warmed considerably at the surface by the sun. Let us take a concrete example from a warm clear summer day with a gentle wind on a heathland in western Europe. 2 July 1976, 14.00 h. Clear, virtually no wind. Yellow sand with some humus at the edge of a woodland.

Surface temperature	51.6 °C
At 2 cm depth	41.8 °C
At 4 cm "	37.4 °C
At 8 cm "	29.4 °C
Air temperature at 1.50 m	29.4 °C.

These elementary data immediately allow some conclusions to be drawn. Solar energy is received and in part absorbed since the surface is yellowish-grey and not white. Part of it is reflected because the surface is not black. Also, the surface radiates heat; this is electromagnetic radiation, just like solar radiation, but with a greater wave-length. Then there is long-wave radiation from the woodland edge and finally, long-wave radiation from the atmosphere. It is obvious that the surface gives off heat to the air, because it is 22 °C warmer. Through conduction some heat is transported down into the soil, as the downward decrease in soil temperature indicates.

After all these parameters have been measured, it is possible to make up the energy balance of the surface, or, if this is easier to understand, a very thin layer at the top of the soil with a negligible heat capacity. This energy balance (Fig. 3) is a state of ins and outs with the Joule per second per m^2, J s^{-1} m^{-2}, or the Watt per m^2, W/m^2 as a unit. Because of the zero heat capacity of the surface, the incoming and outgoing energy fluxes must balance.

All the items in this energy balance are determined by measurement except the amount of heat given off to the air, which is difficult to estimate directly; thus this factor makes the balance. For the sake of simplicity we have chosen an example here with a negligible evaporation (the small arrow H_{ev} in Fig. 3). In the case where there is significant transpiration or evaporation, which costs a lot of energy, this would be an extra item on the balance. Now the various energy fluxes must be discussed in more detail.

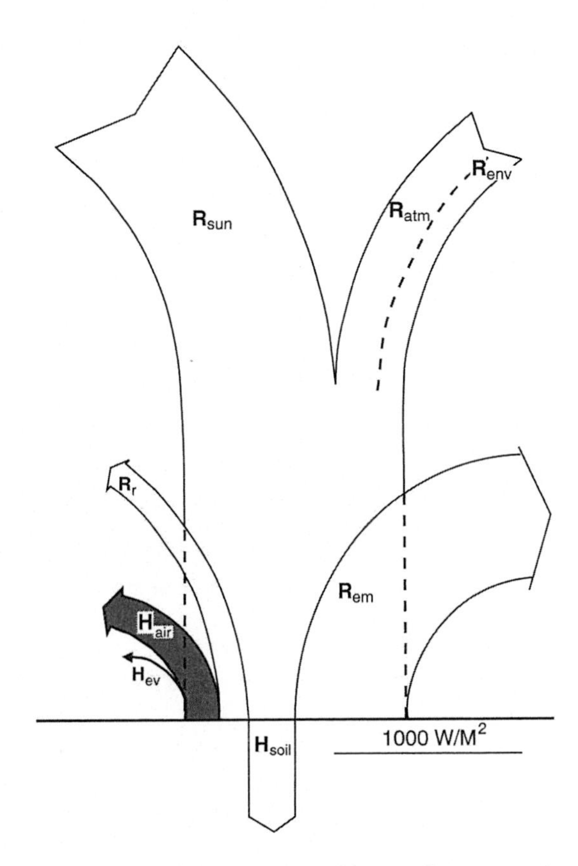

Fig. 3. Energy balance of a dry surface with strong solar radiation: R_{sun} = 832 W/m^2; atmospheric long-wave radiation: R_{atm} = 252 W/m^2; long-wave radiation from the surroundings: R_{env} = 119 W/m^2. Total incoming radiation = $R_{sun} + R_{atm} + R_{env}$ = 1203 W/m^2. Amount of heat emitted by radiation: R_{em} = 629 W/m^2; reflected solar radiation R_r = 154 W/m^2; heat transferred to the air through convection: H_{air} = 168 W/m^2; evaporation heat: H_{ev}; heat transferred to the soil: H_{soil} = 252 W/m^2. Thickness of the arrows indicate the relative amounts of energy flux. H_{air} was not measured, so this post makes the balance.

2.2 Solar radiation

> *"... for discerning bee-masters now find that their hives should not in the winter be exposed to the hot sun, because such unseasonable warmth awakens the inhabitants too early from their slumbers."*
> *Gilbert White (1783): The Natural History of Selborne*

Solar radiation is electromagnetic radiation which forms part of a continuum of radiation types, which are in order of increasing wave-length λ (nm = nanometer = 10^{-9} m; wave-length intervals are approximate):

Cosmic radiation	$< 10^{-3}$ nm;
X-ray radiation	10^{-3} - 100 nm;
Ultra-violet	100 - 380 nm;
Visible light	380 - 780 nm;
Infra-red	780 - 10^6 nm;
Radio waves	1 mm - 10^6 m

All these types of radiation are transmitted in a vacuum, but can be transformed into heat by absorption. Almost all the energy from the sun is in the wave-length range 300 - 3000 nm, but small amounts are X-ray and long-wave radiation.

Within this interval various wave-length intervals can be distinguished and isolated, for instance by means of a prism. The ultra-violet represents only a few percent of the total solar energy, but it has a specific biological impact. Then follow the colours of the rainbow. Wave-lengths greater than 780 nm are not visible to the human eye. Radiation between 780 and 3000 nm is called 'near infra-red', or short-wave infra-red, this is in order to avoid confusion with 'real' heat radiation at greater wave-lengths ($\lambda > 3000$ nm, the 'far infra-red', or long-wave radiation). The term heat radiation does not imply any specific heat effect; other wave-lengths have this warming effect as well. It means that this radiation has a warming effect only. In the meteorological literature this radiation is usually called 'long-wave radiation'.

Regarding photochemical effects of radiation, there is firstly the directly destructive effect of ultra-violet (UV) radiation (λ from 280 to 310 nm) on various components of the living cell, unless special protection has been developed. UV radiation with $\lambda < 280$ nm is completely absorbed by the atmosphere; UV radiation with $\lambda > 310$ nm can have a morphogenetical effect on plants and specific biochemical effects (Lee & Downum 1991).

Light between 400 and 700 nm is used in photosynthesis by green plants. This 'photosynthetically active radiation' or PAR makes up about half of the energy provided by solar radiation (Ross 1975). Photochemical effects at the upper end of the visible spectrum are found both with near red (λ 600 - 700 nm) and far red (λ 700 - 800 nm), especially for seed germination (see Ch. 3.4.4). For purple bacteria (*Rhodobacteriales*) the upper limit for photosynthetically active radiation is at ca. 900 nm. For radiation with wave-lengths over 900 nm perception by the beetle *Melanophila acuminata* has been claimed (e.g. Evans 1966) but not definitely proved.

The total energy flux from the sun is considerable and important, as has also become clear to the general public since solar energy has started to be used as a source of collected energy, used to heat water and generate electricity. The amount of solar energy reaching the earth (measured as joules on a plane of 1 m^2 situated perpendicular to the incoming radiation, outside the earth's atmosphere) is over 80 kJ/min. Earlier, the units calorie and langley (Ly) were used: 1 Ly = 1 cal/cm^2 = 41 868 kJ/m^2. For transforming these older units into presently accepted units of measurements in the SI system (Système International d'Unités) the following conversion factors are useful:

1 J = 0.239 cal = 10^4 erg

1 W = 1 J/s = 0.239 cal/s = 14.33 cal/min

1 W/m^2 = $14.33 \cdot 10^{-4}$ cal cm^{-2} min^{-1}

When describing photosynthesis or other photochemical effects, radiation received is often expressed in quanta or photons. The energy contained in a photon is inversely proportional to wave-length. The number of photons is measured in einsteins (E) or mols: 1 E(mol) = $6.02 \cdot 10^{23}$ photons. Note that this number is Avogadro's number from chemistry. For blue light ($\lambda = 400$ nm) 1 E(mol) represents an energy of 299 kJ, for red light ($\lambda = 700$ nm) this is only 171 kJ ($400 / 700 \cdot 299$).

The average annual value of the radiation intensity (flux density) outside the atmosphere is called the solar constant; its value is 1353 W/m^2 (Coulson 1975). During the year the radiation intensity varies, about 3,5% around the mean with the highest value in January, when the earth is nearest to the sun, and having its lowest value in July, when the earth is farthest away. This means that the difference between summer and winter intensity in the southern hemisphere is greater than in the northern hemisphere, but because of the large masses of water, this is not reflected in a bigger difference between summer and winter temperature.

Part of the solar radiation is scattered by the atmosphere, part of it is absorbed, but the bulk reaches the earth's surface and this can be measured (see Ch. 5.2.4). The intensity of direct solar radiation reaching the earth's surface varies because scattering and absorption vary, notably with the clearness of the air and the thickness of the air layer it passes through. Thus, radiation intensity is higher in the mountains and when the sun is high in the sky. The scattering of the sunlight is mainly caused by air molecules. Shorter wave-lengths are scattered more than longer wave-lengths. Consequently the sky is blue and the setting sun is red: on its long way through the atmosphere direct sunlight is depleted of the short-wave components by scattering and hence the red dominates (Minnaert 1972).

Part of the scattered sunlight reaches the earth's surface as indirect radiation (light from the sky), but since scatter occurs in all directions there is loss of incoming radiation. With a clean atmosphere this light from the sky is a purer, more saturated blue and of a lower intensity than with a polluted atmosphere. The blue is also more dominating at higher altitudes (seen from an aeroplane the sky above us is very dark indeed). Absorption is responsible for another loss of incoming radiation. Absorbing agents are first of all water vapour, but also ozone and to a lesser extent oxygen.

Aerosols are a further agent in both scattering and absorption of sunlight. Aerosols can develop naturally, for instance minuscule clay particles blown up from the surface of the earth, of dust from volcanic eruptions. Pollution from burning fossil fuels, oil refineries and many chemical industries has dramatically increased aerosol concentrations in many parts of the world with considerable effects on the radiation climate (Table 1).

The amount of scatter of sunlight by bigger aerosol particles is largely independent of wave-length. At high aerosol concentrations the blue sky turns into milky white. Natural clearness of the air is dependent on the origin of the air masses. Cold polar or arctic air is clear, i.e. with a low aerosol content. In a high pressure area the sky is clear but often a somewhat whitish blue, which points to aerosol (dust) scattering. Very clean air may occur at lower latitudes after a cold front with much rain has passed: the rain has removed most of the aerosols and polar air is blown in.

Table 1 shows the large variation in intensity of direct and scattered solar radiation as occurring in natural and urban-industrial environments. There is a clear effect of the elevation of the sun, i.e. the path length of the radiation through the atmosphere, and the cleanness and turbidity of the atmosphere.

The properties of the atmosphere as regards absorption and scattering of solar radiation, are conveniently expressed by a turbidity factor T^*; this factor gives a good impression of cleanness or dirtiness of the atmosphere.

The theoretical minimum value of $T^* = 1$, referring to a theoretical absolutely clean and dry atmosphere at sea level. For $T^* = 2$ the absorbed and scattered radiation would be equal to radiation which had passed through twice the standard atmosphere. Characteristic values for the Netherlands are $T^* = 2.75$ for relatively clear dry air in winter, $T^* = 3.25$ for clean polar air in May and $T^* = 5$ as an average summer value (Slob 1982). High mountains in the winter have $T^* = 1.9$.

Table 1. Solar radiation received (W/m²) on a horizontal surface at different elevations of the sun and turbidity values (T^*) of the atmosphere Dir = direct solar radiation; Dif = diffuse solar radiation; Gl = global radiation; Dir ⊥ = intensity of direct solar radiation (irradiance) measured perpendicular to the direction of the incoming radiation. After Schulze (1970).

Environment	Radiation	Elevation of the sun				
		5°	10°	30°	60°	90°
High mountains	Dir ⊥	500	670	990	1120	1160
$T^* = 1.90$	Dir	42	117	500	970	1160
	Dif	25	39	59	60	61
	Gl	67	156	559	1030	1221
Lowland	Dir ⊥	300	490	850	1030	1060
$T^* = 2.75$	Dir	26	85	430	890	1060
	Dif	29	50	84	91	93
	Gl	55	135	514	981	1153
Big city	Dir ⊥	175	330	710	920	970
$T^* = 3.75$	Dir	15	58	360	790	970
	Dif	30	56	109	127	130
	Gl	45	114	469	917	1100
Industrial area	Dir ⊥	86	200	570	800	860
$T^* = 5.00$	Dir	8	36	280	690	860
	Dif	29	59	132	165	170
	Gl	37	95	412	855	1030

On a horizontal plane at 52 °N (e.g. the Netherlands) global radiation is seldom greater than 0.9 kW/m²; on a surface perpendicular to the sun an intensity of 0.95 kW/m² can be reached, even as early as March; with higher elevations of the sun the intensity can increase up to 1.05 kW/m². These values refer to a situation with a clear sky. If white cumulus clouds occur, intensity from reflected light is added and maxima of 1.20 kW/m² can be measured perpendicular to the sun. In the high mountains radiation intensity on a horizontal surface may rise to 1.40 kW/m², which is more than the solar constant of 1.353 kW/m², and 400 W/m² more than what is measured with a cloudless sky (Turner 1961).

Direct solar radiation is reduced by increased scattering. Diffuse solar radiation remains about the same. Thus, global radiation, being the sum of direct and diffuse radiation, is less dependent on turbidity than direct radiation. Diffuse solar radiation increases with higher elevations of the sun, but decreases relative to global radiation.

A total cloud cover reduces the solar radiation considerably, to ca. 75% with a high cirrus cover, ca. 25% with low clouds and only a few percent during a very heavy thunderstorm.

The amount of solar radiation reflected by the soil surface or by vegetation varies considerably, as we know from visual observation (for instance white sand compared with dark soil). The percentage of reflected radiation is called *albedo* (the short-wave reflection coefficient).

Reflection can be selective as to wave-length. Green plant cover reflects little of the visual spectrum but much above 700 nm (near infra-red). Reflection spectra may differ greatly for different types of vegetation and ground cover so that analysis of the reflection spectrum is now used to judge the stage of development and the general health of crops. A treatment of such remote sensing falls beyond the scope of this book.

Table 2. Albedo (%) of various vegetation types and other surfaces. (1) after Keppens et al. (1980); (2) after Piggin & Schwerdtfeger (1973); (3) after Šmid (1975); (4) after Jarvis, James & Landsberg (1976); (5) after Rauner (1976); (6) after Robinson (1966); other figures after Ph. Stoutjesdijk (unpubl.).

Surface	Albedo (%)
Production grassland	21,3-26,0
Dry grassland	20,4-21,0
Moist grassland on poor soil	17,9
Molinia caerulea dominance (1)	15,2
Calluna vulgaris dominance (1)	10,2-15,2
Dead *Calluna* cover	8,7
Erica tetralix dominance (1)	13,3
Reed, *Phragmites australis,* dominance (3)	18
Reed, yellow (3)	21
Coniferous forest (4)	9-13
Oak forest (5)	16
Wheat (2)	23-24
Barley 2)	26
Barley, ripe (2)	30
Barley, stubble (2)	34
Dark raw humus	5-10
Dark dune sand with humus	14-18,5
Yellow sand	24,2
White (mobile) sand	29-36
Black arable soil, dry (6)	14
Black arable soil, moist (6)	8
Grey soil, dry (6)	25-30
Grey soil, moist (6)	10-12
Snow at 30° elevation of the sun (6)	86
Snow at 25° elevation (6)	95
Water at 50° elevation (6)	2,5
Water at 30° elevation (6)	6-8
Water at 20° elevation (6)	12-15
Water at 10° elevation (6)	32-49
Water at 6° elevation (6)	48-70

Table 2 shows some characteristic albedo values. The albedo of fresh snow is the highest found on earth and approaches the albedo of an ideal white surface.

Production grasslands and low crops with a homogeneous, closed texture and an even canopy have relatively high albedo values, up to 25%. More natural grasslands have lower values. Openings and height differences in the vegetation lower the reflection. The optical characteristics of leaves are also important; leaf reflection varies considerably and it is very strong and specular, i.e. mirror-like, in the case of an over-manured grassland. For bare soil visual observation gives a reasonable impression of the albedo; the albedo of white sand is often over-estimated.

At low elevations of the sun (< 30°) the albedo of some vegetation canopies may be increased (Monteith & Szeics 1961), though not for heathland plants such as *Calluna*, *Erica* and *Molinia* (Keppens et al. 1980). Our own data presented in Table 2 refer to an elevation of the sun of > 40°; this is probably the case for the cited data as well. Water has a low albedo, but at low solar elevations specular reflection strongly increases the albedo. Such reflections can have considerable local effects, for instance it favours the cultivation of vines along the river Main in Germany (Volk 1934). Mosses growing on trees on the shores of Lake Lugano in Switzerland clearly profit from extra light (Jaeggli 1943). Reflected light from the sea is one of the aspects of the favourable conditions that the Kullaberg Peninsula (SW Sweden) offers to plants and animals with a southern distribution (Ryrholm 1988).

2.3 Heat radiation

2.3.1 General

Heat radiation (also called long-wave radiation) is part of our everyday life; think of a radiator. A cold radiator radiates heat as well, which is less obvious. A body with a surface temperature T (in K = Kelvin, or °C + 273) emits heat proportional to the fourth power of the absolute temperature of its surface:

$$R_{em} = \varepsilon \sigma T^4 \qquad (2.1)$$

R_{em} is the radiation emitted from the surface in W/m^2, $\sigma = 5.67 \cdot 10^{-8}$ Wm^{-2}K^{-4} (the Stefan-Boltzmann constant); ε is the emissivity of the surface. At $\varepsilon = 1$ we speak of a black surface or perfect radiator. For most natural surfaces ε is close to 1, for polished metal it is very low, e.g. polished nickel has $\varepsilon = 0.02$. Because the emission is complementary to the reflection, such a surface reflects 98% of the incoming long-wave radiation. Thus, a nickel-coated heating element radiates only 2% of the heat a normally painted heating element radiates (the latter having an ε close to 1), but it reflects 98% of the heat radiation from its surroundings. A hand kept close to this element receives hardly more radiation from the element than from any other object in the room. Still, the element warms the room, viz. through conduction and convection.

These relations only hold for a particular wave-length. Thus, the emissivity of long-wave radiation has nothing to do with the colour of an object, because the colour is related to completely different wave-lengths. For instance, the white polar bear and the brown bear emit the same amount of heat. The best known example is snow, which reflects the visible radiation very well, but hardly any heat because its emissivity for heat is close to 1.

Up to 95% of the long-wave radiation emitted at normal temperatures falls into the wave-length interval 6 - 60 µm with the maximum intensity at 10 µm. The radiation emitted by a body has a wave-length distribution which is related to its surface temperature. The peak of the wave-length distribution $\lambda_{max,T}$ is related to temperature by the Wien displacement law: $\lambda_{max,T} = 0.288/T$ cm, where $\lambda_{max,T}$ is the wave-length and where radiation intensity is maximal at T K.

As an example, for a surface temperature of 80 °C (353 K) $\lambda_{max,T} = 8{,}2$ µm and for 0 °C $\lambda_{max,T} = 10{,}5$ µm. At the very high surface temperature of the sun, 6000 °C, $\lambda_{max,T}$ is only 0.48 µm. Thus, the wave-length intervals of solar and long-wave radiation hardly overlap, which is important for radiation measurements (see Ch. 5.2.4).

Because of the relation with T^4 the emitted long-wave radiation increases rapidly with surface temperature, even over the relatively small interval found under natural conditions (Eq. 2.1). The energy balance of the warm sandy surface (Fig. 3) shows that the long-wave radiation of the not excessively warm sand already approaches the intensity of the solar radiation received.

The conceptual difficulty with long-wave radiation is that we never measure absolute intensities but always the difference between emitted and received long-wave radiation. For an object in a closed room heat loss through radiation follows from:

$$H_s = \sigma\ (T_1^4 - T_2^4) \qquad (2.2)$$

where H_s is the heat flux per unit surface (W/m^2), T_1 the temperature of the object and T_2 the temperature of the walls of the room. If $T_1 = T_2$ we do not observe anything, although the energy fluxes may be considerable. In everyday life we also speak of cold radiation, for example with an open deep-freeze box. In fact this is a shortage of long-wave radiation.

As we saw, the emissivity ε for natural surfaces is close to 1. Dirmhirn (1964) found values between 0.94 and 0.99 for leaves, with the lower values for shining, leathery leaves (Gates & Tantraporn 1952). A surface with a complex structure such as a vegetation canopy will always behave like an ideal radiator with $\varepsilon = 1$, since a cavity radiates like a black body, irrespective of the material. Rocks may have a low emissivity in a small wave-length interval. Quartz, for instance, has an average ε of 0.7 between 8 and 12 μm, beyond this range it is almost 'black' for long-wave radiation. A surface can have a certain 'colour' in the infra-red, i.e. reflection and absorption are dependent on wave-length.

One may speak of the effective radiation temperature of a surface, meaning that the surface emits as much heat as a surface with $\varepsilon = 1$ with the same temperature. For $\varepsilon < 1$ the effective radiation temperature is lower than the real temperature.

Most material is impervious to long-wave radiation, even in thin layers. Glass, for instance does not transmit any long-wave radiation but nearly all sunlight. Only some plastics are an exception. Polythene transmits very well in the far infra-red and is used for windows in radiation meters. Well-known for their high transmission are the halogenids such as NaCl. Windows and prisms of rock-salt are used in spectral instruments for the far infra-red.

A substance can also be opaque, i.e. impervious to light but transparent to heat radiation, for example germanium. A glass mirror with a silver coating at the back reflects light but not heat. Surfaces have been constructed for sun collectors which strongly absorb light but have a low emission of long-wave radiation (Robinson 1966). This also explains why an owl cannot 'see' a mouse by its long-wave radiation, as is sometimes said. This radiation would have to pass through the eye, but that consists mainly of water. A rattlesnake, however, can observe a heat-radiating prey because it has temperature-sensitive cells in the epidermis covering the open cavity in the head (Bullock & Cowles 1952).

2.3.2 Atmospheric long-wave radiation

In addition to solar radiation the earth receives much long-wave radiation emitted by the atmosphere. In the example of Fig. 3, long-wave radiation was 252 W/m^2, while the intensity of the total solar radiation was 832 W/m^2. So, long-wave radiation is an important item in the energy balance. The heat radiation of the atmosphere lies in the same wave-length interval as the long-wave radiation of the earth. Interestingly, this important long-wave radiation of the atmosphere is due to radiation by relatively small amounts of specific molecules, water vapour, CO_2 and to a lesser extent O_3 (ozone). The most important components of the air, O_2 and N_2, do not produce any long-wave radiation and do not absorb it either. In the spectrum of atmospheric long-wave radiation there are discrete bands with strong radiation as well as bands with weak radiation. Because emission and absorption are equal, long-wave radiation emitted by the earth, in certain wave-length intervals, is almost completely absorbed by a thin layer of air. In other wave-length intervals hardly any

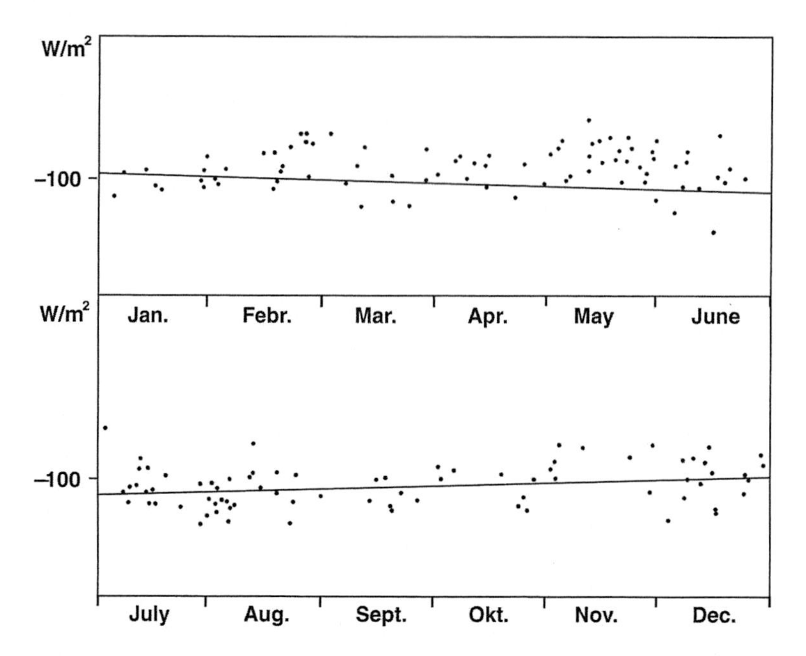

Fig. 4 . Difference between atmospheric long-wave radiation received and emission from a horizontal surface at air temperature, with a clear sky. The line indicates the average expected ($107 - t_{air}$) according to the Swinbank-Monteith formula (2.3 and 2.4).

absorption occurs and heat radiated by the earth in these intervals disappears almost unchanged in interstellar space. An important 'window' is found between 8 and 12 µm where the atmosphere is entirely transparent; as a coincidence the heat radiation of the earth is maximal in this interval.

Thus, in the atmosphere a complicated interaction is found between emission and absorption of long-wave radiation in all directions. These processes help determine temperature distribution in the atmosphere. A side-effect is that due to the small quantities of water, CO_2 and O_3 in the atmosphere, the earth has a much higher surface temperature than would have been possible without these gases, viz. + 15 °C on average instead of –20 °C.

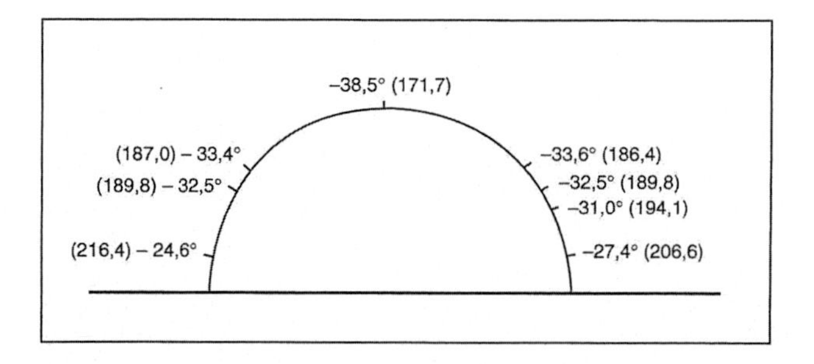

Fig. 5. Effective radiation temperature of the sky dome in different directions. The figures between brackets indicate values of σT^4 in W/m² for each temperature. Place: Arnhem, the Netherlands. Date: 11 November 1978, 11.00 h; clear sky, air temperature 0.5 °C.

The atmospheric long-wave radiation, as it is received at the earth's surface, is mainly derived from the lower air layers. Sellers (1965) gives an example of 58.9% of the atmospheric radiation received being emitted by the lower 100 m of the atmosphere. Clearly, a fair estimate can be made of the total atmospheric long-wave radiation received on the basis of temperature and humidity at a height of 1.50 m. Heat radiation emission by gases follows the same proportion with T^4 as for a solid surface. Although water vapour is the most important contributor to the long-wave radiation in the atmosphere, Swinbank (1963) showed on theoretical grounds that its influence on the intensity of atmospheric long-wave radiation is only limited in the naturally-occurring interval of variation of the water vapour concentration. Swinbank (l.c.) gives the following relation:

$$R_{atm} = 1.214 \cdot \sigma T^4 - 171 \qquad (2.3)$$

where R_{atm} is the atmospheric long-wave radiation (W/m^2), T the air temperature in K, and $\sigma\ T^4$ the emission of a black body at air temperature.
Monteith (1973) derived the following approximation for this equation:

$$R_{atm} = \sigma T^4 - 107 + t_{air} \qquad (2.4)$$

where t_{air} is the air temperature in °C and T the air temperature in K. This expression can be read as follows: if the air temperature is, say 10 °C, the atmospheric radiation received is 107 – 10 = 97 W/m^2 lower than the emission of a black body of 10 °C. Brunt (1932) presented an equation where the water vapour in the atmosphere is taken into account:

$$R_{atm} = (0.53 + 0.06\sqrt{e})\, \sigma T^4 \qquad (2.5)$$

where e is the water vapour pressure in mbar. This equation should give a better approximation than Swinbank's one for very low and high humidity values.

Of course, these equations will only give approximate averages; they do not take into account the fact that temperatures in higher layers of the atmosphere can be much higher, or lower than the temperature at 1.50 m. Wartena, Palland & van der Vossen (1973) have tested the equations of Brunt and Swinbank and found that the Brunt equation gives the best results if different of the values of the constants are used for different months.

Monteith (1973) has shown that the intensity of atmospheric radiation with a clear sky can be estimated by the effective radiation temperature as measured at an angle of 37 °, in other words by replacing the atmosphere by a hemisphere with this radiation temperature.

In Fig 4. results of the measurements are presented taken with a radiation thermometer as described in Ch. 5.2. The straight line indicates the expected value according to Swinbank's formula, while taking into account the average air temperature at midday in the respective month. The agreement is quite reasonable, although considerable deviations (many of them positive) occur. For example, in May, when cold and clear polar air is blown into temperate Europe, the temperature of the sky dome is much lower than predicted by the formula. We should add that the above-mentioned approximations are based on measurements during clear nights, when the temperature near to the ground is lower than in higher layers. In the daytime this temperature relation is reversed and then the radiation intensity of the atmosphere will be somewhat lower than predicted, as is also shown in Fig. 4 which refers to the daytime situation.

As regards radiation, we might imitate the situation for a clear sky with a huge glass hemisphere letting through all the sunlight, but with a temperature 20 - 30 °C lower than the air temperature. The temperature of the dome is not equally low everywhere. The lowest temperatures are found in the zenith; down the dome temperature rises, first slowly, later rapidly. Fig. 5 presents some measurements on a cold November day. The indicated temperature of the dome is also called the effective radiation temperature of the sky in this direction. One can apply this expression also to the total long-wave radiation of the atmosphere. The total radiation received by a horizontal plane is, according to the above, equal to the value indicated for an elevation angle of 37°; the effective radiation temperature of the atmosphere as a whole is –32.5 °C: the total atmospheric long-wave radiation is as strong as if the hemisphere had a uniform temperature of –32.5 °C (Fig. 5). The warmer parts of the sky dome, situated near the horizon, contribute relatively little to the total atmospheric long-wave radiation received on a horizontal plane (cf. Ch. 2.3.3).

This all holds valid for a clear sky. For a cloudy sky the situation is different. A thin, high cirrus veil still has little influence on the atmospheric long-wave radiation, but a thick cloud cover emits almost like a black body. Especially with low, thick clouds the temperature at the base of the clouds does not differ much from that of the air at 1.5 m, usually it is only 2 - 3 °C lower.

2.3.3 Long-wave radiation in spatially complex situations

On plain, open areas the sky dome is the only source of long-wave radiation for the soil surface, or for any other horizontal plane. With obstacles such as hedges or walls screening part of the sky dome, part of the long-wave radiation is intercepted and replaced by the long-wave radiation of the obstacle which has a much higher temperature than the clear sky.

At the base of a wall or hedge the situation is simple: half of the total radiation from the sky is screened. In the situation of Fig. 5 this would mean that the amount of atmospheric long-wave radiation received is $0.5 \cdot 189.8\ W/m^2 = 94.9\ W/m^2$. In the case where the wall has the same temperature as the air, $0.5 \cdot 317\ W/m^2 = 158.5\ W/m^2$ is received from the wall. Thus, there is a gain in long-wave radiation of $63.6\ W/m^2$, which may be important at night. With increasing distance from the wall this influence diminishes rapidly. If we consider the sky dome as half of a sphere, we are dealing with that part of the dome that is screened and the projection of that part on the base (Fig. 6).

For an infinitely long wall seen under an angle of 45°, the surface area A of the projection is:

$$A = (1 - \cos 45°) / 2 = 0.15 \tag{2.6}$$

with the area of the base of the hemisphere put at 1. A is called the view factor. The projection of the non-screened part, $(1 - A)$ is sometimes called the effective radiation loss, a somewhat misleading term. If the wall has the same temperature as the air, point C receives long-wave radiation from the wall according to: $0.15 \cdot \sigma T^4\ W/m^2$; for $T = 0.5$ °C this would be $0.15 \cdot 317 = 47.5\ W/m^2$. The intercepted radiation from the sky is less easy to estimate; if we assume that the sky dome in this segment has an average temperature of –30 °C, then the intercepted atmospheric radiation is $0.15 \cdot 198 = 29.7\ W/m^2$. The 'gain' provided by the wall is $0.15\ (317 - 198) = 17.8\ W/m^2$. For 'wall' we may also read hedge or forest edge, if these have a closed structure.

Another frequent situation is an open site, 'gap', in a forest (Fig. 7). From the centre of the gap, which has a theoretical circle form, one exactly faces the coldest part of the sky. If the trees have a height h and the gap has a radius r^* and $h = r^*$, the free visual field from the

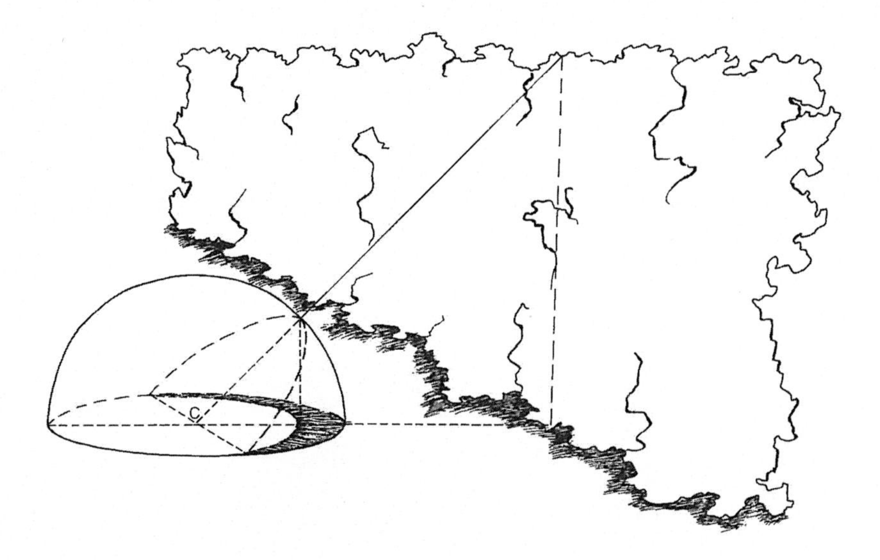

Fig. 6. Long-wave radiation around a hedge in an open landscape. The drawing indicates which part of the long-wave radiation emitted by the hedge is received by a horizontal surface at the centre of the circle and which parts of the atmospheric long-wave radiation are screened and received respectively. These fractions are proportional to the black and white parts of the circle respectively.

centre is a cone with an aperture of 90°. The projection of the gap has a relative area of $(\sin 45°)^2 = 0.5$. The rationale presented here can be applied to many different situations; the essential step is to create a hemisphere over the surface under study and to project the screened part onto the base (Eckert 1959). A vertical surface, for instance, receives half the atmospheric long-wave radiation + half the long-wave radiation emitted by the soil surface.

This can also be applied to the reflected solar radiation, as long as the reflection is diffuse. A vertical surface, for instance a tree bole, receives half the solar radiation reflected by the soil surface. A slope of 30 ° receives $(1 - \cos 30°) / 2 \cdot 100\% = 7\%$ of the radiation reflected by the flat land around, and also 7% of the emitted long-wave radiation.

2.3.4 Net radiation

"Das Gute dieser Satz steht fest,
ist stets das böse was mann lässt"
Wilhelm Busch

The above-mentioned energy balance can be simplified by taking all radiation terms together. The sum of these radiation terms can be positive or negative and is called net radiation. Net radiation can be measured directly with a net radiometer (see Ch. 5.2.4). Continuous measurements with this instrument are rare. The net radiation is a good starting point for various considerations of the energy balance, especially for larger homogeneous areas. Because the net radiometer has an aperture of 180° both upwards and downwards; an average value is obtained for a fairly large area. Measurements of this type only make sense

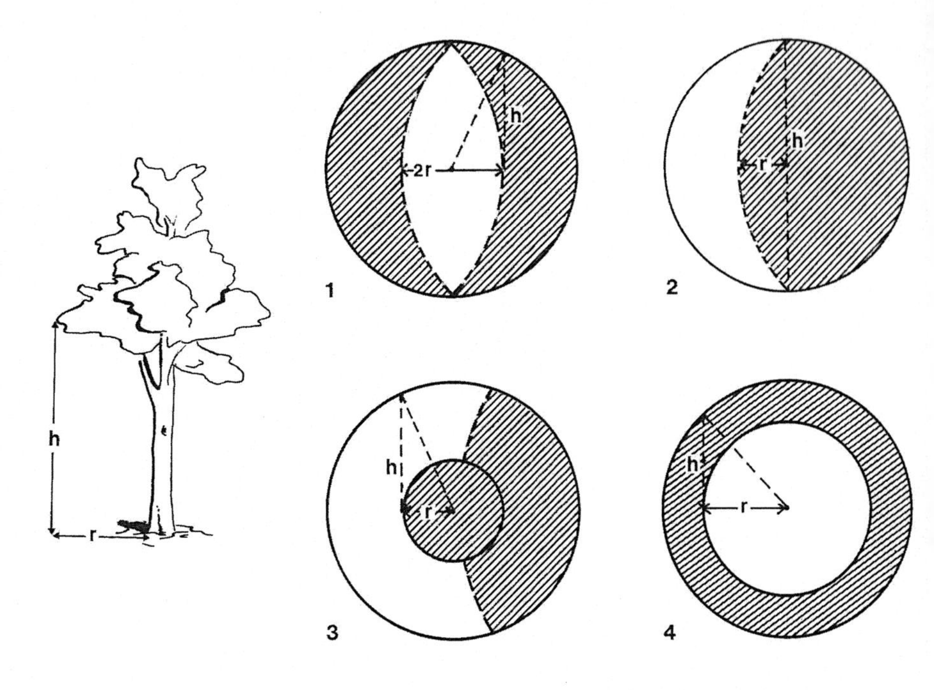

Fig. 7. Long-wave radiation in some forest microsites. 1. The middle of a fire lane; 2. A forest edge, under overhanging branches; 3. The foot of an isolated tree; 4. A forest gap. The hatched surface is the projection of the part of the sky which is screened off (compare Fig. 6).

if one is interested in broad averages. One could speak of net short-wave radiation if one has the difference between received and reflected solar radiation in mind; this can be measured directly (see Ch. 5). Similarly one can speak of net long-wave radiation.

The net radiation is the amount of energy available for the warming of soil and air and for evaporation. Clearly, the net radiation will differ markedly from point to point; a bare surface which is greatly warmed by the sun will receive a low net radiation and, conversely, a strongly transpiring crop with the same albedo as the bare surface, remaining cool, will have a high net radiation. Also, when there is a strong wind, the net radiation received by bare sand will be greater than under still conditions, when the surface is considerably warmed up. In the example of Fig. 3 the surface temperature would easily decrease, maybe to only 10 °C above the air temperature if the wind were stronger. In that case the heat emission of the surface would decrease from 629 to 541 W/m^2. This means that the net radiation would increase by 88 W/m^2. If there had been grass on the same spot, with the same reflectivity as the sand and with a canopy temperature of only 2 °C above the air temperature, the heat emission would have been only 488 instead of 541 W/m^2 and the net radiation would be 141 W/m^2 higher than with warm sand.

For the heat relations of the atmosphere as a whole the net radiation is a very important parameter and maps of its distribution have been made (Budyko 1974). Human activities may have a definitive influence on the net radiation of the earth's surface, e.g. when the vegetation is destroyed evaporation decreases and both the surface temperature and the albedo change. From a plane, in early spring, we see the northern pine forest as a dark green expanse, but clear-felled areas are pure white. In the first case the solar energy is transformed into heat and partly used to melt the snow that has fallen from the trees on the forest floor. In the second case it is reflected 'unused' to outer space.

The net radiation R_{net} in the daytime follows from:

$$R_{net} = R_{sun} - R_r + R_{atm} - R_{em} \tag{2.7}$$

where R_{sun} is the solar radiation, R_r the reflected solar radiation, R_{atm} the atmospheric long-wave radiation and R_{em} the emitted long-wave radiation. R_{sun} and R_{atm} are energy gains here, because they are always directed towards the surface of the earth, R_r and R_{em} are always energy losses because they are directed away from the earth.
R_{atm} is almost always lower than R_{em}; it may be higher when warm air under heavy clouds is situated above a cold surface. R_{sun} is always greater than R_r, but in the case of a surface with a high albedo (snow), the difference may be so small that, even with strong sunshine, R_{net} will be negative. At night $R_{net} = R_{atm} - R_{em}$.

For the total energy balance we can write (Fig. 3): $R_{net} + H_{air} + H_{soil} + H_{ev} = 0$, where all terms can be either positive or negative. Frequently occurring combinations are:
Strong insolation: R_{net} positive, H_{air} and H_{soil} negative, H_{ev} negative or zero.
At night: R_{net} negative, H_{air} and H_{soil} positive, H_{ev} positive or zero.

The sign is determined in relation to the soil surface; so, H_{air} is negative if the surface transfers heat to the air, H_{soil} is negative if the heat flow is directed downward into the soil, H_{ev} is negative if, through evaporation, water is transferred to the air, and H_{ev} is positive with condensation at the surface (dew, hoarfrost).

2.4 Heat transfer to the air

"The air temperature predominating near the ground ... is in fact a return of heat from the soil surface"
G. Kraus (1911): Boden und Klima auf kleinstem Raum

The process of heat transfer from the soil surface to the air becomes clear from Fig. 8, where the temperatures of different layers above and below the soil surface are shown. The temperature at the soil surface is 65 °C and fluctuates little, only 4.5 °C during one minute of observations. The above-ground (as well as the below-ground; see below) temperature drops rapidly with increasing distance from the surface. At 1 mm above the surface the average air temperature is 53 °C, 12 °C lower than at the surface, while the extremes (again one minute of observations) differ by no less than 15 °C. At 1 cm height above the surface the average temperature is again considerably lower, only 41 °C, with the extremes differing by 12 °C. Between 1 and 10 cm the temperature decrease is much less, only 5.5 °C; the extremes, at 10 cm still differing by 10 °C, overlap with those at 1 cm.

On the basis of the general literature on heat transfer (Geiger 1961) this temperature profile can be elucidated as follows. Near the soil surface we find a thin layer of air without any air movement. Heat transfer occurs through conduction, i.e. through collisions of air molecules. The temperature decreases greatly in this thin layer. Above it occurs turbulence; small parcels of hot air are rising and are replaced by colder parcels from above. As a result the temperature fluctuates a great deal here. The temperature amplitudes at 1 mm and 1 cm do not overlap, which means that no air parcels are transported unchanged between the two levels. Between 1 and 10 cm the temperature drop is much less and the amplitudes overlap. Apparently, air parcels mix much better here. From 10 cm to 1 m the temperature lapse is roughly the same as from 1 cm to 10 cm. The effect of the warm soil surface is no longer seen at 1 m height.

The biological implication of this temperature profile is that the steepest part of the gradient is very near the soil surface. This also holds true for air humidity above a wet soil surface. The air in the lowest few cm has completely different properties as far as turbulence, temperature and humidity are concerned than at a higher level. A very small organism living on the ground experiences very large temperature differences already on the scale of its own dimension. Hence it is not easy to characterize the temperature and humidity of its 'environment'. The beetle *Cicindela hybrida* obtains access to heat in the early morning by pressing itself against the warm sand, already warmed up by the sun. As soon as the temperature at the soil surface rises above 40 °C, the beetle avoids heat problems by raising itself up on its legs; its body moves up to 8 mm above the surface (Dreisig 1980). We have observed similar, though less successful attempts by the caterpillar of the moth *Tyria jacobaeae*.

In the Namib desert the beetle *Stenocara phalangium* avoids the adverse conditions near the soil both by stilting and by climbing isolated rock fragments, whereby its white abdomen is directed towards the sun. These thermal refuges are defended against other beetles (Henwood 1975). In the same desert the burrowing spider *Seothyra* spec. makes a sticky mat, flush with the sand surface. The prey is caught where conditions are most severe and it soon succumbs (Lubin & Henschel 1990). Though the authors did not measure this, it seems quite possible that the temperatures on the mat are higher than those on the surrounding sand.

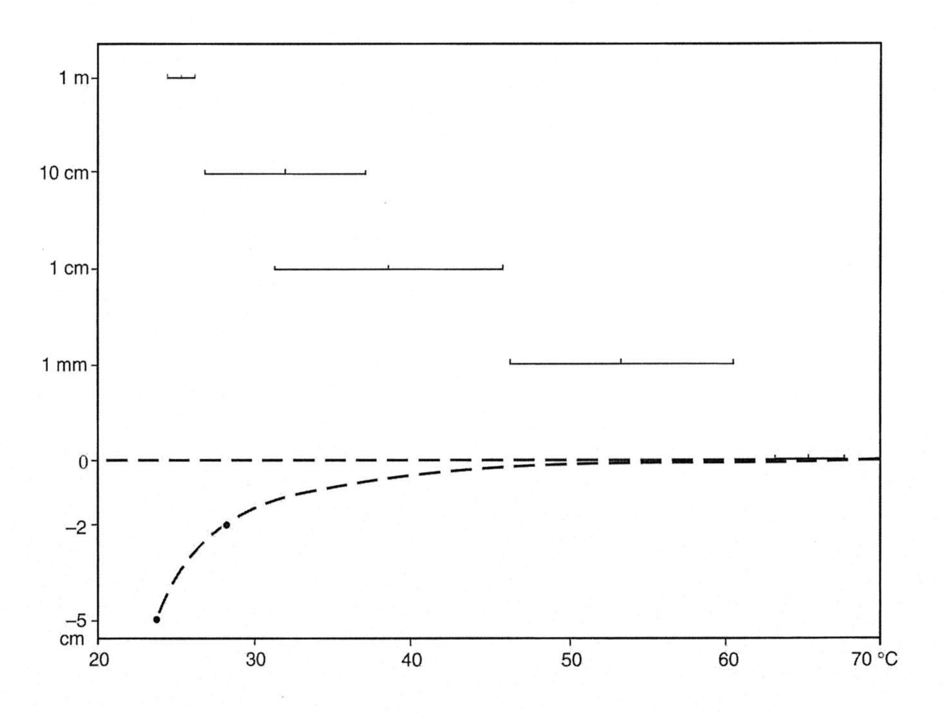

Fig. 8. Temperatures above and below a soil surface round midday, warmed up by the sun. Horizontal lines indicate temperature fluctuations observed within one minute of observations.

Above the level of 1 m, the transport of warm air upward and of cold air downward continues on larger and larger scales, until columns and bubbles with a diameter of a few to hundreds of metres are reached. During a sunny day the upward warming of the atmosphere proceeds and can reach heights of hundreds of metres, and uplifting winds with a speed of several m/s may arise (Elkins 1983). Of course, the upward movements are compensated for by downward ones. On sunny days with little wind such downward movements may be noticed near the ground as squalls which lower the temperature. Uplifting air masses arise on fixed points in a landscape, for instance at the edge of dry warm soil and a moist, cool area. On larger scales uplifting air masses are more common above land surfaces; descending air masses are more common above seas and large lakes.

The reality of uplifting air masses is nicely demonstrated by soaring birds of prey using such thermals. Hankin (see Scorer 1954; Cone 1962; Brown 1976) pointed out as early as 1913 how different birds of prey choose their own suitable time to lift, first the smaller kites, later the bigger vultures. Large birds, e.g. storks and many birds of prey, since they are able to use thermals, choose migration routes over land.

Lifting air masses on a small scale are used by spiders, which on sunny days push out their threads from some high point in the surroundings, for instance a dead stem, and release themselves as soon as warm air lifts them. Gilbert White wrote as long ago as 1783: "Small spiders which swarm in the fields in autumn and have a power of shooting out webs from their tails so as to render themselves buoyant and lighter than air". Vugts & van Wingerden (1976) found that such aeronautic behaviour occurs especially in sunny weather and light winds when the atmosphere has reached a certain level of instability. These conclusions are based on careful observations and measurements during this 'ballooning'. In Missouri, USA, numbers of ballooning spiders caught on stick traps during one week were negatively correlated with sunshine during this week (Greenstone 1990). Here longer periods of fine weather may have a negative effect on the ballooning, although sunshine is needed for the actual take-off. There is also the possibility of updraughts in a turbulent wind field.

As said before, the warming of the atmosphere by the soil surface can be extended to hundreds of metres in the course of a sunny day. The temperature gradient becomes less steep and becomes equal to the so-called adiabatic gradient of 1 °C per 100 m. Then the situation of the atmosphere is neutral, meaning that air from lower layers no longer has a tendency to move further upward. Because the air pressure decreases upward, an air mass moving upward without exchanging heat with the surroundings, would expand and therefore cool down at a rate of 1 °C per 100 m (Oke 1987).

Naturally, thermal convection is dependent on the intensity of solar radiation (at night there is no convection at all) and on the characteristics of the landscape: a strong convection above dry ground, but less above a crop using a large part of the energy received for evaporation (cf. Fig. 31).

Thermal convection is important but is not the only cause of the mixing of air masses. The other process here is dynamic convection through wind. Wind is not a laminar flux with movements only parallel to the soil surface, but turbulent, with both horizontal and vertical movements. Thus lifting and descending air masses are the result of winds as well, but then they are less regular, weaker and fluctuate more. A bird cannot use such air movements for gliding. Gliding is still possible, for example when air movements are controlled by boundaries between two landscape elements, for instance between coastal dunes and the beach, or a forest edge, but in these cases there is no turbulence involved.

Generally, the strength of the air mass exchange increases with wind speed and therefore also with height above the earth. At similar wind speeds turbulence may vary considerably depending on the origin and history of the air mass. Strong winds may still have little turbulence, as we may sometimes observe them above dry sand, moving few sand grains despite the wind force. In this case the wind is less gusty, i.e. direction and velocity vary only a little.

Thermal and dynamic convection may reinforce each other, but usually one process dominates. During the warming up of the earth's surface there is a rather sharp transition from dynamic to thermal convection. If the thermal convection dominates, wind has little influence on the intensity of the exchange process. At some height above the surface the effectiveness of the turbulent exchange process is a hundred times greater than the processes of molecular diffusion and heat conduction, which determine the transport of water vapour and heat in stagnant air. Turbulent exchange is responsible for most of the vertical transport of air masses. Without this transport mechanism heat and water vapour would not be carried upward; moreover, carbon dioxide, needed for assimilation, would not be supplied from higher layers of the atmosphere.

This exchange is not only important for the transfer of heat and gaseous components of the air, but also that of heavier objects such as seeds, spores and small insects. Particularly lighter diaspores may happen to be caught in uplifting air masses of thermal or dynamic origin (Geiger 1961). An interesting example of a higher plant is that of *Senecio congestus*, from the *Asteraceae* family, which during the 1950s and 1960s, spread over large areas of some of the newly reclaimed IJsselmeer polders in the Central Netherlands. This was particularly so during the dry summer of 1959, when seeds of this species were distributed by winds both southward and eastward over many km (Bakker 1960). Even as far as in W Poland, 700 km away, plants were found (E. van der Maarel, pers. comm.). The landing velocity of the seeds of this species is 0.16 m/s. This means that a seed released at the height of the plant (ca. 1 m) would only be carried 20 or 30 m in a laminar air stream. If much clearly visible seed is produced, such as in the case of *Asteraceae*, the upward movements of seeds can often be observed directly. For the much smaller spores of mosses, liverworts and ferns the vertical transport is much stronger, though difficult to assess (Söderström & Jonsson 1989). There are indications that they can be found at thousands of km distance from the source (van Zanten 1978; van Zanten & Gradstein 1988).

The amount of heat dissipated by a heated soil surface to the air, H_{air}, is dependent on the temperature difference surface - air, Δt, and the wind velocity. The intensity of the heat transfer process can be characterized by the heat transfer coefficient α in $Wm^{-2}K^{-1}$:

$$H_{air} = \alpha \cdot \Delta t \qquad (2.8)$$

H_{air} is the amount of heat (W/m^2) transferred to the air and Δt the temperature difference between soil surface and air at a level from where the temperature no longer changes much with increasing height, for example at 1.50 m. In the example treated in Fig. 3 we have: $\alpha = 168\ W/m^2 / 22.2\ K = 7.57\ Wm^{-2}K^{-1}$.

Wind has a great influence on the transfer of heat from the soil surface to the air: "the boundary layer is wiped out" (Geiger 1961). Raman (1936) found a relation between the heat transfer coefficient and the wind velocity at 1.20 m for bare ground (Fig. 9). In addition to Raman's measurements some of our own data are presented here, where the heat transfer coefficient was calculated from the energy balance of the soil surface. Methods for direct measurement of the heat transfer from the soil surface are given by Schuepp (1977) and Stoutjesdijk (1978). With still weather α reaches a value of about 5.6 Wm^2K^{-1}. In that case we speak of free convection: the temperature difference is the driving force of the heat transfer; here data from the technical literature are available. For a horizontal surface that is not too small (diameter > 1 m) we have: $\alpha = 1.78 \cdot \sqrt[3]{\Delta t}$ (see Eq. 2.13). For Δt = 40 °C

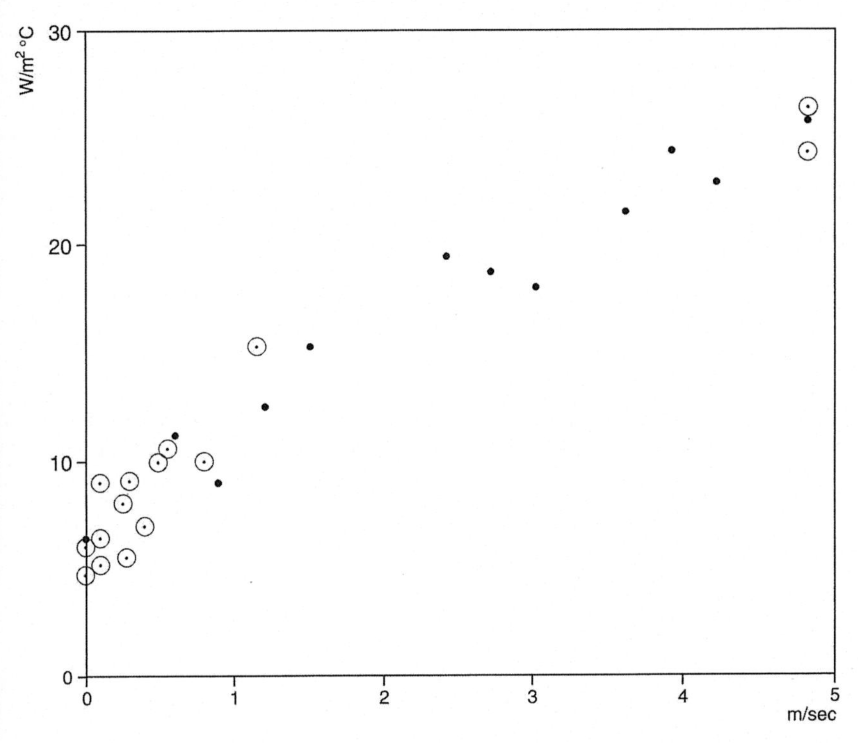

Fig. 9. Heat transfer coefficient for a warm soil surface as dependent on the wind velocity at 1.2 m height. Dots after Raman (1936), open circles after Stoutjesdijk (unpubl.).

$\alpha = 6.2$ Wm^{-2}K^{-1}, which agrees with field observations, if we take into consideration that low values of α are always measured together with high values of Δt. (See Eq. 2.13.)

There is an extensive technical literature on the heat transfer of small objects (Monteith 1973, 1981; Eckert 1959; Gates 1980). The following data serve as a starting point. If air in a laminar flow moves over a flat object, for instance a leaf, this current can remain laminar with zero velocity at the surface and increasing velocity with height above the surface, with the result that in a relatively short distance from the surface air velocity is again equal to that of free movement. The same applies mutatis mutandis to the temperature if the object is warmer or colder than the air in free flow. The thickness of this boundary layer increases gradually from the edge of the object in the direction of the air stream. The heat transfer through this laminar boundary layer occurs through conduction, as through a layer of stagnant air. Air layers with varying speed shift over each other and do not mix. For heat transfer through the boundary layer we have:

$$H_{air} = \frac{k}{d} \cdot \Delta t \qquad (2.9)$$

k is the heat conductivity coefficient of the air, 0.0258 Wm^{-1}K^{-1} and d is the thickness of the boundary layer in m. If, e.g., $d = 3$ mm and $\Delta t = 10$ K, then $H_{air} = 86$ Wm^{-2}. The value of the heat transfer coefficient α is $k/d = 8.6$ Wm^{-2}K^{-1} and Δt is the difference in temperature between the surface and the free moving air.

At a wind speed that is not too low, the flow at some distance from the edge of the object may change from laminar into turbulent. However, there is a very thin boundary layer where the flow is still laminar; this is the so-called laminar sub-layer. Above this layer the air movement is turbulent and above the turbulent layer the free stream occurs where wind speed does not change any more, and usually neither does the temperature. In the laminar layer the speed of the air, temperature and other parameters change rapidly with increasing distance from the surface, while in the turbulent layer above the change is much less. In the turbulent air the thickness of the laminar boundary layer is at most some tenths of a mm, while the turbulent layer is some mm. Eckert (1959) provides instructions for calculating the character (laminar or turbulent) and the thickness of the boundary layer from the wind speed and the dimensions of the object, and from that the heat transfer coefficient. This refers to a laboratory situation, which is rather different from that in nature; hence we will mention only a few details.

If a laminar air flow moves over a flat surface with velocity u, the heat transfer coefficient at a distance l from the edge and measured in the direction of the current, follows from

$$\alpha = 2.15 \sqrt{\frac{u}{l}} \qquad (2.10)$$

Thus, the heat transfer coefficient decreases with distance to the edge, whereas the thickness of the boundary layer increases.

By integration one finds that the average value of α along the entire strip parallel to the edge with breadth l is twice this value: $4.30\sqrt{u/l}$. Note that for very small objects equation 2.10 does not hold (see Ch. 4.1.3).

For a sphere similar values can be mentioned, for instance:

$$\alpha = 0.0516 \,/\, 2r^* + 3.86 \sqrt{\frac{u}{2r^*}} \qquad (2.11)$$

where r^* is the radius of the sphere.

For any object, if the product of u and l exceeds a certain critical value, turbulence may appear in a laminar flow, with the effect that α becomes higher. The main flow can also be turbulent, especially near the surface of the earth and in vegetation. In that case α can be considerably greater than in a laminar flow. Kowalski & Mitchell (1976) found that for a sphere at a height above the ground equal to the diameter of the sphere, the value of α is on average 1.8 times higher than in a free stream with the same wind speed; at a height of 10 times the diameter, α was 1.3 times as high.

In completely stagnant air the driving force of heat transfer is the temperature difference Δt between the surface and the air (free convection). Then, for a horizontal surface of moderate size l, and with a temperature not too different from the surroundings, so that a laminar flow arises:

$$\alpha = 1.44 \sqrt[4]{\frac{\Delta t}{l}} \qquad (2.12)$$

If the flow is turbulent, which is the case for larger surfaces (at least 1 m^2) and higher values of Δt:

$$\alpha = 1.78 \sqrt[3]{\Delta t} \qquad (2.13)$$

These values are averages for the upper side of a surface that is warmer than the air, or the lower side of a surface that is colder than the air. In the reverse case, i.e. the upper side colder or the lower side warmer than the air, the value of α is about half the value indicated by Eqs. 2.12 and 2.13.

For a sphere with radius r^* we have:

$$\alpha = 0.0516 / 2r^* + 1.550 \sqrt[4]{\frac{\Delta t}{2r^*}} \tag{2.14}$$

Always, u is expressed in m/s and l and r^* in m. The above values are only approximate as far as the field situation is concerned. (For further details, see Eckert 1959; Monteith 1973, 1981 and Gates 1980.)

2.5 Air humidity

The discussion above on the thermal budget of soil and biological objects was simplified by leaving out the role of water and water vapour. But, of course, we should now include water. All soils, all air on earth and all organisms contain water and for these organisms water is essential. However, water also influences the thermal budget: it has an impact on the albedo, on the heat capacity, the thermal conductivity of solid objects and thus also on their thermal budget.

Moreover, water may evaporate and there is also negative evaporation, i.e. condensation (haze, mist, fog, dew, hoarfrost). Both positive and negative evaporation have a big influence on the temperature because water has a high heat of evaporation. Evaporation itself depends on temperature. With heat addition to a moist surface not only the temperature but also the evaporation will increase, which counteracts the warming of that surface.

Evaporation also depends on air humidity. This factor will not only determine the speed of evaporation, but also the ultimate water content of organisms, if they cannot compensate for water losses through water uptake from the soil (plants) or drinking (animals). Both evaporation rate and water content of organisms are important ecological factors.

Air humidity can be expressed in different ways. For us, two parameters are important, the amount of water vapour, symbol c, unit g/m^2, and water vapour pressure, symbol e, unit mbar or pascal. Pressure units are related as follows:

1000 mbar = 1 bar = 755 mm Hg = 1 atmosphere = 10^5 Pa (pascal)

Both c and e are absolute units, but in meteorology the notation 'absolute humidity' refers to c only. There are also relative units, the relative humidity RH and the saturation deficit SD (also called vapour pressure deficit VPD). These are always based on e. At a particular temperature a space can contain a maximum amount of water vapour (Table 3).

The vapour pressure related to the maximum amount is called saturation pressure, symbol e_{max}. Now we have the following relations:

$$RH = \left(e/e_{max}\right)100\% \tag{2.15}$$

$$SD = e_{max} - e \tag{2.16}$$

These new parameters could also be based on c, but in practice this makes little difference because there is an almost linear relation between c and e. If $e = e_{max}$, we say that the air is

Table 3. Relation between water saturation pressure e_{max} (mbar) and temperature t (°C) above ice and above water (including supercooled water).

Ice		Water													
t	e_{max}	t	e_{max}	t	e_{max}	t	e_{max}	t	e_{max}	t	e_{max}	t	e_{max}	t	e_{max}
−19	1.14	−19	1.37	0	6.11	20	23.37	40	73.78	60	199.3	80	473.7
−18	1.25	−18	1.49	1	6.57	21	24.86	41	77.80	61	208.7	81	493.2
−17	1.37	− 17	1.62	2	7.05	22	26.43	42	82.02	62	218.5	82	513.4
−16	1.51	− 16	1.76	3	7.58	23	28.09	43	86.42	63	228.6	83	534.5
−15	1.65	− 15	1.91	4	8.13	24	29.83	44	91.03	64	239.2	84	555.8
−14	1.81	− 14	2.08	5	8.72	25	31.67	45	95.9	65	250.2	85	578.1
−13	1.98	− 13	2.25	6	9.35	26	33.61	46	100.9	66	261.6	86	601.1
−12	2.17	− 12	2.44	7	10.01	27	35.65	47	106.2	67	273.4	87	624.9
−11	2.38	− 11	2.64	8	10.72	28	37.80	48	111.7	68	285.7	88	649.5
−10	2.60	− 10	2.86	9	11.47	29	40.06	49	117.4	69	298.4	89	674.9
− 9	2.84	− 9	3.10	10	12.27	30	42.43	50	123.4	70	311.7	90	701.1
− 8	3.10	− 8	3.35	11	13.12	31	44.93	51	129.6	71	325.4	91	728.2
− 7	3.38	− 7	3.62	12	14.02	32	47.55	52	136.2	72	339.6	92	756.1
− 6	3.68	− 6	3.91	13	14.97	33	50.31	53	143.0	73	354.4	93	784.9
− 5	4.02	− 5	4.21	14	15.98	34	53.20	54	150.1	74	369.7	94	814.6
− 4	4.37	− 4	4.55	15	17.04	35	56.24	55	157.5	75	385.6	95	845.3
− 3	4.76	− 4	4.90	16	18.17	36	59.42	56	165.2	76	402.0	96	876.9
− 2	5.17	− 2	5.28	17	19.37	37	62.76	57	173.2	77	419.0	97	909.4
− 1	5.62	− 1	5.68	18	20.63	38	66.26	58	181.5	78	436.6	98	943.0
0	6.11			19	21.96	39	69.93	59	190.2	79	454.8	99	977.6
												100	1013.2

saturated with moisture. This is a somewhat misleading expression because both e and e_{max} are almost independent of the air pressure and even of the presence of air. The parameter e_{max} increases rapidly and progressively with temperature (Table 3). At constant vapour pressure *RH* will decrease with increasing temperature and *VPD* will increase. Also, at constant *RH*, *VPD* will increase with temperature. If there is water in a closed space and $e < e_{max}$, there will be evaporation until saturation is reached. We may also say: evaporation will occur if $RH < 100\%$, i.e. if $VPD > 0$. This holds only, however, if the evaporating surface and the atmosphere have the same temperature.

A moss reaching a surface temperature of 50 °C in the sun ($e_{max} = 123$ mbar), dries up equally briskly in saturated air with a vapour pressure of 10 mbar ($t = 7$ °C, $RH = 100$ %) as in a desiccator with a constant humidity of 8% ($e / e_{max} = 10/123 \cdot 100$ %). A water surface with a certain temperature is in equilibrium with the e_{max} belonging to that temperature. Whether evaporation or condensation will occur, depends on whether this e_{max} is smaller or larger than the vapour pressure of the air. Evaporation can also occur in saturated air, while condensation can occur in unsaturated air.

If we take an ideal salt solution, the equilibrium vapour pressure is a certain percentage of e_{max}, dependent on the salt concentration; this percentage is almost independent on the temperature. If the air above the solution has the same temperature as the solution itself, the relative humidity of the air is at equilibrium equal to the percentage as indicated above. Each salt solution has a certain water potential P (equal to its osmotic value, measured in bar). For solutions that are not too strong we have (Zanstra & Hagenzieker 1977): $P = 14\,(100 - RH)$. However, anomalies occur fairly often (see Solomon 1951). Thus, salt solutions in contact with the atmosphere are able both to dissipate water vapour and to take it up, as organisms do, until an equilibrium is reached, either by increase or decrease of the salt concentration (and in a closed space also the *RH* of the air).

If an object is cooled down condensation will start as soon as the temperature of its surface has reached a level whereby $e_{max} = e_{air}$. This is called the dew point temperature t_{dew}. This can be used as an indirect way of measuring air humidity.

A direct method for the determination of air humidity is to treat air in a closed space with a strongly desiccating substance and to measure the decrease in pressure. Because evaporation cools down an object, the humidity of the air can also be measured by comparing the temperature of a sufficiently ventilated wet-bulb (t_w) and a dry-bulb thermometer (t_{air}) (see 5.2.3). The vapour pressure e follows from:

$$e = e_W - \gamma\,(t_{air} - t_W) \qquad (2.17)$$

with $e_w = e_{max}$ at t_w. The psychrometer constant γ is usually given as 0.66, when e is in mbar, which is somewhat too high for many psychrometers (cf. Ch. 5.2.3).

In a very dry period (July 1976 in the Netherlands) a t_w of only 14.3 °C was measured while the air temperature was 29.2 °C. Since at 14.3 °C e_{max} is 16.3 mbar, $e = 16.3 - 0.66$ $(29.2 - 14.3) = 6.4$ mbar. And, since at 29.2 °C e_{max} is 40.5 mbar, $RH = 15.8\%$ and $VPD =$ 34.1 mbar. The dew point is 0.6 °C, with $e_{max} = 6.4$ mbar. (Table 3).

The difference $t_{air} - t_w$ is called the wet bulb depression or 'psychrometric difference'. At midday in summer its value is usually between 5 and 10 °C; in very humid air under clouds it may be only 2 °C; on the other hand it can be as much as 25 °C in deserts. Evidently, at low air temperatures the deficit and the psychrometric difference cannot be great, because e_{max} itself is small. In winter the air in a centrally-heated house is always dry, because the cold outdoor air can only contain a little water vapour. If this air is warmed up it will have a large saturation deficit and a low relative humidity, unless there are many pot plants in the room.

2.6 Evaporation

All objects that contain water can dissipate water vapour. In the case of soil and water we speak of *evaporation*, which is a purely physical process. With regard to plants and animals we speak of *transpiration*, which is in part a biological process. In the case of vegetated soil it is difficult to separate evaporation and transpiration. So the joint water dissipation of soil and vegetation to the atmosphere is therefore called *evapotranspiration*.

Evaporation has two aspects, an energetic and a diffusion aspect. The evaporation of 1 g water needs about 2.5 kJ. In addition there is a diffusion, or more generally a transport aspect: the water vapour formed must be carried away.

Evaporation is a process which occurs via a boundary layer, just like heat transfer. If this layer is laminar, the water vapour transport is through diffusion, comparable to the conduction of heat. Diffusion is proportional to the concentration gradient in the boundary layer. In a somewhat simplified way, but still realistically as a basis for further consideration, we can say: the boundary layer governing the transport of heat via convection also governs the transport of water vapour via diffusion.

Heat transport in a stagnant or laminar moving air layer with the temperature difference related linearly to the thickness d, is described as:

$$H_{air} = \alpha \cdot \Delta t \text{ or } H_{air} = \frac{k}{d} \cdot \Delta t \text{ or } \alpha = \frac{k}{d} \text{ and } d = \frac{k}{\alpha} \quad (2.18)$$

The amount of heat dissipated, H_{air}, is expressed in W/m^2, α in $Wm^{-2}K^{-1}$ and k in $Wm^{-1}K^{-1}$.

For the diffusion process we have:

$$E = (D/d)(c_s - c) \quad (2.19)$$

and if the diffusion coefficient is defined differently:

$$E = (D'/d)(e_s - e) \quad (2.20)$$

where E is the mass transport ($gm^{-2}s^{-1}$), in this case the evaporation, D the diffusion coefficient (m^2/s), d the thickness of the boundary layer (m), c_s the maximum concentration of water vapour at the temperature of the evaporating surface (g/m^3) and c the concentration of the water vapour outside the boundary layer (g/m^3), while e_s and e express the water vapour pressures at the surface and outside the boundary layer respectively. On the basis of equation 2.18 we also have:

$$E = \frac{\alpha \cdot D}{k} (c_s - c) \quad (2.21)$$

The factor $\alpha \cdot D / k = D / d$ can be considered as conductance, analogous to the terminology in electricity, with E analogous to the strength of current and $(c_s - c)$ to the potential difference. Conversely, d / D is analogous to the resistance r with dimension sm^{-1}. The terms resistance, conductance and current strength are considered here per unit area. Of course, we can also write:

$$E = \frac{\Delta c}{r} \quad (2.22)$$

in which r has the dimension sm^{-1}, in all cases where water vapour is transported along a concentration gradient Δc.

Often the resistance notation is used for heat transfer as well. Instead of $H_{air} = \alpha \cdot \Delta t$ we can write $H_{air} = 1/r \cdot \Delta t$, in which r now has the dimension m^2KW^{-1}. In order to give r the dimension sm^{-1}, r is often defined in such a way that $H_{air} = (\rho Cp / r)\Delta t$, or $r = \rho Cp / \alpha$, where ρCp is the specific heat of air by volume: 1200 $Jm^{-3}K^{-1}$. The resistance notation can be used for heat transport via a boundary layer, for fully turbulent transport in free air, and for transport by conduction in solids and fluids. To express r in sm^{-1} usually has no advantages, only where the transport is fully turbulent does r have the same numerical value for heat and mass transport (cf. Ch. 3.2).

Resistances in series have the advantage that they can be added. Then it is very important to realize to which surface the resistance refers, especially in complicated situations, for instance with several concentric spherical surfaces.

Returning to equation 2.20, the diffusion coefficient D or D' is inversely proportional to the air pressure p. The thickness of the boundary layer d is inversely related to the square root of the air pressure and the square root of the wind velocity u (Eckert 1959); d is also strongly influenced by the roughness and hairiness of the surface of leaves. It follows then that E is proportional to $1/p$ and to $\sqrt{p \cdot u}$, and thus to $\sqrt{u/p}\,(e_s - e)$. The evaporation rate thus increases with wind speed and decreases with air pressure. In the high mountains evaporation is higher than in the lowlands. Moreover there is higher radiation, so that the temperature of the evaporating or transpiring surface (soil, plants) will differ more from the air temperature than in the lowlands. Thus e_s will be higher and also $(e_s - e)$. Essential to this is the ease with which water vapour moves, dependent on u and p, and on the difference in water vapour concentration or vapour pressures, and thus on the saturation deficit. As far as the equilibrium situation is concerned, i.e. the amount of desiccation reached in the end, the appropriate measure is not the deficit, but the relative humidity, if corrected for temperature (thus related to e_{max}), of the evaporating object.

The vapour pressure of the evaporating surface, and thus also the evaporation itself, are determined by the surface temperature, but evaporation also influences the temperature. As said, evaporation of water costs about 2.5 kJ/g. The temperature of an evaporating surface adjusts itself so that the transport and the energy aspect are in balance with one another. Below we will see how the evaporation of a wet surface can be expressed as the sum of two terms one of which is the net radiation and the other the product of α and the saturation deficit of the air.

Some additional remarks will now follow on the numerical values and mutual relations of some of the parameters used above. If we write evaporation in energetic units, we get (see also Eq. 2.21):

$$H_{ev} = \frac{D \cdot \alpha \cdot V}{k} \cdot \Delta c \qquad (2.23)$$

H_{ev} is the energy used for evaporation (W/m^2), V is the evaporation heat of water in J/g. The value of D is inversely related to the air pressure; it is also somewhat dependent on the temperature. With an air pressure of 1000 mbar and a temperature of 0 °C, D takes the value $0.212 \cdot 10^{-4}$ m^2/s; at 40 °C, $D = 0.272 \cdot 10^{-4}$ m^2/s. At 0 °C $k = 0.0243$ Wm^{-1}K^{-1} and at 40 °C $k = 0.0270$ Wm^{-1}K^{-1}. For our purposes we may consider k independent of the air pressure. The evaporation heat of water V at 0 °C is 2501 J/g and at 40 °C $V = 2406$ J/g.

As the vapour pressure e is almost linearly related to the vapour concentration c (cf. 2.25), it can be understood that we can write instead of Eq. 2.23:

$$H_{ev} = \frac{\alpha}{\gamma} \cdot \Delta e \qquad (2.24)$$

The factor γ is equal to $k \cdot \Delta e / (D \cdot V \cdot \Delta c)$ (cf. Eq. 2.23). Thus, γ gives, in energetic units, the relation between heat and water vapour transport in a laminar boundary layer. For turbulent transport (Ch. 3.2) we can define a similar factor which, somewhat confusingly has about the same numerical value. That γ is the psychrometer 'constant' will be understood from Eq. 5.2. The relation between c and e is given by:

$$c = 217 \frac{e}{T} \qquad (2.25)$$

with c in g/m^3 and e in mbar.

Given the relation with D it is also easy to understand why γ is inversely proportional to the air pressure. This holds for a turbulent boundary layer as well, albeit for other reasons (see Ch. 3.2). The temperature dependence of D, k and V as well as the relation between e and c make it clear that γ is somewhat dependent on the temperature.

At an air pressure of 1 bar the factor γ has a value of about 0.57 mbar/K (van der Held 1937). This holds for a laminar boundary layer. In the case of exclusively turbulent transport $\gamma = 0.66$ mbar/K (see Ch. 2.7). For a turbulent boundary layer, where not only turbulent transport processes occur, but also conduction and diffusion, we may expect an intermediate value.

We will now discuss the question of what determines the temperature and the evaporation of a wet surface, for instance a wet soil or the moist skin of a frog. Obviously, this has something to do with the humidity of the air as well as with the energy it takes.

The evaporation of a wet surface is on the one hand determined by Eq. 2.23, but on the other hand the energy balance of the surface should fit. These two conditions can be combined in the following energy balance:

$$H_{ev} = R_{net} - H_{soil} - H_{air} \tag{2.26}$$

Here R_{net} is the net radiation received by the surface, H_{soil} is the amount of heat carried off by conduction, H_{air} is the amount of heat dissipated into the air and H_{ev} again the amount of energy spent on evaporation. In another notation:

$$R_{net} - H_{soil} - \alpha(t_s - t_{air}) - \frac{\alpha}{\gamma}(e_s - e) = 0 \tag{2.27}$$

Later, an example will be given showing how H_{ev} and H_{air} can be calculated from this, if R_{net}, H_{soil}, $\Delta t = t_s - t_{air}$, and e_s are measured.

In order to calculate the temperature of a wet surface from more general meteorological parameters, we need a few changes in notation. Instead of R_{net} we write $R_{neta} - \Delta t \cdot \alpha_{rad}$ where R_{neta} is the net radiation received by the surface as if it had the temperature of the air. The latter can be calculated from the global radiation, the heat radiation received, the albedo of the surface and the air temperature. If the surface is Δt °C warmer than the air, R_{net} is approximately $\Delta t \cdot \alpha_{rad}$ lower than R_{neta}. The value of α_{rad} can be calculated from the increase in radiation from a black surface with a temperature increase of 1 °C. This is $4\,\sigma T^3$, the differential quotient of σT^4. For example, at 20 °C $\alpha_{rad} = 6\ \mathrm{Wm^{-2}K^{-1}}$.

Further, for e_s, which is the maximal vapour pressure at the temperature of the surface, we write (after Penman 1948):

$$e_s = e_{max} + s\Delta t \tag{2.28}$$

Here, s (in mbar/K) indicates how much the maximal vapour pressure increases per degree of temperature rise, and e_{max} is the saturation vapour pressure at air temperature. The value of s increases with air temperature (Table 3).

We can now rewrite Eq. 2.27 as follows:

$$\left(R_{neta} - \Delta t \cdot \alpha_{rad}\right) - H_{soil} - \alpha \cdot \Delta t - \frac{\alpha}{\gamma}\left(e_{max} + s \cdot \Delta t - e\right) = 0 \tag{2.29}$$

Or:

$$\Delta t = \frac{R_{neta} - H_{soil} - (\alpha/\gamma)(e_{max} - e)}{\alpha_{rad} + \alpha + s(\alpha/\gamma)} \tag{2.30}$$

Because $R_{neta} - \alpha_{rad} \cdot \Delta t - H_{soil} - \alpha\,\Delta t = H_{ev}$, we obtain after substituting Δt:

$$H_{ev} = (R_{neta} - H_{soil})\left(1 - \frac{\alpha_{rad} + \alpha}{\alpha_{rad} + s(\alpha/\gamma) + \alpha}\right)$$

$$+ \left(\frac{\alpha_{rad} + \alpha}{\alpha_{rad} + s(\alpha/\gamma) + \alpha}\right)\frac{\alpha}{\gamma}(e_{max} - e) \tag{2.31}$$

If we assume that R_{net} of the wet surface is known and $H_{soil} = 0$, we similarly derive from Eq. 2.27:

$$H_{ev} = \frac{s}{s+\gamma} \cdot R_{net} + \frac{\alpha(e_{max} - e)}{s+\gamma} \tag{2.32}$$

Or, because $e_{max} - e = s\,(t_{air} - t_{dew})$:

$$H_{ev} = \frac{s}{s+\gamma}\left(R_{net} + \alpha(t_{air} - t_{dew})\right) \tag{2.33}$$

This is Penman's (1948) well-known formula. In both cases the evaporation is determined by the sum of the radiation component and a component which is determined by the saturation deficit of the air and the value of α, which is dependent on the velocity of the wind (see Ch. 2.3.4).

If the saturation deficit is zero there may still be evaporation, if only R_{neta} is positive. If R_{neta} is zero there can also be evaporation if the air is unsaturated; in that case Δt is negative and the energy for evaporation is taken from the air.

We can also write α/γ (in Eq. 2.29 and 2.31) as $1/r$. This has advantages in those cases where the resistance is partly in the boundary layer and partly under the surface. In that case r can be split into r_i and r_b, the internal resistance and the resistance of the boundary layer respectively. Then, $1/r = 1/(r_i + r_b)$. This is the case, for instance with a leaf, an insect and a superficially drying soil. The value $1/(r_i + r_b)$ is always smaller, or even much smaller than α/γ, and r_i is independent on the air velocity and often large relative to r_b (see Ch. 4.1.3).

Thus, for a wet surface, temperature and evaporation can be calculated from surrounding factors in all cases where the heat conduction H_{soil} is negligible. For a dry surface α/γ becomes 0 ($r = \infty$).

In that case we can write for Δt:

$$\Delta t = \frac{R_{neta} - H_{soil}}{\alpha_{rad} + \alpha} \tag{2.34}$$

After this exercise we are now certain that the saturation deficit ($e_{max} - e$) and the difference between air and dew point temperature $t_{air} - t_{dew}$ are good measures of the dryness of the air, if we are dealing with water loss from a free water surface or from a living organism. Furthermore, here we have an approach that is applicable to the energy and water budget of all kinds of living and non-living subjects. Note that the resistance r here has the dimension mbar m^2W^{-1}.

Table 4. Energy balance of a wet mud surface (in W/m^2). Air temperature 21.0 °C, wet bulb temperature 17.2 °C, vapour pressure 17.0 mbar, surface temperature 33.5 °C.

In		Out	
Solar radiation	839	Reflected solar radiation	126
Heat radiation, sky and surroundings	321	Heat radiation	501
		Heat flux into soil	119
		Subtotal	746
		Evaporation heat and sensible heat dissipated to air	414
Total	1160	Total	1160

2.7 Thermal properties of the soil surface

2.7.1. Temperature and energy budget of a wet surface

While discussing the heat relations of a dry sand surface we mentioned the term evaporation pro memoria. We will now look at the opposite situation, i.e. a water-saturated soil and the soil surface covered with a closed film of water. We start again with the example of wet mud in the sun, where we measure the surface temperature, the components of the energy budget and other relevant parameters (as specified in Table 4).

We may immediately draw some conclusions. Of the net radiation received, $R_{net} = 533$ W/m^2; 119 W/m^2 disappears in the soil (H_{soil}). Since the surface is 12.5 °C warmer than the air, we may assume that the remainder, 414 W/m^2, is partly used for evaporation (latent heat H_{ev}), and partly dissipated to the air as sensible heat (H_{air}). In order to calculate how the available energy $R_{net} - H_{soil}$ is divided over the items latent heat (H_{ev}) and sensible heat (H_{air}), we make use of the similarity between heat and water vapour transfer.

The calculation is based on Eqs. 2.18, 2.26 and 2.27. Here t_s is the surface temperature (33.5°C), e_{max} the maximal vapour pressure at this temperature (51.5 mbar) and t_{air} and e_{air} the temperature (21.0 °C) and the vapour pressure (17.0 mbar) of the air respectively. If we take an average value of γ(0.615 mbar/K), we calculate that of the available energy 81.5% is used for evaporation and 18.5% is dissipated into the air as sensible heat. With these figures we calculate from Eq. 2.8 that $\alpha = 6.1$ $Wm^{-2}K^{-1}$, which is realistic for this situation without wind.

One may ask which temperatures we may expect on a wet surface in relation to the amount of solar radiation, wind velocity and temperature and humidity of the air. The value attained by Δt in the given circumstances was derived above (Eq. 2.30). The value of Δt is higher with increasing R_{neta} and lower H_{soil}, SD and α. Also, Δt is higher with lower s and α_{rad}. Both parameters increase progressively with increasing temperature. In other words, at lower temperatures, and if the other parameters remain the same, Δt can be higher than at higher temperatures. In view of the above it can be expected that with higher elevations of the sun the surface temperature of the mud can be 10 - 15 °C higher than the air temperature. To mention an ecological implication: Reed seed (*Phragmites australis*), requiring relatively high germination temperatures (van der Toorn 1972) will germinate early in the spring because the mud surface temperature is already high enough. If filamentous green algae such as *Vaucheria* form a thick layer at the surface of a water body, the situation can

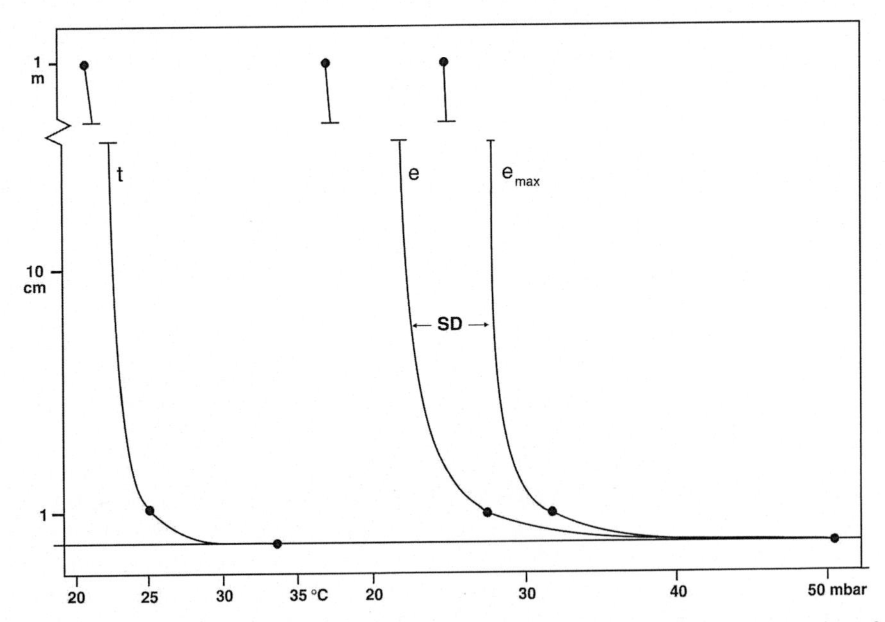

Fig. 10. Temperature (*t*), vapour pressure (*e*) and saturation deficit (SD) = e max - e as a function of the height above a wet surface in the sun.

be compared with that of wet mud. The algae absorb solar radiation and prevent water currents. For the same reasons frog spawn may experience temperatures 10 °C above the air temperature and therefore develop early. Schmeidl (1965) measured surface temperatures on a wet *Sphagnum rubellum* cover of 40.4 °C with an air temperature of 23.0 °C.

On the basis of the analogy of heat and water vapour transport there will be a strong decrease in water vapour pressure in the first mm above the surface, just as with temperature. Fig. 10 shows what can be expected on the basis of measurements at 1 cm. Here it was assumed that the temperature fall and the vapour pressure fall follow the same trend as the temperature fall in Fig. 8, as can be expected in situations without wind.

As soon as the top layer of the soil starts to dry up, the situation becomes more complicated. Heat dissipation still proceeds through the boundary layer. Transport of water vapour occurs first via the top layer of the soil and then via the air layer above; in other words, an extra resistance is included which is not dependent on wind velocity. This resistance soon becomes larger than that of the air above the surface. In this layer the resistance of the air is maximally equivalent to that of a still air layer of some mm thick; a dry soil layer of about 1 cm thickness will certainly have a much greater resistance than an air layer of the same thickness. Consequently, bare soil evaporation during dry periods is small after the superficial soil layers have dried up. We can also say that heat- and water vapour transmission are no longer connected. This is illustrated as follows: even after long dry periods the soil in an inland dune area is still moist at some cm depth. For example, after a dry period of three weeks in an inland dune area the following water content values were measured (volume%):

0 - 3 cm	3 - 5 cm	5 - 7 cm	7 - 10 cm	10 - 17 cm
0%	6.0%	6.4%	7.7%	8.0%

Only in the extremely dry summer of 1976 did the bare sandy soil of many places in the central Netherlands dry up to at least 10 cm deep: eggs of *Lacerta agilis*, laid at 10 cm depth, did not develop at all (H. Strijbosch, pers. comm.).

2.7.2 *High surface temperatures*

"In the hills the rocks will be so hot you could light your cigarette from them"
G. Durrell: My Family and Other Animals

Not only out of curiosity but also from a biological viewpoint it is interesting to know the highest temperatures which may occur at the soil surface under natural conditions. For many organisms the surface conditions are important and small mosses, lichens and seeds even form part of the surface.

An energy budget makes clear where and when the highest temperatures may be expected. In Table 5 two examples are given: on dark heath humus a temperature of more than 47 °C above the air temperature was measured whereas the difference with grey sand was only 22 °C. The energy budget shows that the level sand surface receives somewhat less radiation than the south-exposed humus, but this explains only part of the differences found. There is also a stronger reflection of solar radiation and a much greater heat transfer to the soil. Also, as one can easily calculate, the heat transfer coefficient for the sand is higher than for the humus. Of course, the error in calculating α is relatively large, because it is calculated from the heat released to the air, which is the factor that makes the balance in the energy budget.

Thus, high surface temperatures can be expected with strong irradiation in the absence of wind and on a poorly conductive dark substrate. In the example given solar radiation is far from maximal, suggesting that still higher surface temperatures may occur. Indeed, surface temperatures of 75 to 80 °C are regularly reached and occasionally over 90 °C is measured (Stoutjesdijk 1977a; Barkman, Masselink & de Vries 1977).

The influence of heat conductivity on surface temperatures also follows from a laboratory experiment (Fig. 11). Dry pine needles and a 10 cm thick slab of concrete are heated up by a radiation source, so that the radiation absorbed by both surfaces is equal.

The figure indicates how the absorbed radiation is divided between heat flow towards the substrate, radiation, and convection. Because in both cases, needles and concrete, the deeper layers of the 'material' become warmer and thus the vertical temperature gradient

Table 5. Energy balance (in W/m²) of a sand surface, and a bare heath humus. Surface temperatures 51.6 and 77.0 °C respectively; air temperature 29.4 °C. Still weather. Hum 60: energy balance estimated in W/m² for the humus at a surface temperature of 60 °C and α = 13.7 W/m² (more wind).

In	Sand	Hum	Hum 60	Out	Sand	Hum	Hum 60
Solar radiation	832	909	909	Reflected solar radiation	154	91	91
Long-wave radiation				Long-wave radiation	629	853	698
Atmosphere	252	252	252	Heat transfer to soil	252	112	72
Surroundings	119	119	119	Dissipated to the air	168	224	419
Total	1203	1280	1280	Total	1203	1280	1280

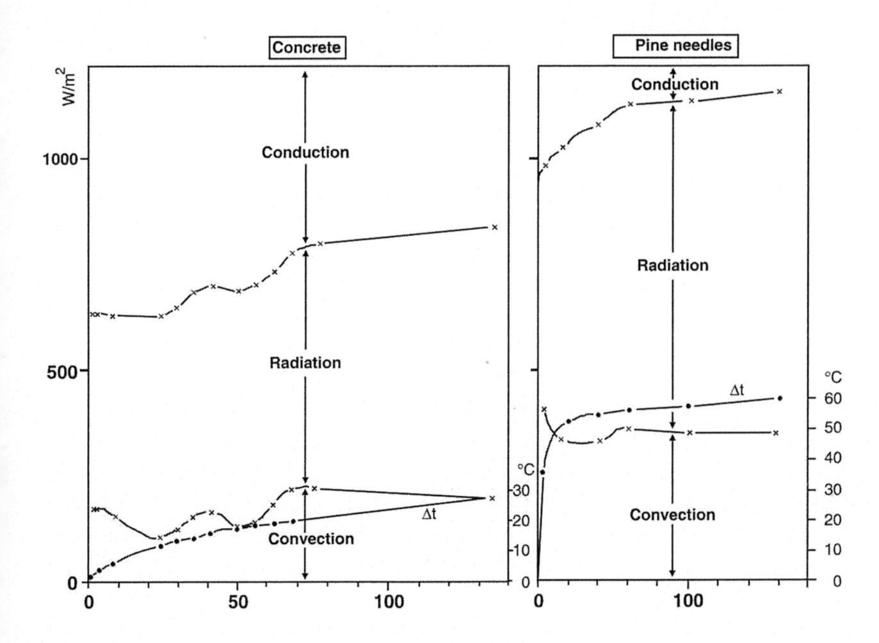

Fig. 11. Partition of absorbed radiation energy and temperature increases as function of time (in min.) for concrete and pine needles. Partly after Stoutjesdijk (1975).

becomes smaller, the heat flow will decrease with time, but with the needles this heat flow is always much less and moreover decreases faster. As a result more energy goes to radiation and convection and the surface temperature can be higher. After a short period of irradiation the relative differences are largest: after 3 min. the temperature rise on the needles is already 36 °C against only 4 °C on the concrete; after 20 min. the temperature excesses are 52 °C and 11 °C respectively. The relative differences become smaller with time but are still considerable. We can understand that under natural conditions solid rock (with similar properties as concrete) only shows a slow rise of temperature in the sun and that excessively high temperatures are not reached (Rejmánek 1971). The properties of sand are between those of these extremes, viz. concrete and needles (Stoutjesdijk 1975).

The influence of wind on the surface temperatures is exerted via the heat transfer coefficient α. It is instructive to estimate the surface temperature under the same conditions as those discussed above (Table 5), but with more wind, so that the heat transfer coefficient is three times higher. For heathland humus this means that α is not 4.7 but 14.1 $Wm^{-2}K^{-1}$. Clearly, the surface temperature has to be lower in the latter case, in order to even up the energy balance. If Δt decreases to 1/3 of the original values, a surplus to the energy budget would arise because long-wave radiation decreases and also because less heat flows into the soil. If we assume that the latter is proportional to Δt, we can approximate the resulting temperature by interpolation. With a Δt about halfway between the two extremes, the energy budget tallies roughly; the surface temperature is then 60 °C. (see Table 5). With a light wind such values are quite frequent, also with a somewhat stronger wind if there is some shelter, for example bare humus spots in between *Calluna* plants.

From the discussion of the energy balance we may also conclude that the effect of wind on surface temperatures is less if the heat flow to the substrate is stronger, especially with a brief gust of wind. If the conductivity of the material (substrate) is high, the temperature at greater depth is relatively high in such a case and if the surface cools off the heat flow may become negative, i.e. heat is leaving the substrate. In other words, the heat reservoir in the substrate acts as a buffer regarding temperature fluctuations at the surface. If we compare rock which is a good conductor with poorly conductive organic material, we see that the rock surface has both lower temperatures and less temperature fluctuation.

The highest values Δt can attain are over 30 °C in winter (Fig. 28) and may be as high as 60 °C in spring and summer. As a rule Δt may reach 50°C on dark organic material on slopes exposed to the sun from early spring to early autumn. On level ground values up to 40 °C may be expected during the summer. For greyish and yellow sand, maximal values of 35 and 30 °C on slopes and of 25 and 20 °C on level ground are found respectively.

High surface temperatures are often found on microsites such as ant-hills, where we measured over 90 °C, south-exposed margins of shrubs (up to 75 °C in front of a juniper shrub) and peat hummocks, where Schmeidl (1965) measured temperatures up to 77 °C (Δt = 47 °C). Vaartaja (1949) mentioned a surface temperature on bare, raw humus in a level forest clearing of 63.5 °C (Δt = 35 °C) and pointed to the influence of substrate type on surface temperature. Vaartaja cites N. F. Olofsson who measured a surface temperature of 72 °C against an air temperature of 32 °C on raw humus in a Swedish forest clearing at a latitude of 69° N. (Since he used a mercury thermometer, the real surface temperature must have been some degrees higher.)

The absolute record value of Δt we are aware of is found in the work of Chrenko & Pugh (1962) in Antarctica: a surface temperature of 61 °C with an air temperature of –7.5 °C! Although the surface was artificial, a black sweater, the high surface temperature effect is evident. Note that the effect is relatively strong at higher latitudes; this is because high values of Δt occur more easily at lower air temperatures as a result of the progressive relation between radiation and temperature.

A similar relation can be expected at high altitudes. Indeed, high values of Δt are known in the Alps. Turner (1958) measured a temperature of 80 °C (Δt = 52 °C) just below a raw humus surface on a south slope near the tree limit. It is known that such slopes cannot easily be colonized by plants (Aulitzky et al. 1961). Temperature maxima of 80 - 90 °C have been reported from desert areas, but usually without an indication of the air temperature. Convincing data on a desert surface are provided by Cloudsley-Thompson & Chadwick (1964): 82.5 °C at an air temperature of 42 °C. Remarkably enough there are also relatively low surface temperatures reported from deserts. Kessler (1974) for instance found no more than 20 °C difference between surface and air in the Sahara with the sun at 60 ° elevation. In the Negev desert we found maximal values of Δt of 19 and 26 °C on yellow loess and sand respectively, on dark rock only 16 °C with an air temperature of 28 °C and the direct rays of the sun about perpendicular to the flat substrate. In view of the favourable exposition and low wind speed the values measured should approach the maximally possible Δt values. These low values in deserts must be related to the high albedo here and the relatively high heat conductivity of sand and loess. The results obtained on the dark rock may have been inaccurate (some degrees too low) because it was impossible to get the thermo-couple used positioned perfectly on the rock surface; still the low Δt values must be considered real here and could be explained by the good heat conduction in the rock. The fact that sand or rock feel much warmer than organic material, despite the much lower surface temperatures, can be explained by the higher conductivity and heat capacity. (See also the quotation from Durrell!)

At northern latitudes very high temperatures have often been reported. This can easily be explained by the slow decomposition of litter there, leading to extensive dark, peaty surfaces. In arid environments we have the opposite situation. On the other hand, the

Table 6. Thermal parameters for diffe[illegible] volumes. Data on rocks after Rejmá[illegible] Eckert (1959), for pelts after Schola[illegible] $C = C_p$, the specific heat with cons[illegible]

Material	Pore vol.%	[illegible]	[illegible]	[illegible]	[illegible]
Copper	[illegible]	[illegible]	[illegible]	[illegible]	[illegible]
Basalt	[illegible]	[illegible]	[illegible]	[illegible]	[illegible]
Limestone	[illegible]	[illegible]	[illegible]	[illegible]	[illegible]
Granite	[illegible]	[illegible]	[illegible]	[illegible]	[illegible]
Quartzite	[illegible]	[illegible]	[illegible]	[illegible]	[illegible]
Sand	[illegible]	[illegible]	[illegible]	[illegible]	[illegible]
	33	[illegible]	[illegible]	[illegible]	[illegible]
	28	[illegible]	[illegible]	[illegible]	[illegible]
	23	20	[illegible]	[illegible]	[illegible]
	13	30	[illegible]	[illegible]	[illegible]
	0	43	[illegible]	[illegible]	[illegible]
Clay	43	0	1.19	[illegible]	[illegible]
	38	5	1.40	[illegible]	[illegible]
	33	10	1.61	1.10	[illegible]
	28	15	1.82	1.43	0.147
	23	20	2.03	1.57	0.145
	13	30	2.45	1.74	0.140
	0	43	2.99	1.95	0.134
Peat	90	0	0.25	0.033	0.052
	80	10	0.67	0.042	0.041
	60	30	1.51	0.130	0.049
	40	50	2.35	0.276	0.057
	20	70	3.19	0.421	0.0060
	10	80	3.61	0.478	0.0061
	0	90	4.03	0.528	0.0060
Flax fibers	98		0.048	0.042	0.158
Fur, various				0.031	
				0.052	
Air			0.00121	0.026	0.781
Water			4.19	0.586	0.0062
Ice			1.93	2.24	0.179
Snow	89		0.192	0.059	0.092
	78		0.384	0.142	0.101
	67		0.577	0.289	0.117
	56		0.769	0.490	0.133

ecologically-significant difference between desert and tundra or forest is not found in the maximum temperatures attained, but rather in the omni-presence of high temperatures in the desert which cannot be escaped by animals, or plants, whereas in boreal and arctic environments high temperatures occur only locally and during the less frequent sunny periods.

Until now we have discussed temperature relations on well-defined, closed surfaces such as rock, sand, peat, a compact dead grass cover, and compact needle litter. However, there are also more open, loose surfaces such as an open cover of dead leaves, or the loose dead biomass of bunch grasses such as *Molinia caerulea*. In these cases the absorption of solar radiation is spread over a thicker layer and the heat transfer to the air is greater; hence

"In almost all latitudes men dig into the earth
for an equable temperature"
Henry Thoreau: Walden

As we have seen, warming and cooling of the soil surface is not only dependent on external factors such as latitude, altitude, season, time of the day, weather conditions, aspect and shadow cast by vegetation, but also on the physical properties of the soil itself, first of all albedo, specific heat by volume ρC and heat conductivity k. The smaller these three parameters, the more the surface temperature will rise during radiation. This does not hold for deeper soil layers, however. Also for deeper layers a low value of ρC is favourable, but, other than for the top soil layer, a high k value promotes the warming up of deeper layers. A similar relation is true for cooling off. From the extensive treatment of the thermal behaviour of soils by van Wijk & de Vries (1963) we summarize the following.

If the temperature course of a surface is known and shows a sine form, roughly as is found during 24 h of clear sky, amplitude and phase-displacement of the temperature wave at a certain depth can be calculated from k and ρC. The concept of thermal diffusivity coefficient (also called temperature conductivity) a is introduced:

$$a = \frac{k}{\rho C} \tag{2.35}$$

Note the difference between temperature conductivity a and heat conductivity k; a has the same dimension as the diffusion coefficient, viz. m^2/s. One also uses the term damping depth, d_D, which indicates how temperature fluctuations at the surface are reduced downward in the soil. For d_D we have:

$$d_D = \sqrt{\frac{a}{\pi} \cdot 86\,400} \tag{2.36}$$

Here, 86 400 is the number of seconds per 24 h, because time is measured in seconds and we are dealing with temperature fluctuations over 24 h. When dealing with fluctuations over a year, we have to multiply by $\sqrt{365}$. Now, for the amplitude of the temperature wave at a depth z, δ_z, we have:

$$\delta_z = \delta_0 \cdot e^{-z/d_D} \tag{2.37}$$

where δ_0 is the amplitude of the temperature wave at the surface; e is the base of the natural logarithm; d_D is the depth where this amplitude has decreased to a fraction 1/e = 0.37 of the value at the surface.

The highest temperature at some depth in the soil is always reached later than at the surface. The time lag in hours of the daily temperature wave is indicated by

$$\frac{24 \cdot z}{2\pi \cdot d_D} \tag{2.38}$$

The time lag in days of the yearly temperature wave follows from:

$$\frac{365 \cdot z}{2\pi \cdot d_D} \tag{2.39}$$

Fig. 12 shows how the process of heat uptake and dissipation is expressed in the daily changes in the temperatures in a stabilized, poorly vegetated sandy soil in a coastal dune area. The sand is greyish and has an albedo of 14 %. As early as 9.00 h there is a strong decrease in temperature with increasing depth, although the temperature at 20 cm depth is still slightly higher than that at 10 cm. The cooling down at 10 cm during the previous night had been stronger than that at 20 cm depth and the warming up during the early morning hours had not yet compensated for this. Thus, between 2 and 10 cm there is a downward transport of heat, between 10 and 20 cm the transport is upward. Soon the temperature at 20 cm is lower than that at 10 cm and on all levels included in the measurement the heat transport is now downward. This should be compared with, for instance, the temperature curve in the soil for 14.20 h in Fig. 13.

The heat transport proceeds through conduction according to the equation:

$$H = k \cdot \frac{dt}{dz} \tag{2.40}$$

where H is the amount of heat passing through a horizontal plane (W/m^2), dt/dz is the temperature gradient (K/m) and k is the heat conduction coefficient ($Wm^{-1}K^{-1}$, or $Jm^{-1}s^{-1}K^{-1}$). In the curve for 14.20 h (Fig. 13) the temperature in the upper 2 cm decreases by 9.5 °C. Thus, the average temperature gradient is 425 K/m. The upper centimetres of the sand are absolutely dry. For dry sand $k = 0.29\ Wm^{-1}K^{-1}$ (Table 6), and thus we find $H = 0.29 \cdot 425 = 123\ Wm^{-2}$. This compares well with the values found by direct measurement (Chs. 2.1 and 2.7). The value found here, which is only meant as an illustration, is an average value for the heat flow in the upper 2 cm of the soil. We see that the temperature gradient decreases with depth. This can be partly explained by an increase of k with depth, because the moisture content of the soil increases with depth. It also means that the heat flow will decrease downward, in other words a certain layer will during the process of warming up, receive more heat from above than it releases downward and consequently its temperature will rise, but also the amount of heat retained is less for deeper layers.

We can now understand that the diurnal course of the temperature at the surface will become both weaker and delayed when transmitted to deeper soil layers. For example, at 20 cm depth the temperature maximum is only reached at 20.00 h. Until then, layers above 20 cm remain warmer. At 30 - 50 cm, the diurnal temperature maximum is reached at midnight, and at 5 - 10 m, the annual maximum is reached in the middle of the winter! Of course, we are dealing here with only very small temperature amplitudes, especially in the latter case.

For the example of the bare sandy soil the damping depth for the upper 5 cm calculated from $\delta_0 = 41$ °C (cf. Fig. 12) and $\delta_5 = 14$ °C, we find $d_D = 4.6$ cm. For the relation between

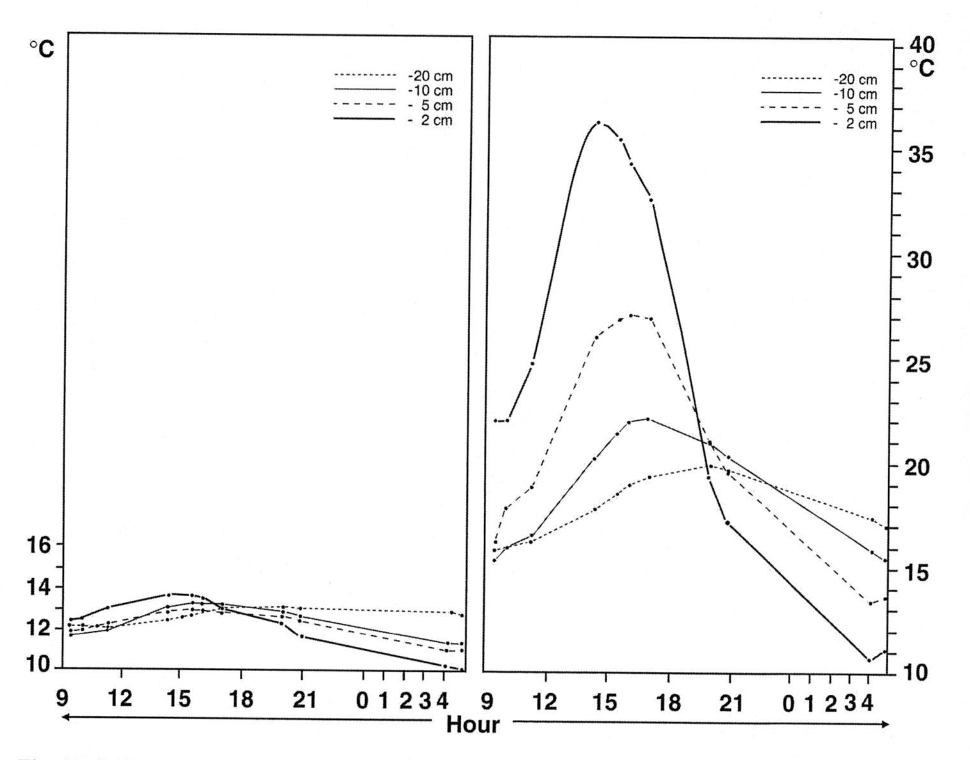

Fig. 12. Soil temperatures from 13 June 1958 at 9.00 h and 14 June 1958 at 5.00 h in unvegetated dune sand and dune sand under a *Hippophae* scrub. After Stoutjesdijk (1961).

the amplitudes of temperature waves at two different depths, e.g. 20 cm and 5 cm, we have:

$$\frac{\delta_{20}}{\delta_5} = e^{-15\,/\,d_D} \qquad (2.41)$$

From this we can derive d_D= 12.0 cm. This higher value can be understood from the higher water content of the soil in this case, meaning a higher value of k, while ρC increases with soil moisture to a lesser extent.

Fig. 12 shows some more details. The average temperature at –2 cm is about 20 °C, at –20 cm slightly lower. The maximum temperature at the surface is ca. 46 °C, the minimum is 4.5 °C, and the average is probably again slightly higher than at –2 cm. The fact that the average value decreases with depth points to a net downward heat transport over the 24 h at this time of the year, i.e. June. The warming up of the soil also becomes clear from a comparison of soil and air temperature: the maximum air temperature at 2 m is 17 °C, the minimum temperature is 5 °C. Generally, the average soil temperature, if taken over longer periods of time, comes close to the average air temperature. However, large differences may occur, especially with bare soil and in times of extensive heating when the net radiation over 24 h is strongly positive.

Fig. 12 also shows the great influence of vegetation. Immediately adjacent to the bare sand an open bush of *Hippophae rhamnoides* was found with the sandy soil underneath covered by a loose 5 cm layer of dead *Calamagrostis epigejos*. The plant material was completely dried out, as was the upper 3 cm of the sand. The depths measured are indicated relative to the sand surface. The temperature curves show basically the same picture as with

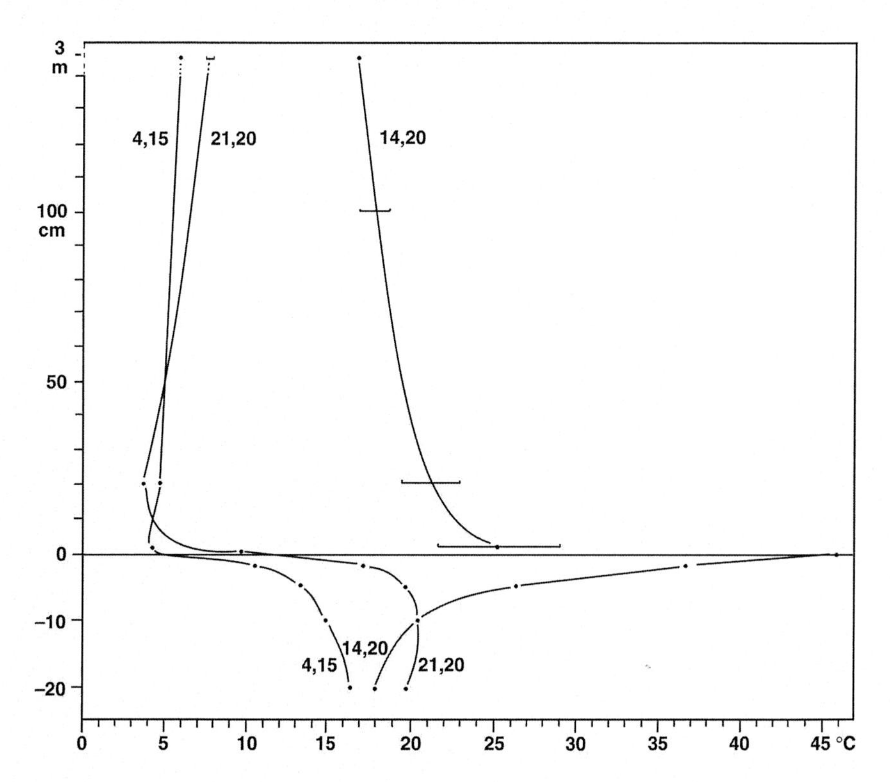

Fig. 13. Temperatures in and above dune sand 13/14 June 1958 at 14.20 h, 21.20 h and 4.15 h. See also Fig. 12. From Stoutjesdijk (1961).

the bare sand, but the amplitudes are much smaller and the average temperature is much lower at all depths.

From the curves indicating the temperature course at various points of time we may calculate how much heat the soil has taken up and released between two points of time. To take an example, the temperature of the top cm of the soil decreases from 42.5 °C to 7 °C between 14.20 h to 4.15 h the next morning (Fig. 13). From Table 7, giving the heat by volume of the sandy soil, we derive that a column of sand with an area of 1 cm^2 and a depth of 1 cm releases 35.5 J. One can calculate likewise for different layers of soil, or, if the heat by volume of the soil does not change much with depth, simply multiply the area between both curves expressed as K · m with the specific heat by volume. The amount of heat released by the soil is important for the temperature near the ground. According to our calculations the upper 20 cm of soil in the bare sand site released 1378 kJ/m^2 between 21.20 and 4.15 the next morning. Note that in the early part of the night some of the heat flow was directed downward but that later deeper layers also contribute to the upward heat flow. Similarly, we can calculate that the heat content of the soil under the *Hippophae* vegetation decreased by 385 kJ/m^2 over the same period and depth. We may state that these figures give a fair approximation of the total heat flow which left the soil overnight (see also Ch. 2.9).

Although less heat is released at night, the soil underneath *Hippophae* remains colder than the bare sand. This is because the bare sand starts the night with a larger heat content, and the measurements were taken at a time of the year where, at least over a longer period

Table 7. Water content and specific heat by volume (ρC) at different depths in bare dune sand and under *Hippophae*. After Stoutjesdijk (1961).

Soil type	Depth cm	Water content vol. %	Specific heat by volume $Jcm^{-3}K^{-1}$
Bare dune sand	0 - 6.5	0.0	1.40
	6.5 - 12	3.2	1.47
	12 - 22	4.2	1.49
Dune sand under *Hippophae*	0 - 3	0.0	1.33
	3 - 20	4.2	1.47

of time, warming up by day prevails over cooling by night. In a season when heat dissipation dominates, bare soils of good conductivity cool off more rapidly than soils of poor conductivity or soils covered by a layer of poor conductivity. This is wel known in horticulture where peat dust is put on top of bare soil.

Up till now water content has hardly been discussed. Still, this factor is very important, affecting the temperature via the albedo, the specific heat by volume and the conductivity, as well as heat used in evaporation and heat set free by freezing. Wet sand and wet peat have a much lower albedo than dry sand and peat, and thus absorb more solar energy. On the other hand these wet substrates evaporate more, and as this effect is greater, they warm up more slowly in the sun. It is also important that both ρC and k increase with increasing water content; consequently the warming of the surface in the daytime and the cooling off at night are much less than on dry substrates. Peat has a lower ρC than sand. Particularly with dry peat ρC and k are very low, so that the surface temperature becomes very high by day, and very low by night. Wet peat has a greater specific heat by volume than wet sand. This can be explained by the much higher water capacity of peat. With peat, and especially *Sphagnum* peat (with a very low ρC and k), drainage has a disastrous effect on the temperature at and immediately above the surface. This may well explain the considerable damage done by night frost to potato crops on reclaimed peat soils in NW Europe.

Together with the water content ρC increases linearly, but the increase in k proceeds more gradually. As a result an increasing water content, though favourable first, is not always favourable for the warming up of the subsoil. In other words, both very dry and very wet soils will warm up more slowly in the summer (in the day time), and will cool down more slowly in the winter, the first because of the low heat conduction, the latter because of the large specific heat by volume.

Table 6 (above) presents thermal data for some types of rock, soil and other material. For sand and clay the damping depth d_D first increases with increasing water content because the conductivity first increases more rapidly than the specific heat by volume, while later on we have the opposite situation. With dry organic material the damping depth decreases with density, this because ρC increases much more than k. For the same reason d_D first increases with increasing water content, but if the water content rises over 10% k rises faster than ρC and d_D increases again.

The low values of d_D for peat, relative to sand and clay at a higher water content may be somewhat unexpected. This is mainly because of the low values of k for peat (irrespective of its water content). With peat the heat conductivity is mainly dependent on the amount of water and water has a low k. With mineral soils the value of k for the parent material is high and the role of the soil water is mainly to promote heat contact between the soil particles. Very loose organic material with a low density, such as the flax fibres in Table 6, which can be compared with a loose layer of dead grass, has a damping depth similar to that of rock.

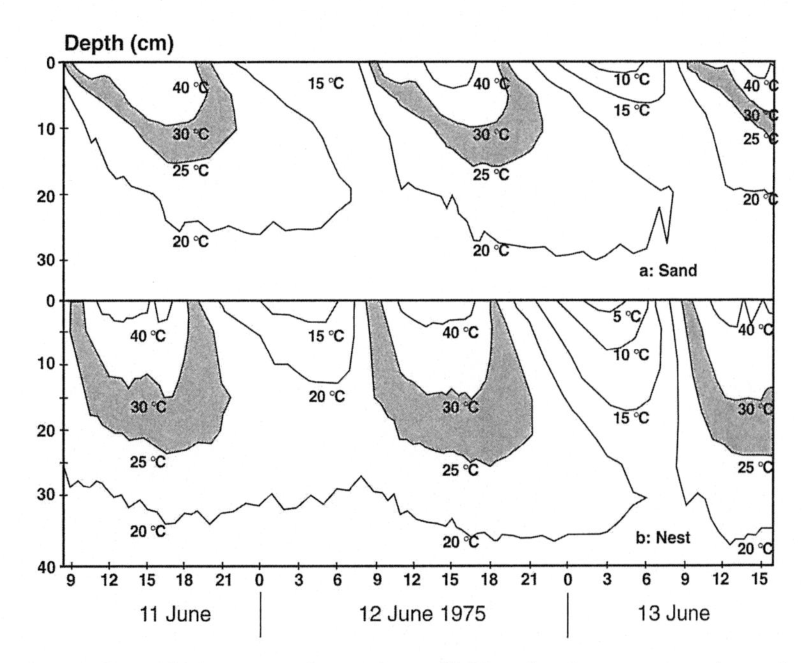

Fig. 14. Isotherms in sand (a) and in the centre of an ants' nest (b). Note that the temperature interval of 25 - 30 °C is larger in the ant hill than in the sand (after Brandt 1980b).

This is because the much lower conductivity of the organic material is compensated for by the much lower specific heat by volume; also note that the organic material layer shows much larger temperature fluctuations at the surface.

The significance of the parameters k and d_D is well illustrated by the thermal behaviour of an ant-hill built of organic material, made by a species of the genus *Formica*. This material conducts heat badly and therefore the surface of the ant-hill is rapidly warmed up by the sun and may reach high temperatures. Frevert (1957) describes how wild pigs (*Sus scrofa*) in Poland make use of ant-hills in winter by using them as an insulating and dry under-layer in their lair. Here it it is the low k which is of importance.

On the other hand, Brandt (1980a, b) points to the considerable damping depth of 20 - 30 cm of an intact ant-hill (here of *Formica polyctena*), which is much more than for sand (ca. 10 cm) or for relatively loose ant-hill material. Apparently, the structure built up by the ants is important here. Regarding its thermal behaviour, the intact ant-hill can be compared with rock and Brandt (1980a) interprets this as an indication of the evolution of the nest of *Formica polyctena* from nests built in cracks or cavities in solid rock. Brandt also points out that the larger damping depth in the nest provides the ants with more space to keep the pupae at the optimal temperature of 25 - 30 °C (Fig. 14).

Van Wijk & Derksen (1963) give a theoretical explanation for the thermal behaviour of soils consisting of two different layers. A frequently occurring combination is that of a moist sandy soil covered by a dry layer of litter or moss. It will be clear without any calculation that the insulating layer does not transmit much heat, which is received by a soil with a relatively high specific heat.

The soil underneath the insulating layer will show only small temperature fluctuations as compared with the bare sandy soil. This can be confirmed by calculations. A similar situation is reached when snow covers a relatively moist soil, which is quite often the case

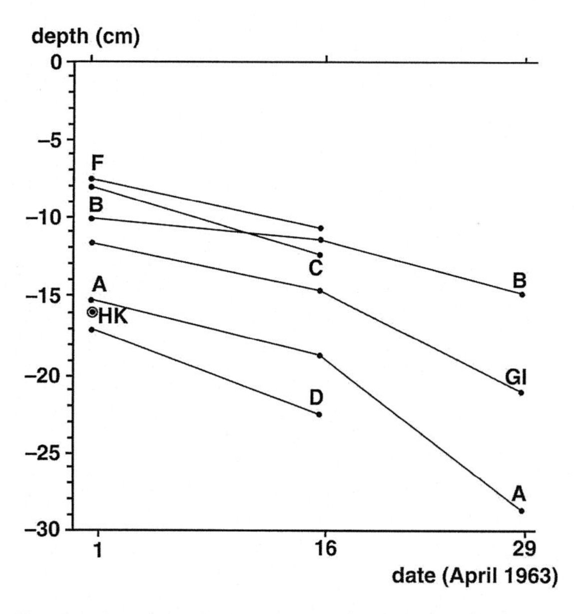

Fig. 15. Frost depths in peat bog in Drenthe, the Netherlands on three dates in April 1963 (J. J. Barkman unpubl.). A. Bare dry turf; B. Dense *Calluna* on dry turf; C. Moderately dense *Molinia* grass cover on moderately dry turf; D. Bare dry turf; F. Very dense *Molinia* vegetation on wet turf; Gl. Moist hummocks in living peat bog; HK. Wet hollows in living peat bog.

at the beginning of the winter.

When freezing starts, the heat conductivity of the soil is decisive for the speed and extension of the frost in the soil. Especially during short periods of frost an isolating organic layer can protect the subsoil from freezing. This means, for instance that some insect life continues in the soil and that birds such as the woodcock (*Scolopax rusticola*) still can probe for food into a soil covered by moss and dead grass, or that moles (*Talpa europaea*) are still active. An example of the differences which may be found is presented in Table 8, comparing the situation in a relatively open sandy soil with a sparse cover of the 1 - 2 cm tall acrocarpous moss *Tortula ruraliformis* and some therophyte rosettes, with a soil covered by litter and a thick layer of the pleurocarpous moss *Calliergonella cuspidata* under a canopy of the dwarf shrub *Salix arenaria*. The measurements were taken on a bright winter day around midday with an air temperature just above zero. In the preceding clear night the temperature had fallen to – 4.5 °C at 1.5 m, to – 10.4 °C at 10 cm and to even –12.0 °C in the *Salix* vegetation. The top 2 cm of the exposed soil thawed, but the deeper layers remained

Table 8. Temperature in dune soils with an open and a closed moss cover respectively (see text). Date 17 January 1957, time 13.00 h; air temperature +0.8 °C.

Depth (cm)	Temperature (°C)	
	Open soil	Covered soil
0 - 2.5	0.4	1.9
5	– 0.2	4.1
10	– 0.2	4.6

Table 9. Average frost depths (cm) in raised bog with different vegetation and water levels.

Vegetation	1 April	16 April
Dense vegetation	9.7	12.7
Sparse vegetation	16.1	18.3
Dry peat	13.1	16.4
Wet peat	11.9	13.8

frozen, whereas the temperature under the thick moss cover was above zero. Apparently, the cold had penetrated deeper in the bare well-conducting sand than under the insulating moss cover. In winter, when the net radiation over 24 h is negative, such temperature differences can be maintained the whole day.

The water content of the soil is a major factor determining the speed of freezing, not only because it influences conductivity and specific heat (ρC), but also because much heat is released with the freezing of water (334 J/g). In wet peat frost penetrates only very slowly, but once frozen thawing also proceeds slowly. It should be noted though that freezing and thawing are not simply reverse processes. Ice has completely different values for ρC and k than water (Table 6). The ρC value for ice is less than half that of water, but the k-value is nearly four times as high. Especially for wet peat, the thermal characteristics of which are determined mainly by its water content, heat is carried away during freezing by a well-conducting layer of ice, whereas during thawing heat is conducted downward through a poorly conducting layer of stagnant water. With mineral soil this effect will be much less because the parent material is more important here, with lower water capacity and larger k, and because even wet mineral soils consist largely of soil particles (Table 6).

The thawing process in a peat bog is very slow because of the high water content and the relatively low k; as a result, peat soil remains cold until late spring, although the temperatures above the surface, and just below, are much higher. As an example, after the severe winter of 1963 in NW Europe, ice was observed in the peat as late as 29 April in the E Netherlands, while the air temperature near the surface had been over 20 °C on several occasions. Consequently, bog plants are exposed to a very high water stress: strong transpiration demands combined with a low water uptake (water uptake through roots of vascular plants nearly reaches zero at temperatures below 5 °C; see also Table 6).

Thus, physiologically, the peat bog is a very dry environment in spring. Moreover, bog plants must be adapted to late night frosts in spring and very early ones in autumn. This implies high frost resistance in the young shoots in spring, or a very short growing season in an environment which is already extreme because of its low nutrient status. Some adaptations observed in bog plants are: *Molinia caerulea* and *Rhynchospora* species develop late in spring; *Drosera* species root very superficially, thereby avoiding the cold soil, while they compensate for lack of nutrients through carnivory; dwarf shrubs such as *Oxycoccus*, *Andromeda* and *Erica* are evergreen with leathery leaves and roots which are insensitive to low temperatures while taking up water (Firbas 1931).

Small differences in the soil temperature regime can thus be important for the microhabitats found in a peat bog: dry hummocks will freeze earlier than wet hollows, at least in temperate regions. In subarctic regions with much snow and frequent winds the snow can be blown off the hummocks and the hummocks can become entirely frozen, while the hollows remain protected by snow cover. Ice may remain in the hummocks throughout the summer, as in palsa bogs described for N Finland and N Sweden (see e.g. Sjörs 1965). Such a snow effect does not occur in the more temperate regions of the Northern Hemisphere, but another complication may arise there: the hummocks are usually covered with a dense plant cover of *Erica*, *Calluna* or *Eriophorum vaginatum*, and the hollows are covered by mosses, usually *Sphagnum* with little vascular plant cover. The dense vegetation restrains heat transport and thus the wet hollows may freeze and thaw earlier than dry hummocks.

Fig. 15 shows the results of the writers' own measurements on frost depth after the hard winter of 1963 in Drenthe, the Netherlands, in both active peat bog and deteriorated bog. It appears that the density of vegetation is more important than the water content of the soil. On April 1 frost depth in open wet hollows was 16 cm, on average, against only 11.5 cm on the dry hummocks. On April 16 no ice could be traced any longer in the hollows. Thawing in the dense vegetation types proceeded slowly, on bare dry turf relatively fast. If we average these data over wet versus dry soils respectively and also over densely vegetated versus open vegetation (Table 9), we see indeed that dry bog thaws faster than wet bog, and open vegetation faster than dense.

As a comparison, in a nearby garden the ice had disappeared three weeks earlier.

2.9 The situation at night

2.9.1 Energy balance, night frost

> *"Moreover, we have often observed that cold seems to descend from above; for, when a thermometer hangs abroad in a frosty night, the intervention of a cloud shall immediately raise the mercury ten degrees; and a clear sky shall again compel it to its former gauge"*
> *Gilbert White (1788): The Natural History of Selborne*

Regarding radiation, the situation during a clear night is a simplification of what happens during the day. Solar radiation disappears and there is only heat radiation left. There is an incoming flux of long-wave radiation from the sky and an outgoing flux of long-wave radiation from the soil surface. The latter is always stronger than the former during a clear night. Consequently the net radiation is negative (Table 10). This negative balance of long-wave radiation is also found in the daytime, but then there is a strong gain through solar radiation (see also Ch. 2.3.4).

In order to obtain a total picture we have to look at the energy balance at the soil surface

Table 10. Energy balance in W/m^2 of a small heathland and bare sand.

In	Heathland	Sand
Long-wave radiation from sky and surroundings	264.5	264.5
Heat flow from the soil	20.2	83.9
Subtotal	284.7	348.4
Withdrawn from the air	25.8	5.4
Total	310.5	353.8
Out		
Radiation emitted	310.5	353.8

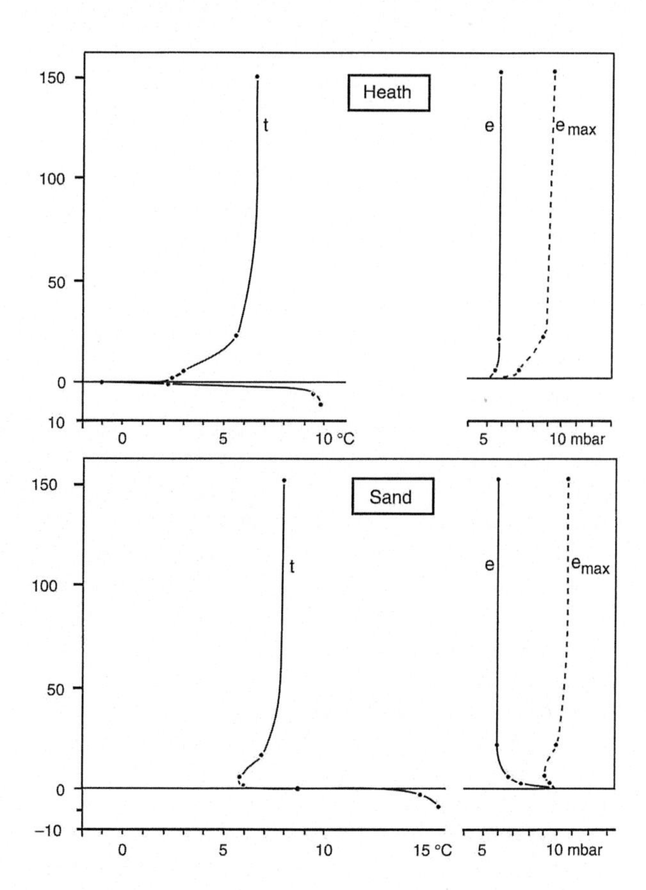

Fig. 16. Temperature and vapour pressure profiles above a heath and a sandy soil on a clear still evening.

and the temperatures in the air above and in the soil underneath. Fig. 16 gives data for a grassy heath where dry grass remains and moss covers a felt-like humus-rich sod.

After a windy, partly cloudy day in the beginning of May, with a maximum air temperature at 1.5 m of 13 °C, the air becomes clear towards sunset. The wind weakens and at 21 h, 45 min after sunset there is hardly any wind left at the soil surface. The temperature at the surface has already fallen to –1 °C; no dew has been formed yet. The air temperature at 1 cm above the surface is 2.6 °C and increases at higher levels to reach 6.6 °C at 1.5 m. The evening temperature profile of the air thus mirrors the daytime profile. Such a temperature profile is called an inversion. As there is gradual warming up of the atmosphere from below to above during the day, there is also a cooling off at night. The cooling extends several hundreds of metres into the air, thus the inversion can also be seen on large-scale temperature profiles of the whole atmosphere.

Downward into the soil the temperature rises rapidly: at –10 cm the temperature is 9.8 °C. The soil temperature profile is thus inverted as well.

The profile of the vapour pressure is almost a straight line. But, as a result of the lower temperatures near the ground the saturation deficit there is much lower than higher up. The temperature at the surface is even somewhat below the dew point of the air above. If the slight increase of the vapour pressure with height is real, it would point to the beginning of dew formation, that is to say a downward water vapour transport, even though dew can not yet be observed visually.

The temperature profiles show that there is a continuing heat transport to the surface, both from the air and the soil. What happens with the energy arriving at the surface? The answer is found in the energy balance (Table 10). Only part of the energy lost by the surface through radiation (246.5 out of 310.5 W/m^2) is compensated for by long-wave radiation of the atmosphere. The remainder (46 W/m^2) must be compensated for by heat taken by the cold surface, from both the air above and the warmer soil beneath. The heat flow from the air is somewhat greater than that from the soil, but small when compared with the amount of heat dissipated in the daytime by a sun-warmed surface. That strong temperature gradients still exist is due to the small amount of turbulent exchange. During the day high temperatures of the lower air layers promote this exchange, while at night the situation is reversed: the colder and heavier air is not easily mixed with higher warmer layers of air and consequently the atmosphere is stable.

Again it is possible to calculate a heat transfer coefficient according to $H_{air} = \alpha \cdot \Delta t$, where Δt is the temperature difference between the air at 1.50 m and at the surface. In this case $\alpha = 25.8 \ Wm^{-2} / 7.7 \ °C = 3.4 \ Wm^{-2}K^{-1}$, a low value compared with what we can find during the day, as we can understand because of the stability of the air at night.

Between the surface and 1 cm above it there is a temperature difference of 3.6 °C. Upwards the temperature gradient becomes less steep, as can be expected because the exchange rate increases higher up. We can calculate how much heat is transported by conduction in the lowest cm: $3.6 \ °C \cdot 2.59 \ Wm^{-2}K^{-1} = 9.31 \ Wm^{-2}$ (the heat conduction coefficient for air is $0.0259 \ Wm^{-1}K^{-1}$. This is much less than the absorbed heat in the energy balance, which indicates that transport through turbulence is still of some importance, despite the still atmosphere. Nyberg (1938) found that at 1 mm above the snow on still clear nights there was an unmistakable effect of turbulence. As stated above, the vapour pressure profile indicates the beginning of dew formation, which has been left out of consideration in the energy balance because it is negligible in energy terms.

During overcast nights with low clouds the downward atmospheric radiation is about equal to the outgoing radiation, and the cooling-off at night proceeds very slowly. Sometimes a cloud cover can form very rapidly during a clear night. At 1.50 m the temperature may then rise 5 °C or so in 10 minutes. Then, the driving force maintaining the inversion disappears and through the turbulent exchange the cold air from beneath is mixed with the warmer air from higher layers.

Immediately after the measurements on the small heath (Fig. 16) the temperature was measured on an area of bare sand of ca. 10 m × 10 m at about 10 m distance from the former measurements. Soil temperatures are higher and the surface temperature is 8.6 °C, almost 10 °C higher than on the heath. The surface is still warmer than the air above it. The air is coldest at 5 cm, though still 3 °C warmer than above the heath, and from here temperature increases again with height; at 1.50 m the temperature was 8 °C, 1.4 °C higher than at the same height above the heath. The simplest explanation of the temperature profile seems to be that cold air from the surroundings flows in and is warmed up from below by the relatively warm sand. However, there are cases where a more complicated and still not completely satisfactory explanation must be given (Berenyi 1967). It is important to know that the temperature profile described has always been observed above bare soil, where a major part of the radiation loss at night is compensated for by the heat flow from the soil. It follows from the vapour pressure profile that the sand, though dry at the surface, releases water vapour to the air. As a result of the higher temperatures of the air the saturation deficit is higher than above the heathland on all levels.

The net radiation of the bare soil is more negative than that of the heathland due to the higher surface temperature. However, this loss is almost entirely compensated by the heat flow from the soil, which is much greater here than in the heathland case. The expression 'almost entirely' is based on the measurements of the various parts in the energy balance (Table 10). If we look at the temperature and vapour pressure measurements (Fig. 16) we see that the surface gives off some heat and latent heat in the form of water vapour to the air. This means that the negative net radiation is 'more than entirely' compensated for. The discrepancy can be easily understood from small errors in the measurement of the components of the energy balance, mainly in the heat flow from the soil. Also when calculated over the whole night, it is quite usual that with a bare soil the heat loss through radiation is wholly or largely compensated for by the heat flow from the soil; the rest is taken from the air. On the other hand, with badly conducting soils only a relatively small part of the heat loss through radiation is compensated for by the heat flow from the soil. Thus, for temperatures at the surface and above, the thermal properties of the soil are of great importance.

On a clear evening in spring or summer, the air above the heath can cool off quickly. For the human observer going from heathland to inland dunes the temperature difference experienced is great, just like when one moves from the countryside into a town. (In both cases the difference at 1.50 m is rarely more than 3 °C.) The explanation is the same in both cases. If the two contrasting types of vegetation at either side of the boundary are extensive, the temperature difference noted can be as high as 1.50 m above the ground. The temperatures above dune sand increase from the edge between the sand and the heath towards the centre of the sandy area. If conditions for a low 'radiation fog' are favourable, we will see this kind of low mist above the heath but never above the sand (Stoutjesdijk 1959).

It will now be clear from the foregoing considerations of energy relations, that the temperature minimum reached on clear nights without wind is strongly influenced by the thermal conditions of the soil. The smaller the heat flow from the soil, the more heat will be drawn from the air and the lower will be the temperature at and above the surface. Sand will remain warmer at the surface than peaty soil, and rock remains warmer than sand (De Felice 1968). In the sequence rock, sand, peat, the surface temperature in the daytime increases and thus the daily amplitude of the surface temperature will increase markedly.

The occurrence of extremely low minima above badly conducting soil has received more attention than the absence of such minima above rock surfaces, although both phenomena are biologically important. The Swedish author Rolf Lidberg compares rock with an old-fashioned tile stove which accumulates heat during the day which it dissipates at night. Stony substrates with a southern exposure in Scandinavia are known for their particular flora, containing southern and continental elements. Examples of such plant species are the grass *Stipa pennata*, occurring in isolated patches in W Sweden (e.g. Fridén 1965) and *Fumana procumbens*, typical of warm stony places on Öland and Gotland (Pettersson 1965; Bengtsson et al. 1988). We should add that it must be biologically significant that the day temperature at the rock surface is not extremely high, i.e. damage by too high day temperatures is not likely.

Well-known soils with very low minimum temperatures are drained peat bogs. Dry organic material is a bad heat conductor and has a low specific heat by volume. Another example is a heath soil with a shallow layer of peaty humus (Stoutjesdijk 1959). Rough grassland with a thick layer of dead organic material and forest clearings with *Deschampsia flexuosa* or *Molinia caerulea* also fall into this category (Geiger 1961).

In such habitats night frosts, i.e. frost at 10 cm above the surface, are frequent. In view of the importance of night frost for agriculture and forestry (plantations) several meteorological stations include in their standard programme temperature measurements with minimum thermometers at 10 cm height above short-cut grass in special instrument shelters.

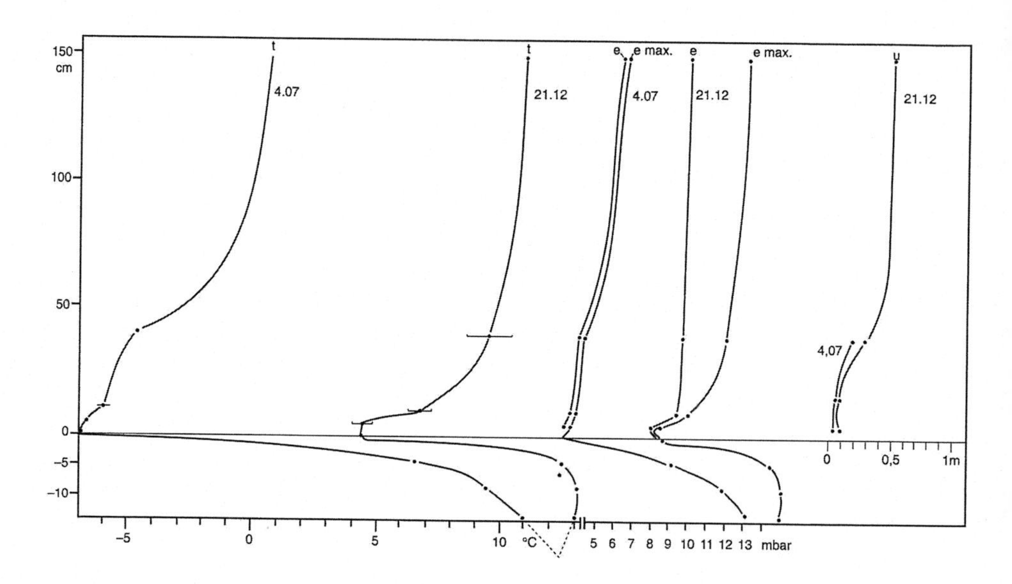

Fig. 17. Temperature and vapour pressure profiles at the beginning and at the end of a clear night in a dune depression in the inner Voorne dunes, the Netherlands, 11 June 1958, 21.12 h and 12 June 1958, 4.07 h. After Stoutjesdijk (1961).

Such measurements are often carried out on mineral soils and most probably minimum temperatures above other soils with a low conductivity as mentioned above, will be still much lower. In the Netherlands minimum temperatures at 10 cm in natural habitats with poorly conducting soils are known to be at least 5 °C lower than at the standard stations of the State Meteorological Service. For example, in a dry reed bed (*Phragmites australis*) in the polder Zuidelijk Flevoland in the central Netherlands the minimum temperature at 10 cm during the period 28 April to 5 May 1976 was as low as –15.0 °C against –9.4 °C at a nearby meteorological station. Note the extreme character of these night temperatures even in the middle of spring. Another case of extreme soil conditions is related to snow cover, which drastically reduces heat flow from the soil. Because a snow cover usually extends over large areas, the impact of snow on minimum temperatures extends to the standard observation level (1.50 m).

Woudenberg (1969) has analyzed data from meteorological stations with near-the-ground minimum temperature measurements available for agricultural sites and found that grassland areas had lower minima than fields.

Figs. 17 and 18 show temperature and vapour pressure measurements in a dune depression with an open grassland vegetation of *Calamagrostis epigejos*. The soil was covered by a 4 cm thick layer of completely dry dead grass. (Measurements in the daytime on the same spot are presented in Fig. 34.) The net radiation loss was 76.9 W/m^2 in the evening and 48.9 W/m^2 the following morning. On average the net radiation loss for the whole night is 1591 kJ/m^2. From the temperature curves and the specific heat of the soil it can be estimated that 569 kJ/m^2 is derived from the soil, or 36%. The remainder is thus taken from the air as sensible heat and as heat released by condensation and freezing, i.e. the formation of dew and hoarfrost. A dew deposition of 0.22 mm was calculated (see Section 2.9.2). As we see the temperature in the depression fell to –7 °C at 0.3 cm height. This was

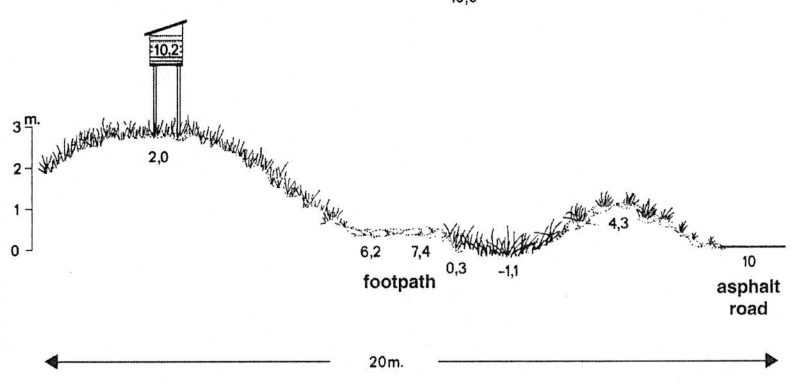

Fig. 18. Surface temperatures in the same dune depression as in Fig. 17 on 23 August, 1972, 21.20 h. Local elevation differences are indicated. Temperature in the instrument shelter was 10.2 °C, the effective radiation temperature of the sky dome was –10.6 °C. The coldest place with –1.1 °C is the place where the measurements for Fig. 17 were done.

a very local effect: on a potato-field at 100 m distance there was no sign of night frost. Two coastal meteorological stations at ca. 30 and 70 km distance had comparable temperatures of +7 and +8 °C. Only at a few inland sites was some night frost (–1 °C) reported during that night. Even in the middle of the summer, night frosts occur in this depression: 17 times in the months of July and August 1956 - 1964, with minima between – 0.1 and –2.7 °C (average –1 °C). Comparable temperatures in two coastal meteorological stations at 30 and 40 km distance were on average 6 and 8.5 °C higher.

Minimum temperatures are not only dependent on the heat relations in the soil but also on the local topography. Depressions are notorious amongst foresters and gardeners for their night frost ('frost pockets'). The common belief is that cold air flows into the depression from the surroundings. Indeed, cold air from the surroundings moves to the depression, but meets air at least as cold near to the ground, which will not be moved by the incoming air. However, above the very cold air in the depression warmer air may be displaced by the incoming cold air and this facilitates the further cooling off of the lower air layers because of a decrease in atmospheric heat radiation. Moreover, if the total cold air layer gets thicker, the exchange with the warmer layers above becomes more difficult. Turbulence and mixing of air layers are also suppressed because depressions are better protected against wind. Furthermore, depressions usually have a thicker humus layer than the surroundings and more litter. Finally, solar radiation reaches the depression for a shorter period, though this will only be noticeable in deeper depressions.

The situation on hill tops contrasts with that of depressions: here cold air will flow downward and the air near the surface will not cool off that much. In the case of mountains the downward movement of cold air on clear nights may be strong enough to be experienced as wind (katabatic wind,'valley wind', German 'Hangabwind').

An extreme form of cold depressions is found in calcareous mountains where deep 'dolines' are found. A famous example is found on the Gstettner Alm near Lunz, Austria at 1270 m a.s.l. At the bottom of the doline temperatures were 30 °C lower than on the edge, only 60 m higher up. The vegetation zonation is reversed here. Higher up, the slopes of the doline are well-developed spruce trees, *Picea abies*, half-way down badly developed,

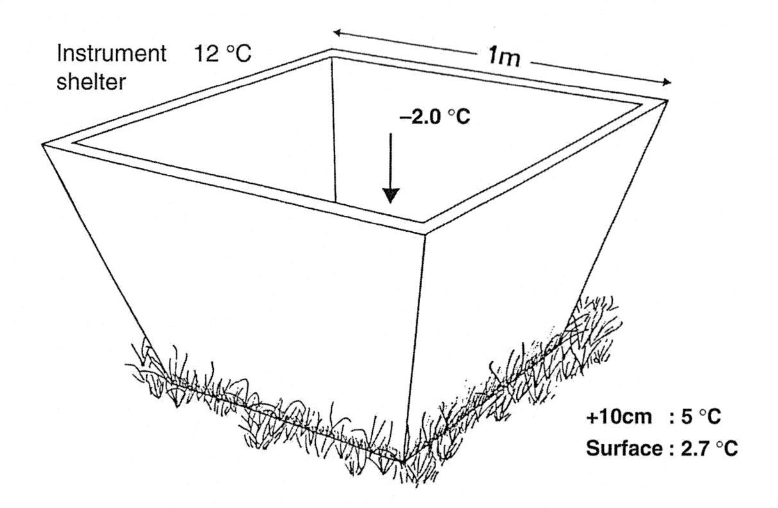

Fig. 19. Simple model of a frost pocket built of polystyrene foam. Temperatures measured on 29 July, 22.00 h; same sites as in Fig. 17.

'Krumholz' spruces and at the bottom an open vegetation of *Pinus mugo*, a small creeping pine which is normally found in the highest zone below the timber line (Sauberer & Dirmhirn 1953). In the Creux du Van in the Swiss Jura, near Neuchâtel a 150 m deep doline is found at 1400 m a.s.l. where typically alpine vegetation without any woody species occurs at the bottom. There is permafrost and in winter temperatures of –50 °C are usual. A reversal of the vegetation zonation is also seen in deep and narrow valleys in the Lunz area. It is not sure whether low minima are the only factor responsible for this phenomenon. Persistence of a snow cover may play a part as well (see also Morton 1963).

Smaller temperature differences can be found in gently undulating terrain. Geiger (1961) reports a difference of 4 °C at 10 cm height for an elevation difference of 1.5 m. Stoutjesdijk (1959) compared the minimum temperatures in a depression situated on a flat heathland. The depression was circa 3 m deep, 50 m wide and 300 m long Here temperatures were 0.2 to 1.8 °C (average 0.7 °C) lower than outside during clear nights. We give this example because it is usually quite difficult to find a situation where a depression can be compared with flat ground with the same thermal properties of the soil.

The reverse effect, higher temperatures up a slope, were observed in heathland as well. On a relatively steep though small hillock (3 m high) minimum temperatures during clear nights were 1.8 - 4.2 °C higher than above the flat heathland. On a bigger hill (30 m) the minima were 5.1 and 6.5 °C greater over two clear nights than on a flat area with heathland (*Calluna vulgaris*). Phenological differences observed on the 30 m hill show that oak scrub (*Quercus robur*) started leafing two weeks earlier than on the flat heathland (Stoutjesdijk 1959). It is also often observed that the young sprouts of beech and spruce are frozen in depressions and undamaged on the upper part of slopes. In Lapland we can see the opposite: fresh green birches (*Betula pubescens* ssp. *tortuosa*) in the valleys and brown-leafed birch scrub higher up. Here the frost kills the larvae (eggs) of *Epirrita (Oporinia) autumnata* (*Geometridae*) which elsewhere may damage and eventually kill vast stretches of birch scrub (Tenow 1975).

One can make a model of a frost pocket from polystyrene foam in the form of a truncated pyramid (Fig. 19). Measurements in such a pocket during ten clear nights showed minima which were on average 6.7 °C lower than in the surrounding low grassland vegetation. In this case the depression effect was of course reinforced by the poor conductivity of the material. In addition to topography and soil cover some macro-meteorological conditions are important for the occurrence of low minimum temperatures, viz. no wind, a clear sky and dry air during a night which follows a relatively cold day. These conditions often occur in a period with sunny, dry and relatively cold weather during the month of May (Northern Hemisphere). From Fig. 4 we see that in May the atmospheric long-wave radiation is lower than average.

Finally, some words on air humidity. This is not only important because atmospheric radiation is dependent on vapour pressure, but also the dew point of the air is reached earlier, i.e. with a higher temperature, at a high vapour pressure. The condensation heat dissipated by dew formation retards further cooling off; this effect is much stronger at higher temperatures, as will be explained in the next section.

2.9.2 Water relations, dew formation and hoarfrost

> *"Behold, I will put a fleece of wool on the floor; and if the dew be on the fleece only, and it be dry upon all the earth beside, then shall I know that thou save Israel by mine hand, as thou hast said. And it was so: for he rose up early on the morrow, and thrust the fleece together, and wringed the dew out of the fleece, a bowl full of water"*
> *Judges 6, verses 37 and 38*

The previous examples were chosen so as to avoid any major influence of evaporation and condensation. However, usually, cooling off at night means that the surface of the soil or the vegetation cover soon reaches a temperature below the dew point of the surrounding air. Then water condenses on the cold surface. This sets free a great deal of heat, about 2500 J/g. With a continuing cooling of the surface, more and more water vapour is withdrawn from the air through condensation. As a result an inversion of both the temperature profile and the vapour pressure profile arises. This is shown in Fig. 20 which also demonstrates how fast not only the temperature but also the water vapour content of the lower air layers decreases. These are measurements above and in dense 50 cm tall vegetation of *Salix arenaria* in a large dune slack with an understorey of herbs and grasses and a moss layer underneath. In particular the measurement at 19.40 h makes clear that the cooling process starts at the canopy level of the vegetation. As shown by the profiles there is a transport of heat and water vapour towards the canopy, both from the moss layer and the air above. Later the cold air sinks and probably the lowest values for temperature and vapour pressure occur between 10 and 50 cm. As can be checked with the temperature and vapour pressure values, the air was almost saturated with water at all levels. On a windy night the temperature inversion is much weaker and the temperature drops much more slowly. There is no clear inversion of the vapour pressure profile (no dew formation) and no decrease with time of the vapour pressure (Fig. 21).

The energy balance at dew formation gives the following picture. The net radiation loss of 69.9 W/m^2 is compensated by (a) 48.9 W/m^2 sensible heat withdrawn from the air + condensation heat; (b) 21.0 W/m^2 heat flow from the soil. In order to determine the dew formation we assume that a similar process of turbulent exchange is responsible for both heat and water vapour transport in the air (see Chs. 2.4 and 3.2). In that case we have:

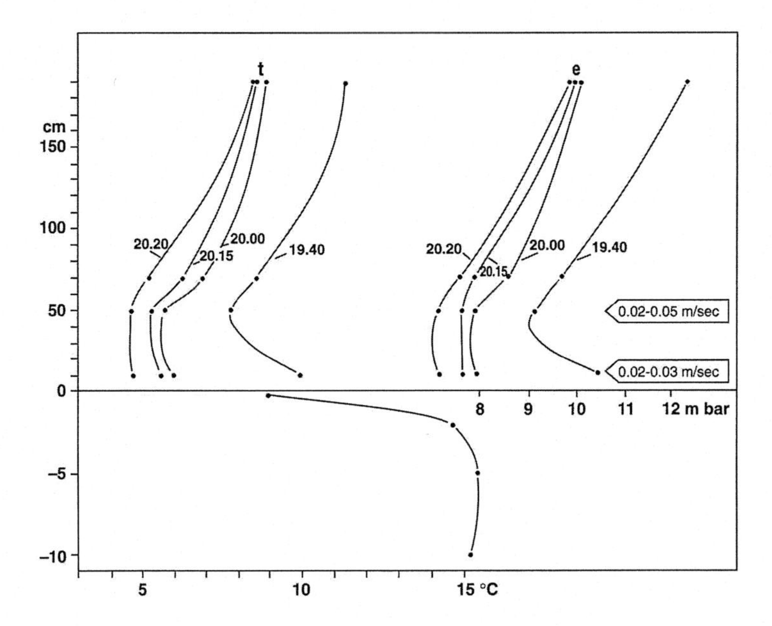

Fig. 20. Temperature and vapour pressure profiles at different times in and above *Salix arenaria* scrub in a dune slack. 22 August 1957, weather clear, very light wind. After Stoutjesdijk (1961).

$\gamma \cdot \Delta t / \Delta e = H_{air} / H_{ev}$, where H_{ev} is the amount of sensible heat transported between two levels and H_{air} the amount of latent heat released as condensation heat. With $\Delta t = 4.0$ °C, $\Delta e = 2.7$ mbar and $\gamma = 0.66$ mbar/°C, we calculate that $H_{air} : H_{ev} = 0.98 : 1$; in other words 24.7 W/m^2 is released as condensation heat and 24.2 W/m^2 is withdrawn from the air. For a night of 10 hours this means 888.2 kJ/m^2, which is equivalent to a water layer of 0.35 mm, provided the conditions remain the same.

Another approximation of the maximum possible dew formation is as follows: as soon as dew formation starts, the air which is touching the cold surface is saturated with water vapour. As further cooling off occurs both sensible heat is withdrawn from the air while condensation heat is released. We see that at an air temperature of 5 °C through the cooling

Table 11. The amount of heat to be withdrawn from 1 l of water-saturated air in order to lower its temperature by 1 °C, for various air temperatures. DW: Decrease in water content (mg/l) with 1 °C cooling; CH: Condensation heat dissipated; SH: Sensible heat withdrawn from the air; C/S: Ratio condensation heat/sensible heat.

Temp. °C	DW mg/l	CH J	SH J	Total J	C/S
0	0.34	0.85	1.29	2.14	0.61
5	0.44	1.10	1.27	2.37	0.85
10	0.59	1.46	1.25	2.71	1.17
15	0.80	1.96	1.23	3.19	1.56
20	1.06	2.60	1.20	3.80	2.13
25	1.38	3.37	1.18	4.55	2.85

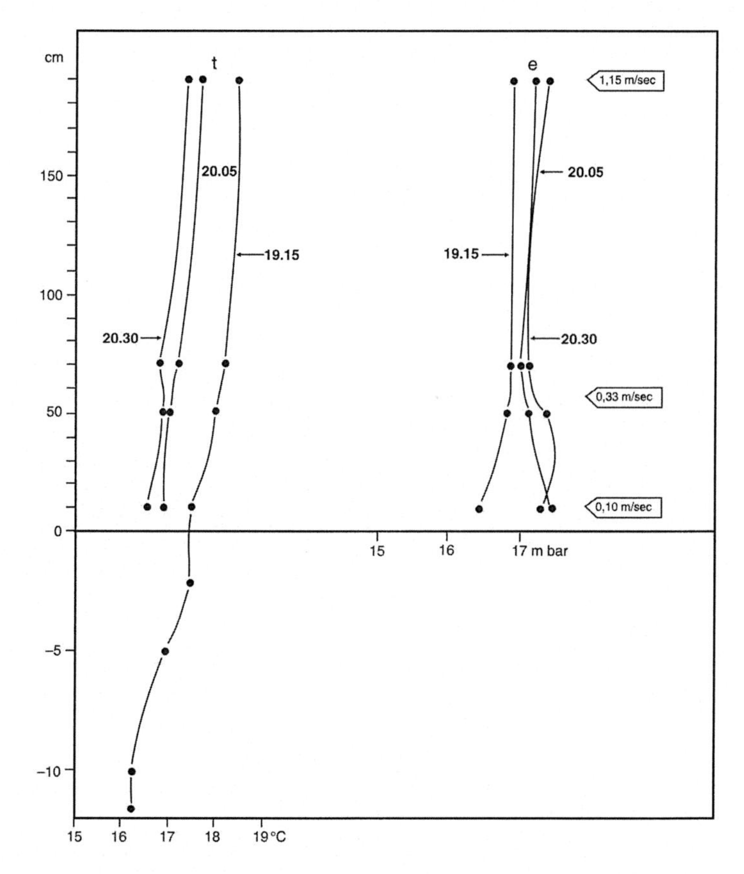

Fig. 21. As Fig. 20, but with more wind. 1 August 1957. No dew formation. Cooling off proceeds more slowly. Small difference between soil and air temperature. After Stoutjesdijk (1961).

by 1 °C of one litre of water-saturated air, 1.10 J condensation heat is released and 1.27 J sensible heat is withdrawn (Table 11). This gives a maximum value for the dew formation. Thus, of the total heat withdrawn from the air ($H_{air} + H_{ev}$) a fraction 1.10 J /(1.10 J + 1.27 J) or 46% can be attributed to dew formation. If the air at some height above the surface is saturated with water, the dew formation should have the value calculated here. If the air at the higher level is unsaturated, the dew formation will be less than the maximum.

If the dew formation already starts at a higher temperature, more water vapour will condense per degree of temperature decrease, because saturated air contains more water vapour at higher temperatures. Dew formation is greater and the ratio between condensation heat set free and sensible heat given off will thus be higher. If one litre of saturated air at 20 °C is cooled by one degree, 2.60 J of condensation heat will be set free and 1.21 J will be withdrawn as sensible heat. This large amount of condensation heat will strongly inhibit cooling off. Therefore, it may be expected that with moister air, meaning that the dew formation starts at a higher temperature, nocturnal cooling will proceed at a slower rate. Wallace (1878) was the first to point this out, in relation to slow cooling off in the tropics.

In summary, we may conclude that the nightly radiation loss is partly compensated for by the heat flow from the soil, while the remainder is compensated for by sensible heat

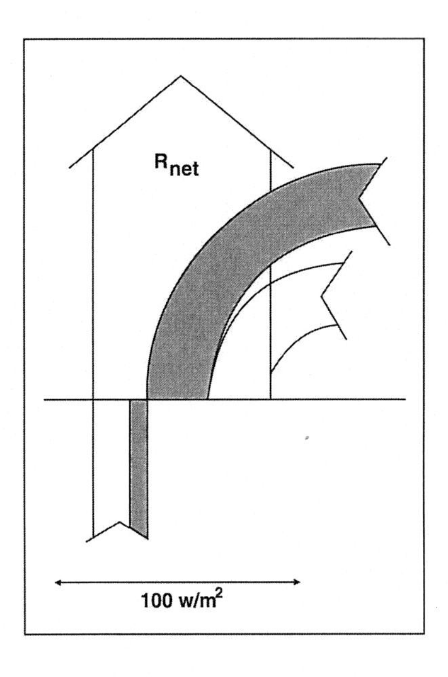

Fig. 22. Heat relations of the soil at night. Stippled: dew formation and heat transport in the soil through distillation. See also Fig. 20.

which is withdrawn from the air, and by condensation of water vapour, withdrawn from the air as well. Clearly, if the heat flow from the soil is great, the energy balance at the surface will not provide much room for dew formation. The latter can even be completely absent, especially with non-vegetated good-conducting soil. In coastal and inland dunes it can be observed that the surface of the dune sand, and scattered mosses and lichens growing there, often remain dry whereas the surrounding vegetation is wet with dew. Thus, Gideon's experiment, described in the citation at the beginning of this section, can be easily explained. The fleece forms an insulating layer locking the heat flow from the soil, and will cool off much more rapidly than the firm loam (soil) threshing-floor. Observations on the spot confirm this. Without looking at the energy budget we understand the absence of dew from bare, compacted soil as the surface temperature hardly falls below the air temperature.

The amount of dew withdrawn from the air can thus be estimated from energetic considerations. There is, however, the possibility that some of the dew formed derives from the soil. Soil temperatures show that the water vapour pressure in the soil is much higher than in the air above; thus distillation occurs of water vapour condensing at the surface of the soil or vegetation. This can be observed on transparent plastic film placed on the soil, where dew is immediately formed on the underside (Fig. 22).

In this connection Monteith's (1954) measurements are of fundamental importance. He placed a 30 cm deep block of undisturbed soil together with its grassy vegetation on a balance. Soil + balance were dug into the ground so that the plant cover of the sod formed part of the surrounding vegetation again. It appeared that often quite a considerable dew formation occurred while the weight of the soil did not increase. Apparently, the observed dew must have arisen from distillation of water vapour from the sod. It should be noted that the grass canopy was cut very short (1 cm) and the soil compacted. The measured heat flow from the soil was about as great as the deficit in the radiation balance; in other words, withdrawal of heat and water vapour from the air was not to be expected. Interestingly, Monteith found that real, albeit weak, dew formation occurred when there was wind: 1 m/s at 2 m height.

On the basis of the energy balance approach, as presented above, Monteith's observations cannot be generally valid, particularly not when the heat flow from the soil is small. The profiles of Figs. 17 and 20 also leave no doubt as to a strong downward transport of water vapour from the air to the vegetation surface.

In summary: real dew is formed mainly at places where the heat flow from the soil is small. In addition, water can be deposited on plants through distillation. Fig. 22 shows the relative importance of the various processes; the intensity of the distillation process was estimated from data by de Vries (1963).

In the cases discussed above there was only condensation on the vegetation. However, with higher air humidity condensation may also occur in the air and then a ground fog, several m high, will develop. This is especially the case in depressions, which lie as lakes of mist in the landscape. Extensive fog formation also occurs above the relatively warm water of ditches and rivers. Here evaporation continues and the water vapour formed condenses in the colder air above the water surface. In the mist formed, there is often an upward stream which is compensated for by laterally incoming air. Through the continued supply of new cold air, condensation will be stronger than can be explained from the heat loss through radiation only.

From energetic considerations it appears that the intensity of dew formation in the strict sense cannot be more than 0.1 - 0.3 mm, even during a night with ample dew. Because the water consumption of vegetation can be as much as 5 mm or more per day, and dew will not occur every night, dew formation will not be very important for higher plants. The exceptions are some deserts with low precipitation and high dew formation, where the vegetation can reduce transpiration and thus diminish the need for water. During prolonged periods of drought in temperate regions air humidity may be so low that at night the dew point is hardly reached, if at all. It seems that when the demand for water is highest, dew can contribute least. In situations with reduced evaporation, for instance on north slopes or in the autumn when the days are short, the relative significance of dew for the total water budget can be much greater.

For many organisms which cannot take up water from the soil the presence of liquid water can be very important, disregarding its origin from real dew or distillation, or from rain. Here actual measurements of the amount of water formed at the surface of objects can be illuminating. For mosses and lichens dew may be essential. These organisms can absorb water up to saturation from a dry starting position and keep this water. In mosses dew is also important for fertilization, allowing the spermatozoids to swim to the archegonia. Rocks and tree boles are a similarly extreme environment for mosses, algae and lichens, as deserts are for higher plants. Some epiphytic species are only found at the foot of boles and do not grow higher up than the nocturnal radiation fog reaches. In depressions and rivulet valleys these epiphytes sharply mark the upper limit of the dew impact (J. J. Barkman unpubl.).

For higher plants the significance of dew is, of course, of a different magnitude than for mosses and lichens. Dew is only a fraction of the total water supply of entire vascular plants. Still, they take up enough dew to profit significantly from it (Steubing 1955). The percentage of the dew formed actually taken up by leaves may vary considerably. In Steubing's experiments the highest percentage was 58.7% in the legume *Trifolium repens*, and the lowest 3.7% in the grass *Setaria viridis*. Patches of *Molinia caerulea* in a heathland can be soaked with water after a night of dew, while the heather has hardly become wet. In the morning after a night of dew, plants can have their stomata open without losing too much water; this is important for plants which cannot take up much water from the soil.

Dew is of special importance for the microflora on the surface of leaves of vascular plants (Ruinen 1961). For many animals dew is an important source of water. Caterpillars of the moth *Philudoria potatoria* really do drink dew drops (as the species name indicates). Lizards are also known to drink dew.

2.10 Wind

"A wind or not a wind. That's the question", Dr. Johnson said.
James Boswell (1785): Tour of the Hebrides

Wind is an environmental factor which varies tremendously from point to point. Elton (1966) writes about a situation similar to the one Dr. Johnson spoke about: "The effect of a little shelter can be dramatic. I have walked in a strong breeze on Pentire Head in North Cornwall and seen nothing visibly active on the bare cliff-top grassland and then come upon a small hollow depression in the ground that by some peculiarity of eddy currents was completely sheltered. Here hundreds of flies and bees were busy collecting nectar from flowers of the autumn Scilla".

At high speeds the most obvious effects of the wind are mainly mechanical, while at low speeds the effects concern transport processes such as heat transfer and evaporation.

Fig. 23 gives an impression of the relation between wind velocity and height above a flat surface with short grass. As shown in Fig. 24, wind speeds at the same heights in a park landscape with strong relief measured at roughly the same time are on the whole much lower.

The measurements above level ground illustrate some general principles. First, wind velocity decreases with decreasing height above the surface. Second, this decrease is exponential. Hooded crows and rooks migrating over the beach against the wind may fly so low that their wing tips almost touch the ground. As with the temperature profile above a warm surface: near the surface differences are greatest.

In the older literature (Sutton 1953) we find the following expression for the relation between wind speed u and height z:

$$\frac{u_1}{u_2} = \left(\frac{z_1}{z_2}\right)^{\beta} \qquad (2.42)$$

When the temperature profile is neutral the value of β is about 0.14 above level smooth ground, but above more broken ground with vegetation the value is slightly higher. The vegetation of the surroundings is also an influence. With an unstable atmosphere lower values of β may occur, with a stable structure (inversion) much higher values are found, up to 0.85 and with a completely laminar air movement β may reach 1.0 in the lowest air layers.

These data fit the following picture of air currents in the lower atmosphere. Wind is not a laminar but a turbulent flow, in which whirlwinds (eddies) occur continuously. This can be seen from fluctuations in both wind speed and wind direction, both horizontally and vertically. The size of these fluctuations may vary. The degree of turbulence is determined by the roughness of the surface, including both substrate and vegetation. Through turbulent exchange there is not only transport of heat and water vapour, but also a continuous exchange of air parcels moving at high speed from higher levels and slower moving parcels from lower levels. Thus there is 'transport of movement' (momentum). With an unstable structure in the atmosphere these exchange processes are stronger than with a neutral

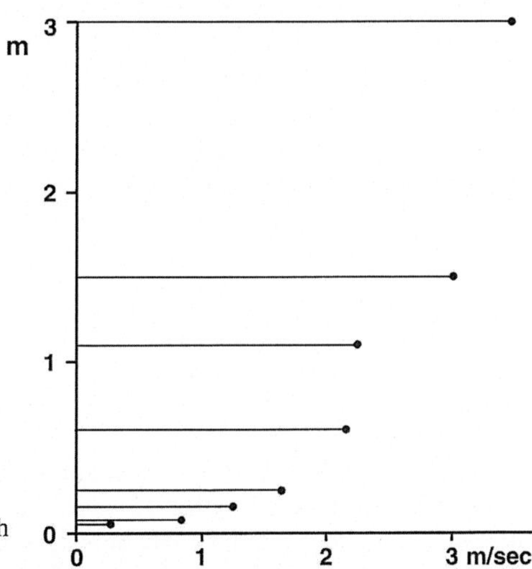

Fig. 23. Wind profile above a flat open area with short grass.

structure. In the former case warm air bubbles rise; these have a low horizontal speed and retard the movement of air at higher levels. Conversely, faster moving colder air masses arrive at the surface as sudden squalls. The gustiness, i.e. the amount of change in speed and direction is greater when there is strong thermal convection.

The opposite case occurs in the evening, when there is a temperature inversion (see 2.9) as a result of which the exchange of heavier cold air near the surface and warmer lighter air at higher levels is hampered. Consequently a less turbulent flow arises with a steeper velocity gradient. The two processes reinforce each other so that it can be completely still near the surface while there is moderate wind at a height of 10 m.

Nowadays the relation between wind speed and height is considered to be semi-

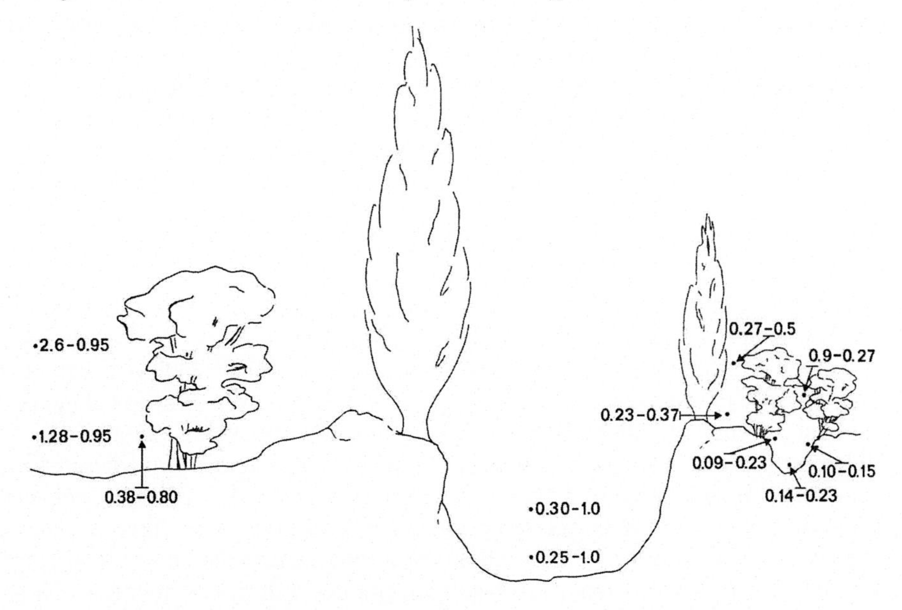

Fig. 24. Wind velocities over broken ground with relief at 20 cm and 150 cm height measured at roughly the same time as in Fig. 23.

logarithmic, both on theoretical and empirical grounds. Thus, $\Delta u / \Delta z$ is considered to be inversely proportional to z. We have chosen the older approach for the sake of simplicity and also because the wind profile and the transport processes in the atmosphere related to it are extensively discussed in physics-oriented texts (e.g. Mattson 1979; Monteith 1973; Oke 1987). The situation as to wind immediately near the ground or just outside or within the vegetation cannot be described with a simple formula (Businger 1975; Grace 1977).

Wind speed is always much reduced within vegetation. This reduction may be gradual in more open vegetation where the wind speed decreases considerably towards the surface, or rather the amount of air movement, because the turbulence can be very great. In vegetation with a closed canopy the amount of air movement decreases abruptly but it may increase again further down where the vegetation under the canopy is more open, and decreases yet again very near the ground. Examples of wind profiles in vegetation are given in Chs. 3.3 and 3.4.

The wind profile as found in open areas is modified by the plant cover, not only within, but also outside the vegetation. This has been investigated on larger scales in connection with the presumed beneficial effect of hedges and other wind screens for agriculture (Geiger 1961; Grace 1977). With a hedge of height h, wind speeds measured at 0.25 h at a distance of 3 h behind the screen can be reduced to 40%, and at 15 h even to 80% of the original velocity. Where networks of hedges occur the effects are greater still. Guyot & Sequin (1975) found, for 8 m high permanent hedges in Brittany (W France) at a height of 2 m and a distance of 3 h from the windward hedge, that there was a reduction to 20% and at 15 h to 60% of the velocity in open terrain. The hedges here were 30 h apart.

Where the screen is closely knit there may also be a local increase in wind speed: gusts of wind may arise, because air masses with a higher speed from higher levels reach the surface through eddies. This can be seen, for instance in the flattening of wheat near forest edges. These effects are still more extreme near tall buildings and towers.

Both at the lee and at the weather-side of an obstacle arises a zone of reduced wind speed. Sand, snow and dead leaves can form a wall, especially at the lee side (Tüxen 1977). Flying insects are often concentrated behind such screens. For weak flyers such as aphids this is mainly a passive effect. They are kept in the air circulation on the lee side of a screen. Stronger flyers such as hover flies often actively seek the shelter (Johnson 1969; Lewis & Dibley 1970). Swarms of mosquitos and *Bibionidae* choose the shelter of a hedge or forest edge. In Lapland mosquitos land on their hosts on the lee side. Butterflies are very sensitive to wind and often seek shelter (Baker 1978). In N Sweden, honey bees are hindered in their collecting activities already at wind velocities >3m/sec (Dr. B. Norin, pers. comm.). The importance of grass tussocks as shelter for arthropods was shown by Bossenbroek et al. (1977a, b).

Shelter from wind also influences heat- and matter-transport and thus surface temperature, air temperature, evaporation, dew formation and night frost (see Grace 1977 for a survey). The effect of shelter on a microscale in arctic-alpine conditions was studied by Wilson (1957). Relatively warm spots in winter are found where maximum exposure to sunlight is combined with maximum shelter from wind. This combination appears to occur in landscapes with a structural gradient, for example from forest edge via scrub to dwarf scrub and grassland. In gaps in the grassland the air near the ground can be completely still even with moderate wind, and the temperature can be relatively high. Temperatures at night in places like this can be quite low, unless a lot of long-wave radiation is reflected back from the shelter. Thus, the highest average temperatures will not be found in the gradient structures described above, but at abrupt forest edges facing south with overhanging vegetation and sunshine penetrating to ground level.

Sometimes mechanical effects can occur even at a low wind speed. Grace (1977) and others found that a wind speed of 3.5 m/s could damage the wax layer through friction between touching leaves and could thereby reduce the resistance to water loss from the leaves via the cuticle. Even moderate shaking can inhibit plant growth (Grace 1981). Visible damage to vegetation can only occur with very strong winds, such as found at the coast. On coasts very exposed to wind, for example on the Hebrides, trees are missing altogether, whereas on coasts with moderate winds wind-pruned trees are common; patches of scrub and wood increase in height in leeward direction.

Mechanical effects on wind-pruned trees are usually combined with more indirect effects, also in the case of wind-pruned trees. The greatest damage to such trees is not the breaking off of branches and the blowing down of whole trees but the desiccation and plasmolysis of young buds by the sea salts carried in the salt spray (Grace 1977; Robertson 1986). This follows from the observation that along the mainland coast of the Netherlands the trees are bent eastward by the prevailing WSW winds, whereas on the West Frisian island of Terschelling the trees are bent southeastward, although the prevailing wind direction is the same; here the effect of the NW winds, carrying salt from the North Sea is stronger than that of the prevailing winds skimming the Wadden Sea, which is narrow and lacks breakers. Barkman (1958) described the change in the branch formation of trees near Medemblik at the former Zuyderzee; after the Wieringermeer polder was constructed salty winds had a much longer fetch over land and consequently trees started to develop branches in all directions. Vulto & van der Aart (1983) found that 50 - 75% of the salt in salt spray is deposited in a zone some hundreds of m wide behind the beach. Hansen (1931, see also Barkman & Stoutjesdijk 1987) found on the west coast of Jutland in Denmark that rainwater contains 12.4 mg/l NaCl, 20 km from the coast less than half, i.e. 5.8 mg/l and much further from the sea, at 110 km from the coast only 4.7 mg/l. In North Holland wind-pruned trees occur at max. 15 km from the coast. An example from Canada shows a similar effect: trees on cliffs in Nova Scotia are bent westward, although the prevailing winds are westerly, but the salt spray comes from the east. Salt spray can affect the plant via the aerial parts or via the root system, and there is little correlation between the respective types of tolerance (Sykes & Wilson 1988).

On Newfoundland, Balsam fir trees (*Abies balsama*) are 'flagging' to the southwest, but the deciduous *Larix laricina* flags in the opposite direction. The northeast winds carry fine ice crystals in winter and they have the greatest effect on the firs, though SW winds prevail throughout the year (Robertsen 1986).

Wind-exposed trees and sheltered trees carry quite different epiphytic assemblages (Barkman 1958). Woodland herbs, e.g. *Teucrium scorodonia*, occur in open coastal dune vegetation, but only in the shelter of dune scrub with *Hippophae rhamnoides* or *Crataegus monogyna*. On exposed sites in the mountains we never find hygrophytes, i.e. plants adapted to moist conditions. Here, plants with leathery leaves and graminoids, xeromorph grasses and *Carex* species, predominate. In the mountains there is an additional effect in that windy sites have little permanent snow in winter and are subjected to relatively heavy frosts.

Another effect of wind is the accumulation of snow and other material behind obstacles (see Ch. 2.11). This can be quite spectacular, e.g. in the flat open landscape of S Öland (Sweden) with a xerophytic alvar vegetation, shallow soils and low precipitation, where snow accumulates behind juniper bushes and stone walls. Such sites may be much moister in spring and may carry a more luxuriant vegetation. Winds carrying drifting snow have a strong scouring effect on plants; buds and leaves can be destroyed. In the mountains of Lapland we find isolated pine trees (*Pinus sylvestris*) with three different layers of branches: near the surface, i.e. under the average winter snow level, there are long, bent or even prone branches, normal branches higher up the bole, and a zone without branches in between. This is the zone above the winter snow where ice crystals blown by the wind damage the buds (Barkman 1951).

If wind loaded with dust meets an obstacle, the dust is deposited, for example on the exposed edge of a forest. Here the trees often have a special epiphyte flora consisting of basiphilous species where the dust is from a limestone quarry, or nitrophilous species where the dust comes from fields (Barkman 1958). A special case is caused by bushes of *Sambucus nigra* on eastern slopes in seaward dune valleys. Winds from the sea loaded with sand deposit much of it as they meet the scrub canopy. Consequently the epiphytic moss flora on *Sambucus* include species of calcareous dunes which do not occur as epiphytes elsewhere (Barkman 1958).

Wind does not only supply particles, it does of course also blow them away. A well-known example is the removal of leaf litter in wind-exposed woodland edges and small woodland lots; this process is known as 'aushagern' in Germany. As a result nutrients are lost and the soil becomes poorer and more acid, and species such as *Anemone nemorosa* are replaced by acidophilous species such as *Vaccinium myrtillus*, and finally mosses such as *Dicranum scoparium* and *Leucobryum glaucum*. Gonschorrek (1977) measured the amount of litter at various distances from the western edge of a woodland. At 3 m distance, only a few g/m^2 were found, at 15 m, 130 g/m^2 and at 30 m, 1700 g/m^2. In relation to this effect a clear vegetation zonation could be observed: from the centre of the woodland to the edge: 1. *Festuca altissima* (basiphilous); 2. *Deschampsia flexuosa - Mnium hornum*; 3. *Deschampsia - Calluna vulgaris* (acidophilous).

Luff (1965) found that new shoots of the grass *Dactylis glomerata* grew mainly on the weather-side of the big tussocks, where dead grass material was blown away; on the lee side dead material accumulated and offered no room for shoot development.

The removal of leaf litter in forests is favourable for terrestrial mosses, which would otherwise suffocate. In an open beech wood in the W Netherlands mosses were observed only at slight elevations in the wood and at the western edge and around trunks, all places where leaf litter is blown away. Tüxen (1977) observed that only in the centre of woods do leaves remain on the surface everywhere; nearer the edge where there is more wind leaves are blown away from tree trunks in whirls around the trees, causing leaf-free zones up to 100 cm wide. Boulders on the forest floor, from which the leaves are blown away, carry a conspicuous green moss cover. Barkman (1979) described an oak wood on a hill, exposed to the wind, where all the leaves are blown away and the soil is covered with terrestrial mosses. In the moss carpet the dwarf shrub *Empetrum nigrum* has established and after some years has formed a close cover. This *Empetrum* carpet acts as a shelter so that oak leaves are now no longer blown away to the same extent as before. As a result the mosses have disappeared. A moss carpet underneath *Empetrum* heath, even in pine woods, is normally quite common. Later measurements by Barkman of the wind speed illustrate the effect of the *Empetrum* carpet:

Date 19 September 1983, *Empetrum* vegetation 11 - 20 cm: wind speed in m/s.

Height (cm)	*Empetrum*	Moss cover
100	1.25 (0.80 - 1.50)	1.30 (0.50 - 1.80)
30	0.95 (0.80 - 1.20)	0.97 (0.70 - 1.40)
4	0.12 (0.05 - 0.18)	0.61 (0.30 - 0.90)

It appears that within the *Empetrum* carpet, at 4 cm height, wind speed is only 20% of the speed above the moss carpet.

Wind is, of course, also very important for all anemophilous and anemochorous organisms. First we report on unpublished research by P. W. Vroege on the pollination of *Salix arenaria.* This shrub is pollinated both by wind and by insects. Female catkins were surrounded by netting to prevent insects from pollinating. In the flowering period April - May 1983 there were 107 6-h periods suitable for wind pollination out of a total of 244, suitable meaning: wind, dry pollen, precipitation less than 1 mm during the period and the period before; 23 6-h periods were suitable for insect pollination, meaning daylight, temperature above 12 °C, no rain, wind not stronger than 8m/s. (Meteorological data obtained from a nearby weather station.) Two *Salix* patches were compared, one on an exposed dune top, one in a sheltered valley. On the dune top 30% of the catkins were pollinated by insects and 70% by wind. In the valley the situation was the other way round: 80% insect and 20% wind pollination.

The biological significance of wind can be summarized as follows: wind can be unfavourable for plants and animals because of (a) the mechanical damage it causes, especially in combination with air-borne salt, sand or ice crystals; increased transpiration; (b) reduction of dew formation; (c) supply of air pollutants; (d) hindrance of insect or bird flight. All these effects vary from organism to organism. On the other hand, wind promotes the dispersal of pollen, spores, seeds, cysts and small animals, as well as bird flights. For many organisms the risk of night frost, as well as the risk of overheating, is reduced. The topic 'wind in relation to plants and vegetation' is treated in detail by Grace (1977).

2.11 Snow

"The food of the migrant birds that are coming from over sea was there dormant under the snow"
Richard Jefferies (1889): Field and Hedgerow

Snow cover changes the thermal budget at the surface of the earth. Fresh snow reflects solar radiation, not only the visible light but also shorter and longer wave radiation, up to 95% (see the albedo data in Table 2). Older snow has a lower reflection because some of the snow has changed into ice and because of dust now in it. Because of the high reflection the radiation balance may be negative even with strong sunshine. However, for long-wave (heat) radiation snow acts as an ideal black surface, with an absorption of 99.5 % and only

Table 12. Energy balance of a snow surface (W/m^2). Air temperature at 1 m –5.8 °C; at 1 cm –7.0 °C; at 0 cm –8.5 °C. Temperature under the snow (14 cm) –1.5 °C; wind 2 - 3 m/s. Date 4 January, clear sky.

In		Out	
Solar radiation	184.6	Reflection	142.0
Atmospheric long-wave radiation	188.8	Emitted long-wave radiation	278.3
Heat flow from snow	2.4	Evaporation	14.7
Subtotal	375.8		
Derived from air	59.2		
Total	435.0	Total	435.0

0.5% reflected.

Table 12 shows an example of an energy balance. The negative radiation balance is also expressed in the temperatures. The lowest temperature is found on the snow surface; despite the occurrence of wind a temperature inversion is maintained. At the soil surface the temperature is just below zero, that is 7 °C higher than at the surface of the 14 cm thick snow layer. The energy taken from the air is partly used for evaporation. The evaporation energy is calculated in a similar way to that of a wet surface (Ch. 2.6). As the difference in vapour pressure between the surface and the free air is small, evaporation is probably less than the hoarfrost formed at night. As wind strength and/or elevation of the sun increases sublimation of the snow can be considerable, so that a thin snow cover will disappear with prolonged frost.

Because of the small amount of heat from the soil during a clear and still night snow can cool off rapidly at the surface, causing very low night temperatures. Moreover, high radiation intensities during the day (both from direct and reflected sunlight) may lead to relatively high temperatures in the middle of the day. As a result, plants growing above the snow may be subjected to very large temperature differences, with a damaging effect. Investigations of this phenomenon were carried out on *Pinus cembra* by Holzer (1959) and Turner (1961) in the Austrian Alps. The low night temperatures cause freezing of the needles, while in the sun rapid thawing occurs. On N slopes beyond the reach of the late winter sun no damage was observed. We have observed a similar situation for tall heather, *Calluna vulgaris*, in the Netherlands when temperatures went down as low as –25 °C and branches projected above the snow.

Snow has low values for k and ρC, especially when the snow cover is new. A particular characteristic of snow is its insulating effect, protecting soil and vegetation from frost. On top of the snow cover it is much colder than on the bare ground; underneath the snow it is much warmer. Many plants in alpine and arctic environments with winter temperatures of –20 to –50 °C, would be killed in temperate regions because of the severe frost with no protecting snow cover, as many gardeners know. The insulating effect of snow is also used by willow grouse (*Lagopus lagopus*) and many other birds in the cold boreal winters (Korhonen 1980, 1981; Shilov 1968). The grouse often stay for the major part of the day in cavities under the snow where the temperature can be as much as 25 °C higher than outside; this temperature is of course partly due to the additional heat produced by the birds. Under such a long-lasting snow cover many animals, both vertebrates (e.g. lemmings and other small rodents) and invertebrates can remain active for most of the winter (Coulianos 1962; Nordman 1962; Merriam, Wegner & Caldwell 1983).

Some observations may illustrate this:
On the night of January 17, 1963 at Wijster, NE Netherlands, the minimum temperature at 1.50 m was –14 °C, at 10 cm above a snow cover –16.2 °C; at the snow surface –22.4 °C, at the snow surface in a small frost pocket –27.1 °C. In the afternoon of the same day at 15.30, with sunny weather and a moderate NE wind, the temperature at the snow surface was –11.7 °C, at 10 cm depth in the snow –9.3 °C, at the soil surface underneath a *Molinia* tussock (47 cm of snow) – 0.1 °C and between tussocks under 27 cm of snow –5.5 °C. Thus, the temperature at the soil surface varied more than 5 °C depending on the vegetation pattern. Such differences can be of importance for organisms hibernating on and in the soil. The situation described here may be deviant in that a thick humus layer had developed in and between the *Molinia* tussock vegetation, as a result of which little heat was dissipated from the soil. Normally heat conduction is so much greater that the temperature underneath a snow cover of at least 20 cm is always above zero.

Even if the soil has been frozen, after a snow cover has developed, the top soil layer can thaw again; the heat supply to the air has stopped, but the heat supply to the soil surface goes on. Skaters know that the ice underneath the snow can be dangerously thin, whereas tracks which have been swept have much thicker ice. Natural interruptions in a snow cover lead to local changes. Luff (1965) described how cold air can reach the soil surface along tussocks of the grass *Dactylis glomerata* where the snow cover has been interrupted. Whether such interruptions occur depends on the structure of the plants. If we compare two heathland plants: *Calluna vulgaris* has an uneven canopy causing interruptions in a snow cover, whereas *Empetrum nigrum* with an even canopy has a closed cover of snow.

Snow offers an effective protection against low temperatures; on the other hand little solar radiation is passing through a thick layer of snow. Another ecological aspect of snow is that after the snow melts the soil may be saturated with water. All these effects show spatial patterns in relation to topography, presence of obstacles, wind strength and direction when the snow falls, etc. Thickness of the snow cover and time of melting are often varied spatially according to fixed patterns, which can be correlated with vegetation patterns found in the summer (Friedel 1961).

In the Scandinavian mountains many plants spend the winter with green shoots under a snow cover several dm thick. The shoots are often formed in the autumn and may even show some growth during the winter. Examples are *Carex bigelowii* and *C. lasiocarpa*, *Deschampsia caespitosa* and *D. flexuosa*, *Eriophorum angustifolium* and *E. vaginatum*. *Vaccinium myrtillus* is an evergreen under the snow, but it is killed by frost when unprotected (Warenberg 1982). This is of great importance for foraging reindeer.

Sheltered depressions and spaces behind obstacles receive a great deal of snow, mountain ridges very little. In the dunes snow remains longer on north slopes than on south ones. Variation in snow cover ranges from very short duration with strong frost and desiccating winter storms, to a relatively long duration, especially in arctic-alpine environments where the growing season is short and the soil soaked with water during much of the season. If the soil is snow-free for less than three weeks, no vegetation will develop. If the snow-free period lasts at least three weeks the moss-dominated *Polytrichum sexangularis* may develop on silicate rock in the Alps. A period of at least two months free of snow may lead to a dwarf-willow scrub vegetation, the *Salicetum herbaceae*, which does not grow taller than 1 to 3 cm. On alpine meadows with snow-free periods of 3.5 - 5 resp. 5 - 7 months the associations *Caricetum curvulae* and *Festucetum halleri* may occur. If the snow-free period is still longer, and the winter is cold, the limiting effect of frost on the vegetation is considerable: an evergreen dwarf shrub vegetation, the *Loiseleureo-Cetrarietum* develops, characterized by small leathery leaves. *Vaccinium myrtillus* tolerates snow-free periods of 8 months, *V. vitis-idaea* up to 10 months. 7-10 months free of snow leads to the development of the *Loiseleureo-Cetrarietum cladinetosum* with *Vaccinium vitis-idaea* and various species of *Cladonia* subgenus *Cladina.* If there is hardly any snow at all in alpine environments, the *Loiseleureo-Cetrarietum alectorietosum* develops with many lichen species, especially *Cetraria* species. Here, the rare lichen *Thamnolia vermicularis* occurs. This species does not only grow in the Alps and in arctic Scandinavia, but also on the island of Öland in SE Sweden. The southern part of this island consists of almost barren or shallow-soil limestone. The strong easterly winds blow away the snow, and there is large-scale frost on exposed sites. In other places high drifts of snow are formed. The many smaller and larger scrub patches of *Juniperus communis* create a pattern in the snow-depth. On the lee side, i.e. the western side of the juniper bushes snow stays for a relatively long time, the soil is much moister in spring here and there is a different vegetation.

Clearly, lichens growing in snow-free places in high mountains and polar areas have to tolerate very low temperatures. In MacMurdo Sound (Antarctica) the average summer temperature is –5.3 °C, the average winter temperature is –30 °C. Lichens occurring here tolerate temperatures of –75 °C, even in wet conditions (according to laboratory experiments). They can assimilate at – 20 °C and have their photosynthetic optimum at low light intensities and very low temperatures of 0 - 5 °C (Lange 1963; Lange & Kappen 1972; Gjessing & Øvstedal 1989).

The influence of snow can be easily observed on trees, stones and rocks in the subalpine and alpine zones. In the birch woods of Lapland the bases of the trees are covered with grey and yellow thallose lichens while shining dark brown foliose lichens are found higher up the trunks. The boundary between the two zones is very sharp and is an indication of the average snow depth in the area.

The best known snow indicator is the foliose lichen *Parmelia olivacea* growing on the trunks of the mountain birch *Betula pubescens* ssp. *tortuosa*. The lower limit of the lichen indicates the average depth of the mid-winter snow cover (Nordhagen 1927). Sonesson & Lundberg (1974) and Verwijst (1988) found that the actual snow depth figures measured in the field were slightly less than indicated by the 'olivacea limit', which Sonesson & Lundberg explain from the tendency for snow to accumulate around trees. The dark colour of lichens above the snow cover may be related to their heat requirement. In the primeval forest of Bialowieza in Poland *Hedera helix* is found creeping up the tree boles, but not higher than the average height of the snow.

A completely different adaptation to snow is found in plants occurring on sites where the snow melts very late in the season (German 'Schneetälchen'; snowbed communities; cf. Ellenberg 1988). Vascular plants have only a short time to develop leaves, flowers, fruits and seed and several species start their photosynthesis when the leaves are still covered with snow. *Soldanella alpina* sends up flowering stalks, which are relatively strong and flexible, through the snow. This mechanism is also found in the forest herb *Anemone nemorosa*, which sends its flexible stalk through the cover of leaf litter. *Ranunculus glacialis* can even spend a whole summer under snow and yet flower the next year (Larcher 1980).

The question is how plants can assimilate under the snow. Older snow partly turns into ice and becomes more transparent. A fresh snow cover of 20 cm transmits only 3% of the solar radiation between 400 and 700 nm, but an older snow cover of the same depth transmits 13% (Geiger 1961). Seedlings of *Pinus cembra* under a snow cover of 15 cm can assimilate at 50% of the maximal possible intensity (Turner & Tranquilini 1961). Under icy snow a sort of greenhouse effect may occur. In Antarctica temperatures of +13 °C were measured on lichens growing under an icy snow cover, with an air temperature of –9 °C (Lange & Kappen 1972; Lange 1972). Warenberg (1982) presented similar data on temperature differences and showed that many plant species maintain green shoots underneath a snow cover, where there is sufficient light.

Solar radiation, because of the high reflection from snow, has little influence on snow melt. As soon as objects appear above the snow, plants or stones, melting proceeds much faster.These objects release heat through conduction and radiate heat which is fully absorbed by the snow. In this way holes arise in the snow around stones and tree trunks.

2.12 Aspect and inclination

The amount of solar radiation absorbed is of great importance for the heat and water budget of plants and animals, and thus of vegetation as a whole and also the soil surface.

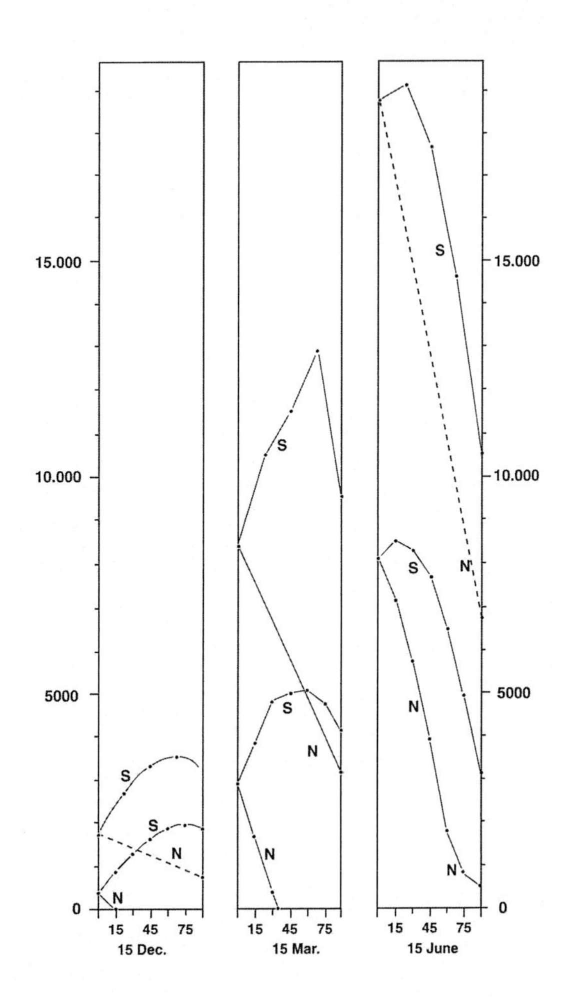

Fig. 25. Climatological data on direct and global solar radiation on a horizontal plane and on north and south slopes of different inclination. For global radiation on north slopes, only data for slopes of 90° are available. Daily sums in kJ/m^2. After Slob (1982).

Many animals and even some plants can adjust to the amount of radiation required by turning to the sun. Animals can simply turn the body axis and expose their broadest side to the sun to receive more radiation. Some species of *Lupinus*, e.g. *arboreus*, follow the sun through the day, at least as seedlings. On the other hand the desert plant *Lupinus arizonicus* folds its leaves when it is dry and exposes the narrow side to the sun in order to lower the amount of radiation received (Forseth & Ehrelinger 1982). *Lactuca serriola* arranges its leaves in a north - south direction in order to avoid maximum exposure to heat and light. Subtropical trees such as the American *Schinus molle* and Australian *Eucalyptus* species have their hanging leaves oriented north-south.

Aspect (compass direction) and inclination (steepness of angle with the horizontal plane) are of direct significance for the microclimate, this in relation to geographical latitude and time of the year. In the tropics and subtropics it depends on the season whether north or south aspects are exposed to the sun. Very steep east and west slopes may even receive more radiation than south or north slopes. Towards the poles the effect of exposure disappears, but not that of inclination.

The effects of aspect depend on latitude and inclination as demonstrated by comparing the difference between a termites' nest in the tropics and an ants' nest in a temperate region. Termites build vertical nests with the broad side north-south oriented, which receives

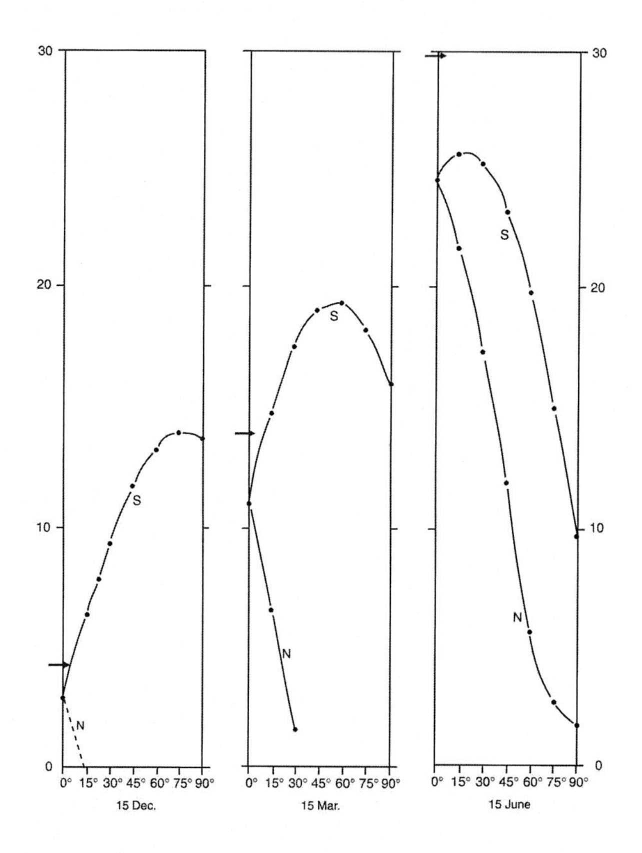

Fig. 26. As in Fig. 25 but for cloudless days and direct solar radiation. The global radiation on a horizontal plane is indicated with an arrow. Values in MJ/m^2.

relatively more radiation early in the morning and in the late afternoon, but little during the hottest hours of the day. Ants of the genus *Formica* build dome-shape nests which become steeper towards higher latitudes, thus adjusting to the generally lower elevation of the sun. According to the Belgian entomologist Raignier some *Lasius* species occurring near the tree limit in the Alps, build their oblong nests NW - SE so as to expose them optimally to the early morning sun in the summer months.

The difference in microclimate between north and south slopes is well-known and well investigated especially in dunes. Differences in the radiation intensity on these contrasting slopes with the sun in the south and a clear sky, can be derived from Table 1 (p. 16). This gives an indication of the maximum differences to be expected. On average the differences over a day are naturally smaller. In temperate regions a north slope even gets more radiation in the early morning than a south slope, round the summer solstice.

With a less clear sky the differences are smaller, because the direct radiation decreases whereas the indirect radiation increases. On a cloudy day the differences are almost negligible. Average radiation sums for slopes of different aspect in different times of the year, which take into account the actual cloud cover and transparency of the atmosphere, are hardly available. It is clear that such data are important for soil temperatures, evaporation and the phenology of plants and animals.

The curves of Fig. 25 give the amount of direct solar radiation on north and south slopes, taking into account the average degree of cloudiness and atmospheric turbidity. As an approximation of the global radiation on north slopes with different degrees of steepness, the values for global radiation on a horizontal plane and the north face of a wall are

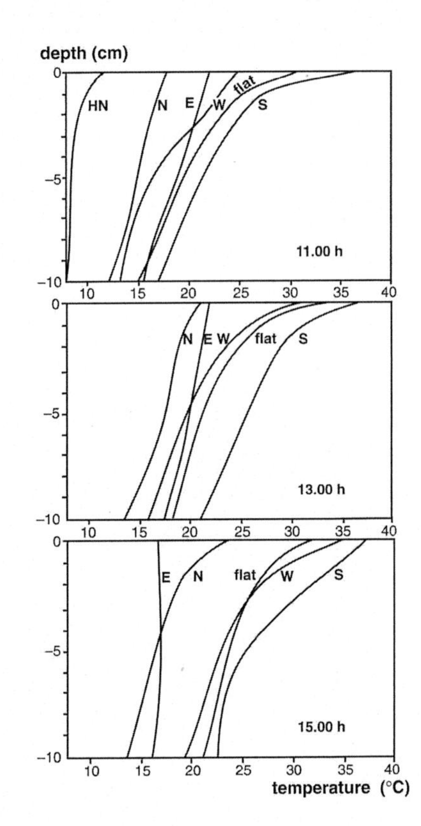

Fig. 27. Soil temperatures at various depths on slopes and on a flat part of an inland dune system on 25 May 1955, at three occasions. Air temperature at 2 m at 11.00 h was 15 °C, at 15.00 h 18 °C. N = north slope of 30 °; E = east slope of 33 °; S = south slope of 25 °; W = west slope of 15 °. HN = north slope with *Calluna* vegetation. (After Stoutjesdijk 1959.)

connected by a dotted line. An important point here is that when the elevation of the sun is high, the difference between a south slope and flat ground is much less than between a north slope and flat ground. When the sun is low the opposite is true for steep slopes (Fig. 25). In Fig. 26 the values for direct radiation on north and south slopes are given for cloudless days. The value for the global radiation on a horizontal plane is indicated with an arrow. The radiation received is now mainly determined by direct solar radiation. In winter it is maximal on a steep slope, in summer on a gentle slope. The difference between these two maxima is much less than in the first case (see Figs. 25 and 26). The ratio is about 1 : 2 when the situation on 15 December is compared with that of 15 June. Where actual values at midday are concerned the differences are still much smaller (see Ch. 2.2).

To show the effect of aspect on soil temperature, Fig. 27 presents measurements on slopes of different expositions in an inland dune area with a sandy soil in the Netherlands on a bright day in late May. Regarding surface temperatures, these were measured with mercury thermometers and may be a few degrees too low. Before midday the gentle west slope is warmer than the steep east slope. Temperatures in the deeper soil layers reflect the integrated effect over a longer period. A measurement on 31 October (see Table 13) shows that the temperature differences between a north and a south slope in the same area are still somewhat higher, and that the north slope is now clearly colder than the air temperature The surface is moist, while on the south slope the top cm of the soil are completely dry.

At low elevations of the sun the maximum radiation intensity is received on very steep slopes. In winter the ideal slope-aspect is an almost vertical plane. So, high surface temperatures may occur on the bark of tree boles even in winter. Fig. 28 shows that on the

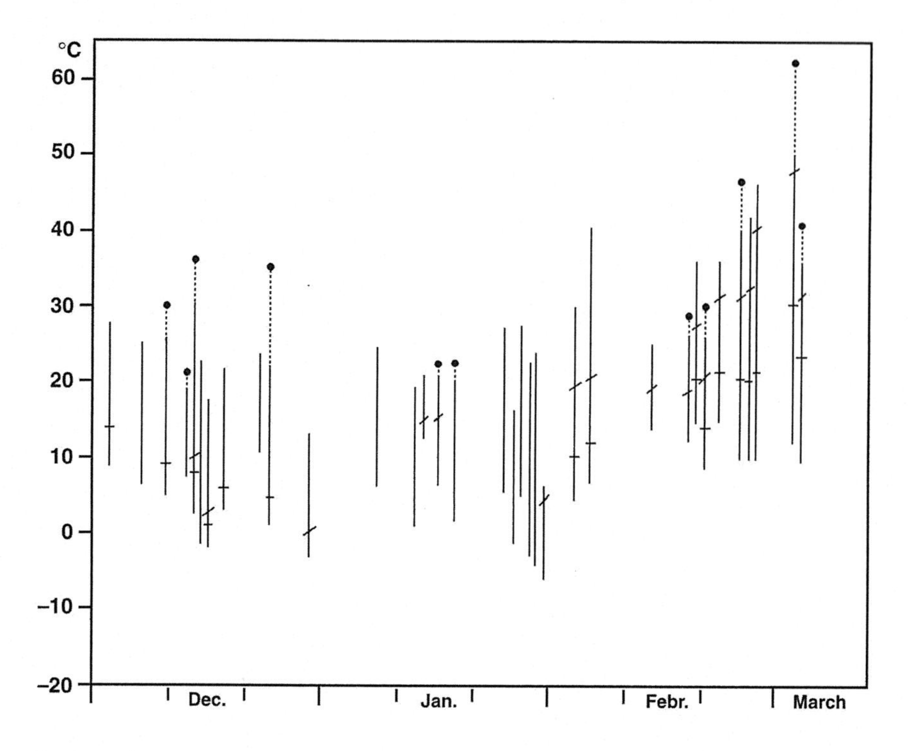

Fig. 28. Air and surface temperatures on tree boles are indicated by the lower and upper ends of the vertical lines in the figure. If higher temperatures were found in the surroundings than on the bole these are indicated with dots. Temperatures on a level surface (raw humus) are indicated by horizontal strokes, temperatures on a slightly sloping surface exposed to the south are indicated with oblique strokes. After Stoutjesdijk (1977b).

bark of *Pinus sylvestris* this temperature may be as much as 30 °C above the air temperature in December and even 40 °C in March. High values of Δt are only found on badly conducting material, for instance on tree stems with a thick bark, as on *Pinus sylvestris* or *Quercus robur*. When the cork layer is thin, as with the beech *Fagus sylvatica* and the conductance much higher, values for Δt are low.

As far as bark temperatures are concerned, similar results were obtained by Nicolai (1986) for a considerable number of trees. Characteristically, cambium temperatures are lower on trees with a thick cork bark (which have the highest surface temperatures) than on those with a thin cork layer. Many arthropods live exclusively on the bark of trees. The

Table 13. Soil temperatures on a north and a south slope in an inland dune area, 31 October 1955, 14.30 h, air temperature 10 °C. After Stoutjesdijk (1959).

Depth	Temperature °C		
cm	N 28°	S 28°	Plane
0	4.4	25.0	13.5
−2	4.2	18.0	10.8
−5	4.2	15.3	9.4
−10	4.2	11.0	7.7

Table 14. Frequency distribution of occurrences of the lichen association *Physcietum elaeinae* in southwestern France and the Netherlands in relation to aspect.

Aspect	S France		Netherlands	
SE	2		11	
S	1	5	28	63
SW	2		24	
W	4		3	
NW	5		0	
N	4	13	0	0
NE	4		0	
E	2		2	

microclimatic conditions on the cork bark of *Quercus robur* probably make it possible for spiders like *Agyneta innotabilis* and *Entelecara penicillata* to reproduce even in early spring and late autumn. This is exceptional as most central-European spiders reproduce in summer (Nicolai l.c.).

Very high temperatures may be reached on small steep edges of peaty or raw humus soil and partly decayed grass, up to 35 °C above air temperature in December and even 52 °C in March. Large temperature differences may also be found on grass tussocks in the winter in relation to aspect. For instance, on 24 December with an air temperature of 1 °C, the surface temperature at the south side of a (largely green) tussock of *Deschampsia flexuosa* was 19 °C whereas at the north side it was –2 °C.

The influence of aspect on vegetation development is well known. South slopes are the most suitable habitat for northern-hemisphere organisms with a southern distribution and north slopes for organisms with a northern distribution. On the south slopes of the Kullaberg Peninsula in southwestern Sweden several arthropod species are found far to the north of their main distribution area, e.g. the moth *Idaea dilutaria*, the beetle *Danacea pallipes* and the spider *Theridion conigerum*. Ryrholm (1988) has given a fine analysis of how macro- meso- and microclimatological factors operate together here. The southern part of this east-west oriented peninsula slopes gently towards the sea. The normal habitats of the southern species are found on short steep parts of the generally gentle slopes. In the winter these steep areas catch not only much of the direct solar radiation available but also reflected radiation from the sea.

The plant geographer S. A. Cain made the remark that a plant species at the border of its distribution range finds itself on the slope with the aspect from where it can look towards the centre of its distribution area. As an example, wild strawberry, *Fragaria vesca* is found in the Mediterranean lowlands where it occurs mainly on north slopes, in temperate Europe on level area and gentle slopes, whereas in N Norway the species is restricted to south slopes. (J. J. Barkman unpubl.). The mediterranean *Frankenia laevis* shows a clear preference for the south side of sandy ant hills of *Lasius flavus* in Norfolk salt marshes (Woodell 1974).The community of the epiphytic, nitrophytic lichens, *Physcietum elaeinae*, is known from trees in S France where it preferably occurs on the north side of tree trunks; in the Netherlands it is mainly found on southern aspects (Table 14).

On the 800 m high Lägern, a limestone mountain range running east-west near Baden in the Swiss Jura, one observes a very sharp transition from a warm microclimate on the south

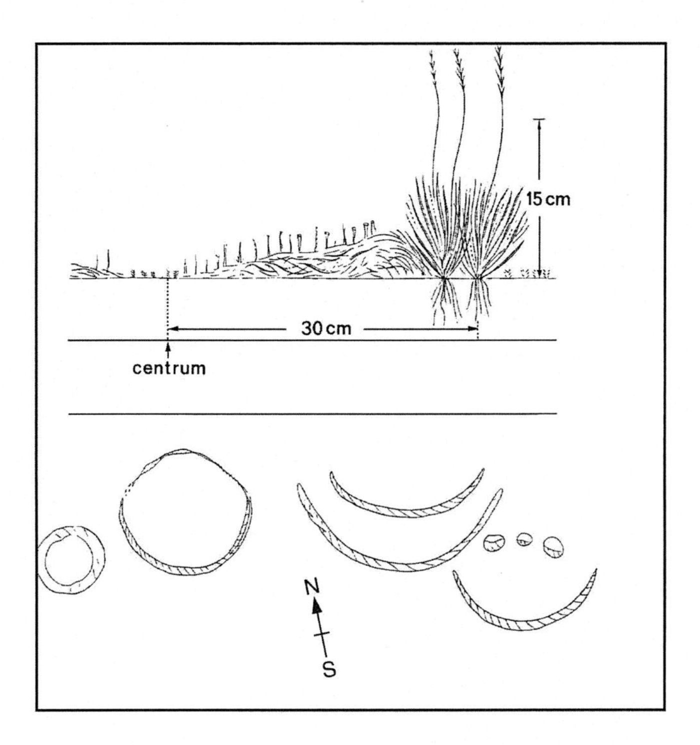

Fig. 29. Cross section through a growth ring of *Nardus stricta* and different stages of development seen from above.

slope and a much cooler microclimate on the north slope, separated by a mountain ridge 50 cm wide. On the south slope we have a thermophilous *Quercus pubescens* woodland with southern elements such as *Orchis pallens*, *Viola mirabilis* and *Melittis melissophyllum*. On the north slope a beech (*Fagus sylvatica*) forest is found with subalpine elements such as *Thlaspi alpestre* and *Asplenium viride*. It is like comparing an altitudinal difference of 1000 m or a latitudinal difference of 1000 km!

In northern Denmark one may find the boreo-alpine species *Trollius europaeus* on the north slope of a steep hill, where the south slope is dominated by the continental thermophilous species *Geranium sanguineum*. The latter species' red colour seems to show its preference for heat. An ornithological example here is the cirl bunting (*Emberiza cirlus*) which prefers south-facing slopes near the northern limits of its distribution area (Simms 1971).

The marked differences between north and south slopes of temperate coastal dunes have interested plant ecologists for a long time, especially in the Netherlands. Van Dieren (1934) in his classical study on organogenic dune building was the first to notice the contrast between an open plant cover with *Corynephorus canescens* and lichens on south slopes and a dense dwarf shrub with the boreal *Empetrum nigrum* on north slopes.

In stable inland dunes the colonisation of the south slopes is a slow process. When blown sand is no longer received by the slope, then the few grass tufts (e.g. *Corynephorus canescens*) present may die as their root system is laid bare by rain water. The dead grass tufts act as centres from which crustose lichens can spread over the sand surface. Gradually podetia of *Cladonia nemoxyna* and *Cladonia glauca* develop. *Polytrichum piliferum* invades these slopes starting at the foot and catching the sand that washes down the slope.

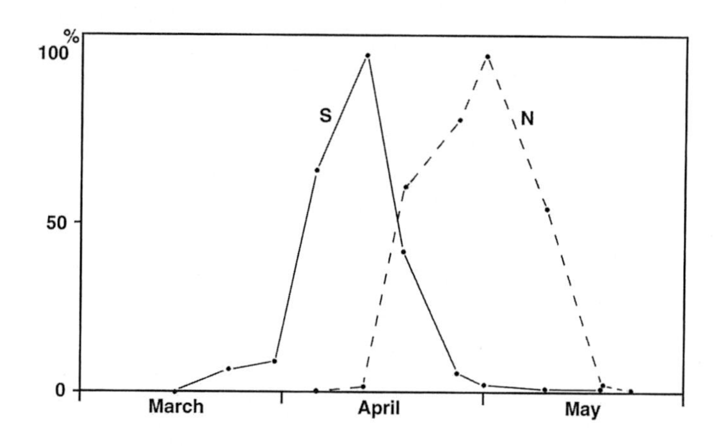

Fig. 30. Flowering frequency of *Taraxacum* spec. (percentage of flower heads) on the south and the north side of a dike. Data from Dr. A. A. Sterk.

Gradually a community is formed of *P. piliferum* with many *Cladonia* species, *Cornicularia aculeata,* a few tufts of *Corynephorus canescens* and *Festuca ovina.* The lichens use the dead *Polytrichum* as a fixing point. Where the first cm of soil under this community has a higher humus content (the colour changes to grey), the first seedlings of *Calluna vulgaris* are found. On the north slope the colonisation by *Calluna* (sometimes *Empetrum*) is rapid, together with some mosses, e.g. *Webera nutans.* A dense heath vegetation may be found here on pure yellow sand without a trace of humus whereas the south slope is still completely bare.

With the colonisation of the south slopes by *Calluna* the differences between north and south slopes are much less but they are still noticeable both on fixed inland dunes and on heathlands of long standing with a well developed podzol profile. We mention the presence of *Erica tetralix* and *Empetrum nigrum* on north slopes on pure *Calluna* heaths. In the moss layer several liverworts and *Leucobryum glaucum* are typical for the north slope (Stoutjesdijk 1959).

Differences between north and south slopes can also be found on a very small scale. In Scandinavia the vegetation on north and south exposed sides of ant hills is different, both regarding total cover (much higher on the north side) and species composition. Another example is the asymmetrical development of tussocks of the grass *Nardus stricta* on fixed inland sand dunes (Stoutjesdijk 1959). The plant grows in tussocks which die off in the centre. New shoots developing at the periphery are more abundant on the southern side of the tussocks. In the end the plant may disappear completely on the north side and a horseshoe pattern arises (Fig. 29).

An old demonstration of the effect of aspect on a micro-scale is that of Kraus (1911) who remarked that the southern side of a willow catkin flowers much earlier than the north side. A more recent animal example is that given by Williams (1981) who found that the eggs of the butterfly *Euphydryas gillettii* are deposited most often on leaves exposed to the east, perpendicular to the early morning sun which accelerates the development of the eggs.

When a plant species grows on both north and south slopes, the development of the same species is often much faster on south slopes. Differences in the flowering period are not only dependent on aspect but also on soil type (van Os 1981). The difference in development time

Table 15. Thickness of soil horizons (cm) and organic matter concentration (weight %, loss on ignition) in humus podzols on two north and two corresponding south slopes. After Stoutjesdijk (1959).

Slope type	Layer	Thickness	Organic matter
North 1	A_0	3	
	A_1	17	9.3
	A_2	13	1.5
	A_3	8	11.2
	B_1	12	3.3
	B_2	25	1.5
	C		0.5
North 2	A_0	4	
	A_1	18	7.4
	A_2	14	1.3
	A_3	8	9.2
	B_1	12	4.5
	B_2	15	1.2
	C		0.4
South 1	A_0 }	1	
	A_1 }	16	3.1
	A_2		
	A_3	10	5.3
	B_1	20	3.0
	B_2	14	1.2
	C		0.4
South 2	A_0 }	1	
	A_1 }	12	5.7
	A_2		
	A_3	4	5.2
	B_1	8	4.1
	B_2	15	1.3
	C		0.5

of *Leucanthemum vulgare* on north and south slopes on löss and marl was only a few days, but on clay four weeks. Fig. 30 shows how the flowering frequency curve of dandelion (*Taraxacum* spec.) on the south slope of a dike is almost entirely completed before that on the north slope, though the last flowers on the south slope flower almost as late as those on the north slope.

De Fluiter, van de Pol & Woudenberg (1963) describe experiments by Evenhuis who worked with pupae of the moth *Enarmonia pomonella*, which were put on vertically arranged pieces of bark with various aspects, and which were compared with pupae which were protected from direct solar radiation. The development on the south side was always the earliest. Some phases in the development could be as much as 33 days earlier in the most favourable situation.

The role of aspect in the population dynamics of the butterfly *Euphydryas editha* in Californian serpentine grasslands is quite complicated (Dobkin, Olivieri & Ehrlich 1987; Weiss, Murphy & White 1988). Weiss et al. state: "The highest survivorship of pre-diapause larvae comes from early flying females which lay egg masses on cool north-facing slopes where host plants remain edible later in the season. The females that fly earliest, however, are those that developed as post-diapause larvae on warmer south-facing slopes, where pre-diapause starvation rate are highest". Furthermore there is interference with development and senescence of the host plants on different slopes. Where there is much variety of aspect on a small scale the larvae are able to switch from one aspect to another.

Alpine marmots (*Marmota marmota*) prefer to make their burrows on south slopes. Yet the high temperatures on the south slopes restrict the time they can feed above-ground (cf. 4.1.5). North slopes would be more favourable in summer. Their preference for south slopes is probable due to better hibernation conditions there (Türk & Arnold 1988).

The differences between north and south slopes are also reflected in differences in soil development. This is clearest on a larger scale when mountain slopes are compared. Aulitzky et al. (1961) found in the Ötztal (Austria) a podzolic brown earth on south slopes and an iron-humus podzol on north slopes. On smaller scales differences are also clear. Podzol development on north and south slopes of sandy ridges in the central Netherlands with a *Calluna* heath was different, with deeper A1, A2 and A3 horizons on the north slopes. The amount of organic matter was also greater there (Table 15).

Vegetation can both reinforce and reduce differences in microclimate. This can be demonstrated using the factor aspect. In dry and warm climates such as in the Atlas Mountains, forest can only develop on north slopes. Already existing differences in microclimate between north and south slopes only become greater. The same is found during the colonization of bare sand on dunes. Original differences between the two aspects are reinforced in this way (Fig. 27). It shows how the north slope overgrown with *Calluna* is much colder than the bare north slope. The same development is found on north versus south slopes on the dunes of the Frisian islands. Because of the maritime climatic constraints no further development beyond the heathland and lichen steppe phases (see above) is possible. However, on the inland dunes a further development towards forest is possible (either spontaneously or after planting) and if this occurs the differences of the vegetation may be reduced again later on. In that case the microclimatic differences near the ground will become smaller as well.

This was investigated in a formerly mobile inland sand dune area in Drenthe, the Netherlands. From 16 to 23 September 1974, measurements of maximum and minimum temperatures at the soil surface and of evaporation at 3 cm height were made on two steep (25 - 35 °) north and south slopes, one slope of each type covered with heathland, one with pine wood. The level at which the measurements were taken are well under the canopy

Table 16. Temperature and evaporation (mm/day) on north and south slopes in heathland and pine wood. (J. J. Barkman unpubl.)

Site	Maximum temperature °C		Minimum temperature °C		Temperature Amplitude °C		Evaporation (mm/d)
	Average	Extreme	Average	Extreme	Average	Extreme	
Heath (N)	17.6	24.5	7.2	2.4	10.4	22.1	0.65
Heath (S)	29.7	43.3	6.5	0.9	23.2	42.4	1.04
Difference	12.1	18.8	0.7	1.5	12.8	20.3	0.39
Wood (N)	17.0	21.7	6.7	2.2	10.3	19.5	0.43
Wood (S)	20.1	33.8	7.3	2.5	12.8	29.7	0.56
Difference	3.1	12.1	0.6	0.3	2.5	10.2	0.13

height of the heather, and compared with bare soil a certain levelling effect may be expected and seems evident from the maximum temperatures on the south slope which are not very high. The effect of the pine wood is still evident, especially regarding temperature amplitude, as is seen in Table 16.

As may be expected, differences in vegetation in the pine wood are smaller. This can be seen by calculating the floristic similarity between sets of vegetation samples. Indeed, the similarity between the heathland samples was considerably less than the similarity between the woodland sets.

There is a clear relation between aspect and inclination. First, inclination influences the effect of aspect such that the difference between steep north and south slopes is greater than that between gentle contrasting slopes. On very steep slopes there is also a difference with respect to precipitation: the total amount of rain falling on a steep slope may be less because the receiving surface is smaller, but where the slope faces the wind, more rain is received. As far as snow is concerned, it slides more easily down a steep slope.

A special case occurs on slopes of > 90 °, i.e. overhanging rocks and overhanging boles of trees. Unless the surface is exposed to wind it will not receive any rain and the vegetation is limited to unicellular green algae and a few thallose lichens. All foliose and most of the common thallose lichens are missing. The lichens occurring here are crustose, bear soredia and have bright white, light blue, yellow or orange colours. Examples are *Lecanora umbrosa, Acarospora chlorophana, Leproplaca xantholita* and several species of the genera *Cetraria, Lepraria* and *Ramalina.* They are entirely dependent on water in the air (Büttner 1971; Creveld 1981) and are not favoured by rain because rain dislodges the air from the intercellular cavities in the plectenchymatic hyphen tissue of the fungus component as a result of which the O_2 - CO_2 exchange is decreased which is detrimental to the symbiotic green algae.

It would seem that plants underneath overhanging rocks receive less light. This is not true for lichens, however, at least not in climates with snow! Lichens are photosynthetically active in winter; in such places they will not be covered by snow and therefore receive more light than lichens elsewhere. Moreover, they receive a great deal light that is reflected by the snow on the ground underneath the rock.

On overhanging granitic rocks in south Norway a special lichen association, *Lecanoro-Acarosporetum chlorophanae* occurs. Creveld (1981) found that this association receives the highest light intensities in winter because of snow reflection and can be photographed in winter without using a flashlight, but in summer a flashlight is needed. She called this adaptation *cheimophotophytic* (cheimoon = winter). Moreover, the moisture conditions for lichens such as *Lecanora* on rain-free spots in winter are more favourable (a higher relative humidity of the air) and photosynthesis is possible at very low temperatures. Therefore we can assume that the main assimilation season for these lichens is in winter and early spring (Creveld 1981).

3. The influence of vegetation on microclimate

3.1 Introduction

Vegetation exerts a range of influences on the microclimate, which will be the subject of this chapter, and the introductory section lists the various aspects of the macroclimate with which vegetation interacts. As to solar radiation, grassland reflects 15 - 25% of the sunlight, which is about as much as a sandy surface does; woodlands reflect only 10 - 15%, which resembles dark soil. Because of transpiration by plants the temperature on a vegetation canopy will be less than compared with bare soil of about the same reflectivity. In addition the absorption of radiation by vegetation is not limited to a sharply defined surface but especially in an open canopy absorption takes place over a layer of some depth, particularly when the leaves grow vertically, as in many grasslands. Finally, heat transport to the air is easier from a loose canopy than from bare soil.

There are vegetation structures which may develop a higher temperature at their surface compared with bare soil, for example the greyish lichen cover found on stabilized dune sand or, even more extreme, the dark green moss cover of *Polytrichum*. Such carpets may dry up completely in the sun; then the transpiration is not only zero but also the specific heat by volume has become very small and the albedo and conductivity are less than those of bare sand. So that we have higher temperatures there than on the bare soil surface. The moss *Tortula ruraliformis*, well-known from dry dunes in NW Europe, is usually in a similar dry state and looks brownish. It needs only a small shower to suck up sufficient water to become green again, a process which takes a few seconds. Willis (1964) studied this process experimentally and showed how under dry conditions the green papillate cells of the lamina are hidden behind the thickened brown cells on the abaxial surface of the mid-rib of the leaves, and how through absorption of water by the cell walls the leaves unfold in a few seconds.

Plants form humus, which has a low albedo, a low specific heat by volume (ρC) and a low conductivity (k). These factors all contribute to the occurrence of high surface temperatures during the day and low temperatures at night by insulation of the subsoil. As with plant cover, the heat flow into the soil is relatively small during the day. Anyway, the humus will then have little influence on the total energy balance. At night the influence of the humus is great, however, especially on heathlands and in peat bogs.

Another important factor is the presence of litter. Here, the size of the decaying leaves is significant. Large leaves form big air cavities which are cool, moist and rich in CO_2, which is important for many soil organisms, favourable for some, unfavourable for others. Small leaves and needles form a loose type of litter which easily dries up, which is unfavourable to most organisms.

The structure of the vegetation has a manifold influence on the energy balance of the soil surface and on the microclimate in the vegetation. It will be discussed in the following sections.

First, we will discuss the influence of vegetation on the energy balance of the surface. Then we will compare the distribution of temperature and vapour pressure values in

Table 17. Energy balance (W/m^2) of a Dutch grassland (13 May 1980). Temperature at 1.50 m was 16.5 °C, at 45 cm 18.9 °C. Water vapour concentration at these heights: 6.0 and 9.1 g/m^3 respectively.

In		Out	
Solar radiation	839.0	Reflected radiation	200.0
Atmospheric long-wave radiation	287.5	Long-wave radiation emitted	421.0
		Heat flow to soil	12.6
		Released to air as latent and sensible heat	492.9
Total	1126.5	Total	1126.5

vegetation types with varying structures. Finally some vegetation types will be discussed in more detail, with different emphases for each type.

3.2 The energy and water balance of the vegetation

It is possible to make up an energy balance for a system of soil + vegetation, in the same way as we have done before, for bare soil with a dry or wet surface. Energy and moisture exchanges will occur in a layer rather than on a surface. These exchanges also determine temperature and moisture at various levels in the vegetation. An extensive treatment of heat transfer and evaporation by vegetation is found in Thom (1975).

We will now concentrate on the energy budget of the vegetation as a whole, viz. a grassland with a closed vegetation of 40 cm height. Table 17 presents the energy balance for the entire system vegetation - soil. Again, the closing entry on the budget is the amount of energy needed for evaporation and for warming up of the air. It follows from the data on temperature and vapour pressure that there is an upward transport of both heat and water vapour. Both the temperature and the vapour pressure just above the vegetation are greater than higher up in the air. The distribution of the available energy over sensible and latent heat cannot be calculated in the same way as we did for a wet surface. Heat is transferred in the same way, i.e. from the surface of the grass leaves to the air, but the water vapour follows a more complicated path via the interior of the leaves and the stomata.

Another possibility is to make use of the gradient in vapour tension and temperature in the air above the vegetation. When there is no lateral transport, or rather, when in a lateral direction the net influx of heat and water vapour in a vertical column of air is equal to the net outflux, we have for the water vapour transport E (in $gm^{-2}\,s^{-1}$):

$$E = \frac{A^{*}.\,(c_1 - c_2)}{(z_1 - z_2)} \qquad (3.1)$$

where c_1 and c_2 are water vapour concentrations in the air (in g/m^3) at heights z_1 and z_2. A^* (in m^2/s) is a parameter indicating the intensity of the turbulent transport process.

For the transport of heat we can follow the same reasoning, and more generally for all characteristics of the air, which we can express as amount per volume unit of air, e.g. for CO_2 concentration or the number of pollen per m^3. This because the turbulent exchange process does not function selectively. In order to apply this idea to heat transport we should not work with temperature differences but with differences in heat content per m^3 of air. Then we have:

$$H_{air} = A^* \cdot \rho C_p \cdot \frac{(t_1 - t_2)}{(z_1 - z_2)} \tag{3.2}$$

H_{air} is the transported heat ($Jm^{-2}s^{-1}$), ρ is the density of the air (= 1225 g/m^3), C_p is the specific heat of air (1.01 $Jg^{-1}K^{-1}$). If we also express evaporation in energetic units (H_{ev}), then, with V = 2465 J/g for the evaporation heat of water:

$$H_{ev} = A^*.V \cdot \frac{(c_1 - c_2)}{(z_1 - z_2)} \tag{3.3}$$

or

$$\frac{H_{ev}}{H_{air}} = \frac{V\ (c_1 - c_2)}{\rho C_p\ (t_1 - t_2)} \tag{3.4}$$

The values of $(c_1 - c_2)$ and $(t_1 - t_2)$ are 3.1 g/m^3 and 2.4 °C respectively (Table 17), so H_{ev} / H_{air} = 2.57.

From the energy balance we deduce: $H_{ev} + H_{air}$ = 492.9 W/m^2. Further we have H_{ev} = (2.57/(2.57 + 1)) · 492.9 W/m^2 = 354.8 W/m^2. Now we can calculate the value of A^*in Eq. 3.1, returning to the evaporation in g/m^2: E = H_{ev} / V = 0.144 $gm^{-2}s^{-1}$. Substitution in Eq. 3.1 gives A^* = 0.049 m^2/s, for $(z_1 - z_2)$ = 1.05 m. A^* has the same dimension as the diffusion coefficient D for water vapour in still air. The value of D is 21.2 · $10^{-6}m^2s$ but the value of A^* is more than 2000 × that of D. Thus, in this case the transport process in the atmosphere is about 2000 × as effective as the diffusion process.

We can write $A^* / (z_1 - z_2)$ as $1/r$, where $1/r = 0.049 / 1.05$ and $r = 21.4$ s/m (cf. 2.6). In this case the numerical value of r is the same for the transport of heat and water. The calculated evaporation is 518 $gm^{-2}h$, which amounts to a layer of water of 0.5 mm thick per hour. Thus, for a sunny summer day in the temperate zone the evapotranspiration for a closed crop canopy is about 5 mm per day.

Let us return to Eq. 3.4. The factor $V/\rho C_p$ is inversely proportional to the atmospheric pressure because ρC_p is proportional to that. The dependence on the temperature is very small. Because of the relation between concentration of water vapour and vapour pressure we can replace Eq. 3.4 by

$$\frac{H_{ev}}{H_{air}} = \frac{(e_1 - e_2)}{\gamma\ (t_1 - t_2)} \tag{3.5}$$

The factor γ for turbulent transport (Ch. 2.6) is just the same as the factor of the same name occurring in laminar transport, proportional to the atmospheric pressure and somewhat dependent on the temperature, albeit for other reasons. It is a coincidence that the numerical values differ little. The thermodynamic psychrometer constant used in large-scale meteorology is identical with γ in Eq. 3.5.

In the case of the grassland the larger part of the available energy $(R_{net} - H_{soil})$ was spent on transpiration, but there was still heat dissipated to the air. Of course, it is feasible that in dry air and with a strong wind, transpiration from a vegetation cover may be strong enough for the heat to be withdrawn from the air. However, this can only be a small amount, because the decline in vapour pressure is less at lower temperatures of the vegetation. The situation of withdra wal of heat from the air with strong radiation can be observed when a relatively small area of an irrigated crop occurs in arid surroundings (the oasis effect). In this case H_{ev} can be up to 70% higher than R_{net} (Linacre 1976). Šmid (1975) found occasionally a weak oasis effect at midday in reed swamps in Czechoslovakia.

Generally, for an extensive area of crops, the net radiation seems to provide a reasonable basis for the calculation of the transpiration, with the crop transpiration H_{ev} always remaining below R_{net}. Šmid found a maximum transpiration of 6.9 mm/day, for a reed swamp, with the energy balance over 24 h as follows:
R_{net} = 19150 KJ/m^2; H_{soil} = 1600 KJ/m^2; H_{air} = 600 KJ/m^2; H_{ev} = 16950 KJ/m^2. Extremely high values have sometimes been calculated on the basis of measurements of plants in pots. Gessner (1956) calculated a transpiration of over 40 mm/day for *Scirpus*. The discrepancy may be ascribed to the fact that transpiration was related to the surface area in the pot, while the sides of the vegetation were left out of the consideration. For an extensive vegetation such a transpiration would mean such a strong cooling off that transpiration would become impossible. Where lysimeters are used, i.e. containers with an untouched column of soil + vegetation placed on the level in the vegetation, good agreement with energy considerations is found (Monteith & Szeics 1961).

Penman (1948) was the first to attempt to calculate the transpiration of a green crop with a good moisture supply from general weather data. He started from the situation of a free evaporating water surface and arrived at the formula for a wet surface named after him (Eq. 2.33). For the step from a wet surface to a transpiring crop Penman used an empirical factor:

$$H_{ev/veg} = 0.8 H_{ev/wet} \tag{3.6}$$

where $H_{ev/veg}$ is the transpiration of a closed green vegetation cover with a good moisture supply. This method for the calculation of the potential evapotranspiration yields undoubtedly good results for agricultural crops, because we are dealing with crops which must assimilate intensively and thus have their stomata wide open with a good moisture supply.

Natural vegetation may have lower water use than the Penman formula indicates, even if there is no shortage of water. Stoutjesdijk (1959) estimated on the basis of the decrease in moisture content in the rooted soil layer, that heathland uses only 40% of R_{net} for transpiration. The amount of heat dissipated to the air is about 3 × as high as with a grassland. This estimation was confirmed by Grace (pers. comm.) through measurements in Scotland (see also Cernusca 1976). The differences are also clear from air temperatures at 1.50 m above the heath. These were 2 - 3 °C higher than the maxima measured at nearby stations. Where vegetation consists largely of mosses and lichens the heat and water budget may be still more extreme. Water losses are prevented almost entirely, as soon as the upper cm's are dried up. It is estimated that for alpine vegetation above the tree limit in Norway, at least two thirds of the net radiation is dissipated to the air as heat (Stoutjesdijk 1970b; Lewis & Callaghan 1976). In peat bogs where the top soil is dried up the same situation arises.

In all these cases great instability arises in the atmosphere, visible to the eye. Such instability may lead to local whirlwinds over heathlands and shrublands with isolated trees, which may carry leaves and even bigger branches up into the air. This phenomenon has been described for various situations in shrub communities above the tree limit, dry grasslands, and raised bogs (Stoutjesdijk 1959, 1970b; Geiger 1961; Warren Wilson 1957). Van Dieren (1934) presented a vivid description of small whirlwinds over the dunes of the island of Terschelling and ascribes the rejuvenation of lichen vegetation on warm south slopes to such local disturbance, which creates bare soil.

Dry vegetation has little transpiration and a heat and water budget which deviates markedly from green vegetation. An example can be seen in *Molinia caerulea* vegetation

Table 18. Energy balance (W/m^2) of a dead heather stand. Uddel, central Netherlands, April 14, 1980, midday. Air temperature at 1.50 m 21 °C and at 40 cm 24.4 °C. Vapour pressure values 6.4 and 6.9 mbar respectively.

In		Out	
Solar radiation	762	Reflected solar radiation	48
Atmospheric long-wave radiation	316	Long-wave radiation	460
		To soil	20
		Subtotal	528
		To air as sensible and latent heat	550
Total	1078	Total	1078

which remains dry and yellow until June. The energy budget for the dry *Molinia* will be similar to that of a 40 cm tall heather vegetation (*Calluna vulgaris*) attacked by the heather beetle *Lochmaea suturalis* (Table 18). The dead leaves, sucked dry by the beetles, remain on the branches. From the energy balance we can calculate that only 19% of $H_{ev} + H_{air}$ is spent on evaporation (104.7 W/m^2) and 81% (445.3 W/m^2) is dissipated to the air as sensible heat.

If we compare the energy balances of the grassland (cf. Table 18) and the dead heathland with that of bare soil (Fig. 31), we notice that the net received radiation for the grass is much higher than for raw humus and sand, despite the lower total amount of radiation received and the higher reflection. Due to lower temperatures of the grass canopy the emitted long-wave radiation is much lower. The total amount of sensible and latent heat ($H_{air} + H_{ev}$) released to the air is 97.5% of R_{net} for grass, because here the heat flow to the soil is small. For sand it is only 40%; 60% of R_{net} flows into the soil (see also 2.7.2).

It should be noted that the amount of sensible heat (H_{air}) dissipated by the grass to the air, though smaller than for bare soil, is still considerable. In other words: the contribution to the thermal convection may still be important for an evaporating surface. Finally, the dead heather has the highest net radiation, although the total radiation (solar plus long-wave) received is low as is the solar radiation. The higher net radiation as compared with the grass is mainly due to the low albedo. H_{ev} and H_{soil} are small and consequently H_{air} is high; this is because the temperature does not become very high and thus the emitted radiation remains low. The effective radiation temperature of the dead heather is only 27 °C.

The fine branches and leaves have a high heat transfer coefficient. They do not reach high temperatures and besides, deeper layers in the vegetation receiving little sunlight contribute to long-wave radiation. Again a value of α can be calculated from $H_{air} = \alpha \cdot \Delta t$, where Δt is the difference between the effective radiation temperature of the vegetation (27 °C) and the air temperature at 1.50 m (21 °C). This gives $\alpha = 74.2\ Wm^{-2}K^{-1}$. This value is about five times as much as the value expected for bare soil with the same wind speed (1.1 m/s). So, by its looser structure, only vegetation causes a change in the thermal budget of the earth's surface.

3.3 Vegetation and microclimate

The simplest case is that of low vegetation covering the soil like a carpet, for example *Loiseleuria procumbens* in the high mountains of Europe (Stoutjesdijk 1970b). Surface temperatures of such a plant cover are in between those of a wet and a dry soil. In

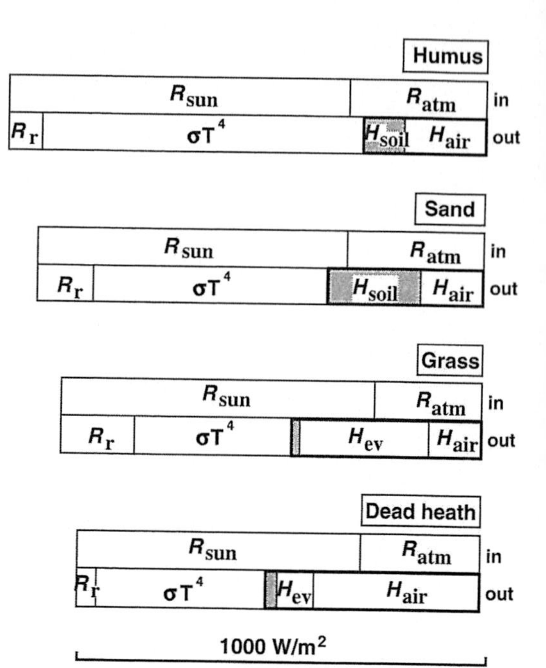

Fig. 31. Energy balance of the earth's surface with and without vegetation. R_{sun} = solar radiation; R_{atm} = atmospheric long-wave radiation; R_r = reflected solar radiation; $\sigma \cdot T^4$ = emitted long-wave radiation; H_{soil} = heat flow to the soil (grey); H_{ev} = evaporation; H_{air} = heat dissipated to the air. The section within the thick line indicates the net radiation.

Scandinavia Δt was over 20 °C in *Loiseleuria* vegetation. If there is taller vegetation, absorption of radiation is more gradual, dependent on the structure of the vegetation. Also some exchange occurs, i.e. transport of heat and water vapour between the various layers of the vegetation and to the soil and the free air above the vegetation. There is also lateral transport. Assuming that the net effect of the latter is small for homogeneous vegetation, because at similar levels temperature and humidity of the air are the same, we can characterize the microclimate in such vegetation by means of profiles, i.e. curves indicating the change of the various parameters with height.

Goudriaan (1977) has also tried to simulate the microclimate in vegetation on the basis of its structure and the physiological characteristics of the participating plants. The vegetation can be divided into a number of horizontal layers and the mutually dependent energy budgets of these layers are used in the simulation. Such a theoretical treatment of the process of interaction between soil, vegetation and atmosphere certainly contributes to a better understanding of the processes involved. However, for a qualitative comparison of microclimates and for an understanding of the microclimate factors limiting plant growth, the simulation approach may be too complicated. Therefore, we present only some typical and extreme measurements in vegetation types of different structures. An extensive bibliography was presented by Dierschke (1977).

We start with a tall closed reed swamp (*Phragmites australis*) on wet soil covered by a thin layer of water. Fig. 32 shows profiles for radiation, temperature, vapour pressure, saturation vapour pressure and wind. The vegetation was 2.40 m tall with isolated stems of 2.80 m, the highest level of the measurements. From these tops down to 1 m radiation intensity decreased almost linearly with height . Below 1 m the decrease was quite small. If leaves had been equally distributed vertically, the decrease would have been more asymptotical. In reality the density of the leaves was between 2.40 and 1.80 m at the maximum, while below 1 m the vegetation was fairly open.

The air temperature was 24 °C and changed little between 2.80 and 2.40 m. From there

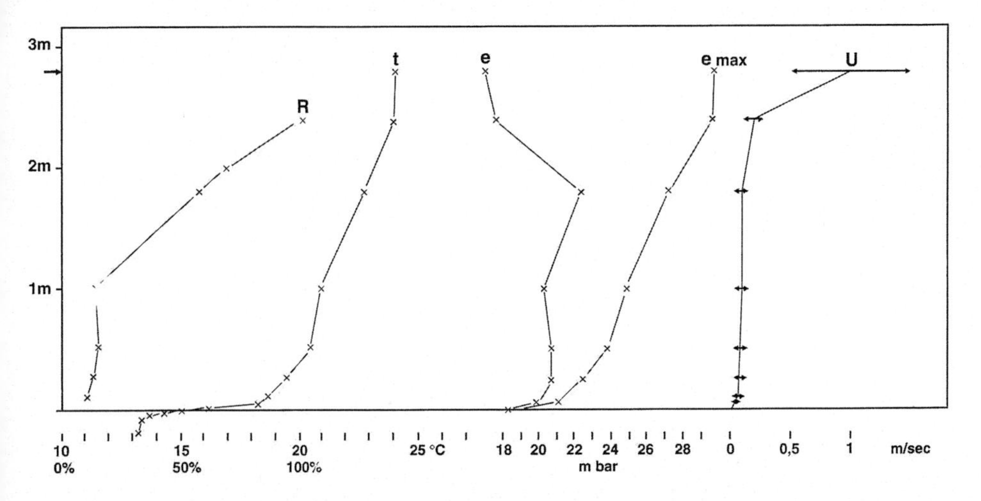

Fig. 32. Profiles in a tall reed swamp. R = Solar radiation; t = temperature; e = vapour pressure; e_{max} = saturation vapour pressure; U = wind speed. Radiation intensity is expressed as % of the radiation above the vegetation. July 1973, 12.00 h, clear sky. After Stoutjesdijk (1974).

it dropped more rapidly, especially so below 60 cm. At the surface of the water the temperature was 15 °C. Below 2.40 m a continuous heat flow toward the cold water surface occurred. Whether the vegetation as a whole released heat to the air above or took it up cannot be decided because of the small temperature changes above 2.40 m.

Vapour pressure showed a clear increase from 2.80 to 2.40 m and a still larger increase to 1.80 m. This may be related to the low wind speed and the resulting low exchange coefficient. Below 1.80 m vapour pressure decreased, but it had a second maximum between 0.50 and 0.30 m. The minimum at 1.00 m would suggest that water vapour was disappearing. This is possible in principle, because the vegetation is three-dimensional and lateral transport may occur (Byrne & Rose 1972). Anyway, there was downward vapour transport in the vegetation, a very uncommon situation, which was possible because of the low water surface temperature. As a result there was continuous condensation at the water surface. This could be demonstrated by putting a piece of aluminium foil on the surface; the foil became fogged immediately. The saturation deficit, i.e. the distance between the curves for e and e_{max}, decreased gradually and approached zero near the surface.

The wind speed showed a strong decrease to ca. 0.1 m/s as soon as the vegetation became denser, and then a more gradual decrease to a few cm /s near the surface. A situation with a similar temperature profile was found in an irrigated sugar cane field, with the difference that the vapour pressure increased continuously towards the surface (Geiger 1961).

An example of a vegetation structure with a dense canopy cover above an almost empty space is provided by a *Ligustrum vulgare* scrub in coastal dunes (Fig. 33). The vegetation reflected 15% of the incoming radiation. Of the remainder, 97% was absorbed by the canopy, which was ca. 30 cm thick. As can be expected the maximum air temperature was found on the canopy, which was 5 °C above the free air temperature. Such a great difference is relatively rare. It is due to the compact structure of the canopy.

At a depth of 1 cm the soil was already much cooler than the air above it. Apparently the

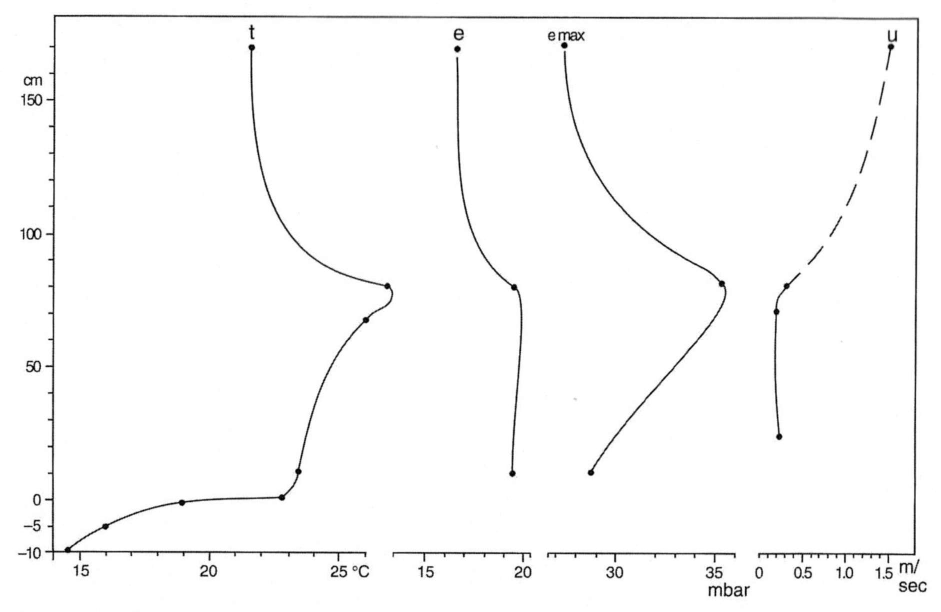

Fig. 33. Profiles of temperature, vapour pressure, saturation pressure and wind in dense *Ligustrum vulgare* vegetation. (See also Fig. 32.) 9 July 1958, 11.45 h; intensity of solar radiation 837 W/m^2. After Stoutjesdijk (1961).

soil takes up heat from the air together with long-wave radiation from the canopy and the small amount of short-wave radiation that was transmitted. Soil temperatures did not rise much during the day, in contrast to the situation with the nearby open dune sand. The lowest wind speed values were recorded in the canopy though they were not much different beneath the canopy. The vapour pressure in the canopy was a little higher than above it (1.70 m) and remained constant down to the soil surface, which indicates that the soil was not a source of moisture. The saturation deficit between the tops of the vegetation was higher than above the vegetation, since, because of the higher temperature, e_{max} increased more than e.

The *Ligustrum* vegetation is a clear example of drastic changes in the canopy because the leaf density of *Ligustrum* absorbed almost all the solar radiation.

In taller scrub without understorey the wind is stronger and is combined with the shady microclimate with small differences in temperature and vapour pressure compared with the open (Stoutjesdijk 1961). Still smaller are the differences in isolated patches of wood or scrub, e.g. *Crataegus monogyna* with a dense canopy but with an open side (e.g. near a path). If the soil is dry the climate, even near the ground, is the same as normally found at 1.50 m: the climate of a Bedouin tent with open sides!

Fig. 34 presents a different situation of a low and open vegetation with scattered tussocks of the grass *Calamagrostis epigejos* and a maximum height of 40 cm. A 4 cm layer of dead, dried-up grass remains covering the soil. This layer was reached by the sun at many places. The temperature maximum was 3 mm above this litter layer where the temperature was on average 22 °C higher than at 1.50 m. There was a remarkable fluctuation in the temperature at 3 mm above the litter,with extremes 15 °C apart, observed over one minute. At 5 cm the average air temperature was still ca. 10 °C higher than at 1.50 m. Fluctuation was still strong, 9 °C, but the interval did not overlap with the interval at 3 mm. The fluctuation interval at 10 cm did not differ much from that at 5 cm, an indication of a better exchange. At 5 and 10 cm the temperature was much higher than above a warm sandy

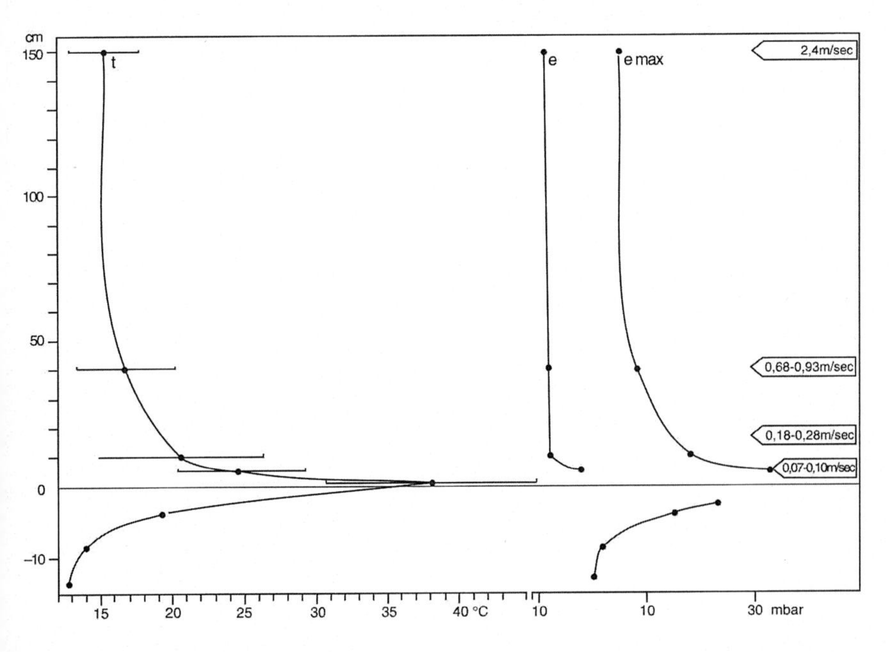

Fig. 34. Profiles of temperature, humidity and wind in open *Calamagrostis epigejos* vegetation. June 11, 1959, midday. After Stoutjesdijk (1961).

surface under comparable conditions. Apparently the vegetation, although thin, reduces the movement of air.

The vapour pressure profile was completely different from the temperature profile. At 10 cm the vapour pressure was little higher than at 1.50 m, and at 5 cm still not more than 2.7 mbar higher. This situation points to a vegetation which dissipates little water vapour and a great deal of heat to the atmosphere. Consequently, the saturation deficit was highest at 5 cm.

3. 4. The microclimate of forests

3.4.1. Temperature and humidity

In principle the microclimate of a forest is not different from that of other types of vegetation. Forests give shadow, change the light climate, slow down the wind, make the air moister through evaporation and temper temperature fluctuations. The main difference, though, is in its size relative to the human observer. Man can walk inside the microclimate and take measurements with classical apparatus. Hence, the oldest measurements we know are from forests and almost 100 yr old.

Through the deeper root systems of the trees short periods of drought are less important, and temperate European oak woods have survived severe summer droughts such as the one in 1976. On the other hand massive plagues, especially of insects can drastically change the

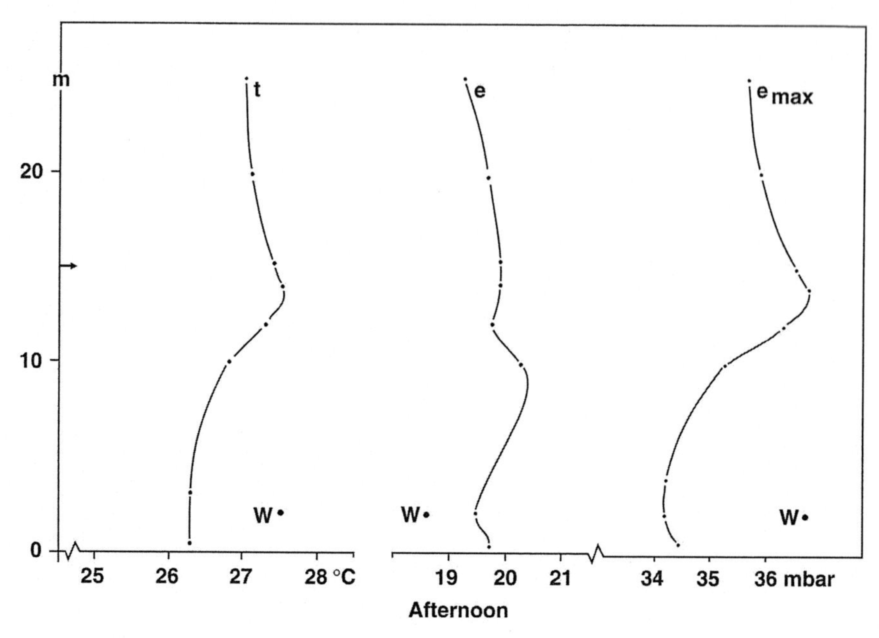

Fig. 35. Temperature and vapour pressure profiles in and above an oak wood, 15 m tall (see arrow). Data from an instrument shelter outside the wood are indicated by W. Data from Heckert (1959).

forest, for example the leaf roller *Tortrix*. In contrast to many temperate grasslands and heathlands and most tropical vegetation, temperate woodlands are deciduous and their microclimate in winter is different from that in evergreen structures. Temperate forests form thicker layers of litter and humus, which influence the microclimate near the ground, especially in sunny patches. Trees can keep a great deal of precipitation in their crowns and the throughfall along stems, branches and openings in the canopy creates an irregular pattern of precipitation on the forest floor.

Natural forests are heterogeneous because of the mixed tree composition, the mixture of different age and size classes and the pattern of smaller and larger treefall gaps. In the vertical direction there is often a continuous development of branches and leaves. As a result a forest develops its own microclimate. Unfortunately there are only a few reliable measurements from primeval forests. Much of what we know about the microclimate in forests is based on woodlands with a structure which may be quite different from natural forests. For instance, in planted woods with a dense canopy and very little understorey there is more wind and exchange of air with air above and outside the wood. This leads to an even microclimate, which is often not greatly different from the microclimate in low dense vegetation. This is illustrated by measurements by Heckert (1959), summarized in Fig. 35.

We are dealing here with a 15 m tall oak wood subjected to silviculture, which has a a crown cover of 90 %; 9% of the solar radiation penetrates down to a height of 1 m above the ground. The profiles of temperature and vapour pressure give the usual picture. The temperature maximum is found just below the canopy, the vapour pressure maximum a few m lower. Vertical differences are small. In tropical rain forest the vertical differences may be somewhat greater (max. 6 °C), but still rather small in view of the height of the forest, 45 m (Chiarello 1984). The temperature interval in the oak wood is only 1.2 °C. For the vapour pressure the extremes are 1.2 mbar apart. The temperature in the wood at 2 m is only 1.2 °C lower than outside the wood, the vapour pressure 1 mbar higher, and the saturation

Table 19. Maximum and minimum temperatures and amplitudes in °C on 4 - 5 August and 30 - 31 August in and above oak wood and heathland, compared with some data from meteorological stations (Ms) in the neighbourhood. After van der Poel & Stoutjesdijk (1959). hm = measuring height (m); t_{max} = maximum temperature; t_{min} = minimum temperature; at = temperature amplitude.

Site	hm	t_{max}		t_{min}		at	
		4-5	30-31	4-5	30-31	4-5	30-31
Oak wood	13.5	21.7	24.5	8.7	11.3	13.0	13.2
Oak wood	7.0	20.2	23.9	9.0	10.0	11.2	13.9
Heath	2.0	22.2	25.1	2.7	8.1	19.5	17.0
Oak wood	2.0	20.0	23.5	9.0	10.0	11.0	13.5
Heath	0.10	25.6	28.6	– 0.2	6.4	25.8	22.2
Oak wood	0.10	19.6	23.0	9.1	12.4	10.5	10.6
Ms Wageningen, NL	2.0	20.3	23.8	8.3	12.3	12.0	11.5
Ms Wageningen, NL	0.10	22.7	26.0	5.0	10.4	17.7	15.6
Ms Deelen, NL	2.0	21.0	23.0	8.0	10.0	13.0	13.0

deficit 3.3 mbar lower.

The small vertical differences in temperature and vapour pressure can be ascribed to the relatively strong movements of air within the wood. At 2 m height wind speed was still 12% of that above the wood. This is much more than in the vegetation types discussed earlier. The widely spaced tree trunks provide less resistance to air movement than the more densely packed stems of, for instance, a reed swamp. Also, the canopy has a looser structure than the canopy of lower vegetation. In a natural lime - hornbeam forest (*Tilio-Carpinetum*) in Poland wind speed at 0.2 m in summer was 7 - 11% of that at 18 m; in winter this was 12 - 19% (Olszewski 1974). The direction of the air flow is very changeable and can even be reversed suddenly (Sidorowics 1959).

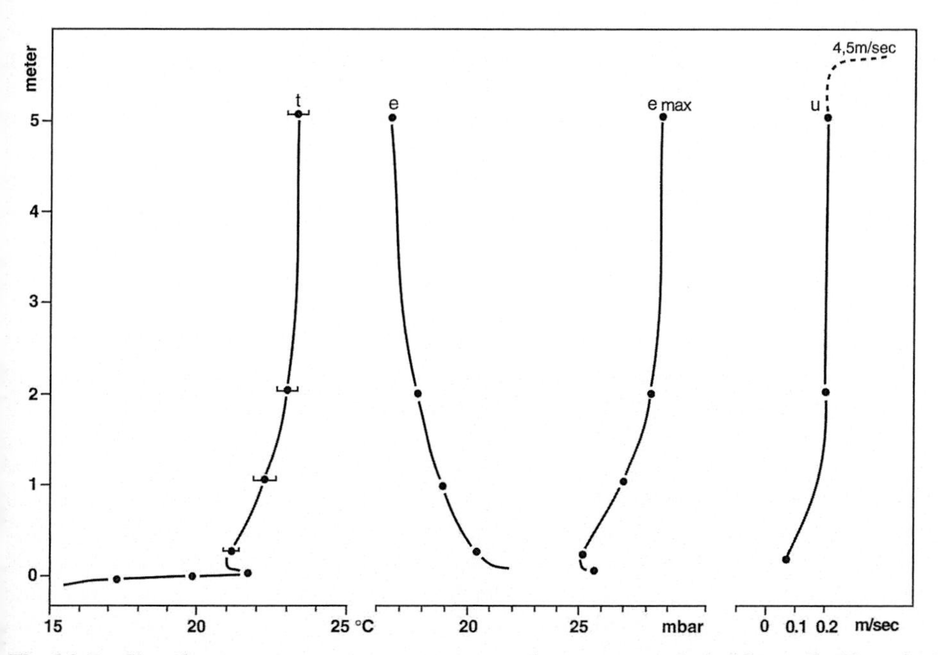

Fig. 36. Profiles of temperature vapour pressure, saturation pressure and wind for a tall willow shrub. September 5, 1958, 14.30 h. Intensity of solar radiation 558 W/m^2. After Stoutjesdijk (1961).

As far as the temperature is concerned, measurements by Kiese (1971) in a 25 m tall beech forest give the same picture as those in the oak forest by Heckert (Fig. 35). In an oak wood which originated from an oak coppice of 13 m in height the maximum temperature in the wood was 0.3 °C lower than at the nearest meteorological station (van der Poel & Stoutjesdijk 1959). The maximum increased regularly from 10 cm above the forest floor to the canopy level (Table 19).

Fig. 36 presents data on temperature, moisture and wind for a willow shrub 5 m in height with a slightly moist soil surface.Of the solar radiation received at the top of the canopy 5% was transmitted to the soil surface. The vertical differences in temperature and humidity were much greater here than in a dry oak forest, despite the fact that the intensity of the solar radiation was rather low. The wind in the scrub was weak, about 5% of the wind speed above the scrub (compared with the 20% in the oak wood).

The temperature falls steadily from the canopy downward while the vapour pressure increases. As a result the saturation deficit decreases strongly from the tops downward. A comparison between the measurements at 1.50 m and in the vegetation on the forest floor of a dry *Quercus* wood and a swampy *Alnus-Populus* wood both in the northern Netherlands gives the following results:

	1.50 m			In herb layer		
	t	*e*	*VPD*	*t*	*e*	*VPD*
Quercus wood	26.9	17.3	18.1	25.5	23.3	8.6
Alnus-Populus wood	23.3	20.9	7.7	20.8	24.0	0.5

As in every type of vegetation the radiation absorption in forests occurs at levels high above the surface. The level where most of the absorption takes place, and where the greatest transfer of energy and the strongest reduction of wind speed occur is called the 'active surface' in microclimatology. On bare soil and in open vegetation the active surface is at the soil surface, for epiphytes it is the stem surface, for epilithic vegetation the stone or rock surface, and for a closed forest the active surface is the canopy. Only open forests have a second surface, though of minor importance on the forest floor and eventually additional surfaces on tree stems and in the open shrub layer, but then the active surfaces are all discontinuous. In extremely diverse forest structures there may hardly be any clear active surface at all. In such open forests several temperature maxima may occur. Whether the major one is in the canopy, or on the forest floor, depends on the structure of the forest, but also on the type of substrate. In a wetland forest the soil surface is relatively cool; in a forest on clay the ground temperature is lower than in a forest on dry peat or sand with a thick humus layer. Especially warm are forest floors on limestone.

The great influence of the canopy on the temperature profile follows from the differences in that profile between a deciduous forest with and without leaves. If the forest is in leaf the temperature at the ground is lower than that of the air above the forest. In the forest without leaves the floor may, on the contrary, be warmer.

One would expect the number of temperature maxima during the day in a forest to be parallelled by an equal number of temperature minima at night. For the oak forest of Heckert this is indeed the case (Fig. 37, compared with Fig. 35). The air among the tree crowns at 15 m is clearly colder than the air between 20 and 25 m, although the differences are small. Downward in the forest the temperature decreases further, but only a little, with not more than 0.8 °C difference between 0.5 and 15 m. The vapour pressure does not show much variation at all.

In Table 19 the minimum temperatures in the above-mentioned oak wood are given

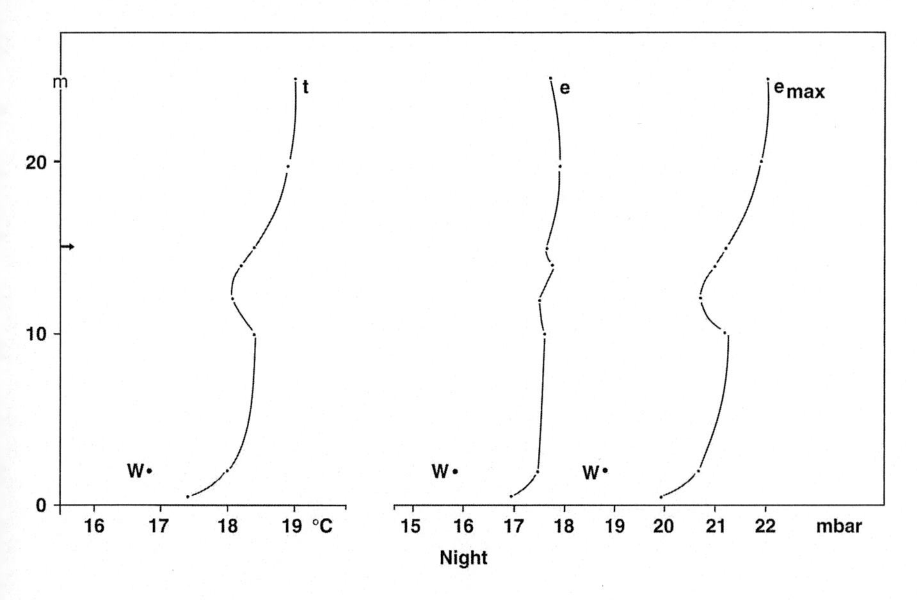

Fig. 37. Profiles in the same oak wood as in Fig. 35. After data from Heckert (1959).

(van der Poel & Stoutjesdijk 1959). The lowest temperatures were found in the tree tops, which is the active surface as far as the nightly radiation loss is concerned, but on lower levels the temperature is not much higher. Outside the forest there is a clear inversion. As a result the temperatures at 10 cm height outside the forest are much lower than inside; on nearby heathland the ground temperature even reached freezing point. The measurements by Baumgartner (in Geiger 1961) in a young, dense, spruce plantation show an almost isothermic situation on clear nights. In the 5 m tall willow scrub the minimum temperatures at 10 cm were on average 6.4 °C higher than in the surrounding dune valley with open *Salix arenaria* cover and a thick moss layer - measured during 23 clear nights between the end of April and early October. In the valley the temperature sank below zero on no less than 20 nights in the same period, against only 5 in the willow scrub.

Generally, the night temperature at 10 cm is not much higher or may be even lower than in the canopy. This is because the cold air formed outside the forest will usually descend towards the forest floor. As a result the forest floor becomes colder than one would expect, whereas the tree tops remain warmer than the ground surface outside the forest. Only in the case of a dense canopy would the cold air remain above the forest. Even then a canopy is a much looser structure than a soil surface with short vegetation, heat exchange with the surrounding air is much easier and the cooling effect is spread over a much thicker layer of air.

The soil contributes little to the nightly energy budget. But what about the above-ground parts of the forest? The heat stored in the trunks and branches cannot compensate for more than ca. 20% of the nightly radiation loss, even in a tall beech forest. The forest soil, exchanging long-wave radiation with the canopy, which has about the same temperature, does not cool off much below the air temperature or may even be warmer. Hence, there is little dew formation, although there is some on the tree crowns. On clear nights there may be some evaporation in the forest but outside there may be heavy dew formation i.e. condensation occurs (Wilmers 1968; van der Poel & Stoutjesdijk 1959). Isolated small woodlands can be warm islands in the sea of cool grasslands and fields around. Even a single large tree has a warmer microclimate at night than the surroundings, as can be experienced during an evening walk on a temperate summer night. Once we observed (van der Poel & Stoutjesdijk 1959) how soon after sunrise the warm air in a wood was suddenly replaced by the cooler air from outside so that the temperature at 2 m height dropped 4 °C. Apparently the first warming up by the sun had put the entire system in motion.

By day the warm air in the canopy will not mix readily with the colder air down between the tree trunks. On the other hand, by night we expect that the cold air in the canopy will sink down between the trunks, i.e. the forest floor could have a 'valve' function similar to that which is found in a lake (Wartena 1968). At first sight we may expect that with bright weather the air in the forest will be colder than outside the forest, when we average over a 24-h period. Reality is more complicated, however. The overall energy balance of the forest and, for example, the grassland with which it is compared may be different. Furthermore the cold air formed by night is spread over a layer at least as deep as the forest is high while with low vegetation it is concentrated near the ground. Then exchange conditions in the trunk space are different from those in the free air. We shall now look at the empirical evidence.

Heckert's measurements between 8.00 and 17.00 h show that the temperature at 2m height in the forest is clearly below the temperature outside the forest and the difference varied between 0.3 and 1.9 °C and the average difference was 1.0 °C. For the rest of this 24-h period the temperature inside the wood was between 0.0 and 1.3 °C (on average 0.9 °C) higher than outside of it. For the whole 24-h period the temperature inside the wood was only 0.1 °C lower than outside of it. In a dense 20-yr old pine plantation in Massachusetts the average summer maximum in the wood, from weekly readings, at 1 m height was 25.6 °C, while it was 4.1 °C higher outside the wood (Spurr 1957). Both in deciduous and pine woods in Japan the average temperature for July was 0.8 °C lower than outside the wood (Yoshino 1975).

In winter there is generally a net loss on the radiation balance of the forest as a whole as well as outside the forest. Between the trunks and on the forest floor the situation is fundamentally different for deciduous forest on the one hand, and evergreen coniferous or broadleaved forest on the other. A leafless deciduous canopy transmits much more solar radiation than when in leaf. Still the interception is quite considerable: 73% on a bright day for Heckert's oak forest and 44% on an overcast day. It stands to reason that the direct radiation from the sun which arrives at a low elevation angle is more affected than the diffuse radiation which comes from all sides. The reduction of the long-wave radiation loss from the forest floor (both by day and by night) is probably even less than the 44% (for diffuse short-wave radiation) because the effective sky temperature is lowest at the zenith.

Besides a 'day-and-night temperature valve' deciduous forests may also have a 'summer-and-winter temperature valve'. However, the average temperature in the deciduous wood in Japan mentioned before (Yoshino 1975) is less than 0.1 °C below that outside the forest. The winter temperature in the pine wood also studied by Yoshino (l. c.) was 0.7 °C below the temperature outside the forest. Spurr (l. c.) found in the pine plantation mentioned earlier an average minimum of –16.7 °C against –18.4 outside. The maximum was 2.7 °C in the pine wood and 5.7 °C outside it. The average winter temperature in the pine wood (–7.1 °C) was just the same as in the Japanese pine wood, somewhat below the temperature outside (–6.7 °C).

In summary: the temperature between the trunks at, say, 2m height is not greatly different from the air temperature outside the forest. Perhaps more significant is that, because of the weak solar radiation, nowhere do surface temperatures on tree trunks and on the forest floor rise significantly above the air temperature; often, they are even somewhat lower. Very important is the effect of a canopy of leaves or needles on the minimum temperatures near the ground (c.f. e.g. Table 19). In this respect there is a big difference between evergreen coniferous and deciduous woods on still, clear winter nights. Thus on a cold clear winter night the following temperatures were measured at 10 cm height: Under dense Douglas fir (*Pseudotsuga menziesii*) –4 °C, pine wood (*Pinus sylvestris*) –9 °C, *Larix leptolepis* –13 °C, *Quercus robur* –14 °C. Deciduous forests leafing early (e.g. beech and hornbeam forests) are important lowland habitats for some species with a predominantly mountanous distribution. So we wonder if there is a relationship with microclimate here. For instance, in the forest of Fontainebleau near Paris (mainly beech, *Fagus silvatica*) five montane and subalpine moss species are found on trees. This was known in the past in forests in the central Netherlands, but here the epiphytic species have disappeared because of air pollution.

In woods of the exotic *Larix leptolepis* in the northeastern Netherlands five boreo-montane mosses are found on the forest floor: *Plagiothecium undulatum*, *Ptilium crista-castrensis*, *Rhytidiadelphus loreus*, *Dicranum majus* and *D. fuscescens*. In pine woods in the same area only two of these species occur, one in juniper scrub and none in woods with the exotic *Pseudotsuga menziesii*. These five species occur optimally in coniferous forests in the Alps and Scandinavia. Why would these species occur mainly in *Larix* wood outside their main distribution area? The hypothesis is that they are not only bound to the humus of coniferous species but also to a relatively cool winter microclimate, which is the case in the deciduous *Larix* woods. Measurements of minimum temperatures do indeed show this (Barkman 1965a). The highest minima are found in *Pseudotsuga* wood; in *Juniperus* scrub minima are on average 0.7 °C lower, in *Pinus* wood 0.8 °C, in *Larix* wood 2.0 °C, and in *Quercus* wood 2.9 °C lower. The extremes are still further apart, see above. It may also be important that deciduous trees have a more favourable light climate in winter, which is especially important for mosses and this is combined with shelter from direct sunshine i.e. avoidance of overheating and extreme desiccation. On the other hand, litter of broad-leaved trees such as oaks may suffocate mosses, and here the fine litter of *Larix* does not have this effect. Some of the five species do indeed occur in oak woods, but only in places where the dead leaves are blown away.

In early spring the temperature on the forest floor in a deciduous forest may rise above that out in the open field. The trees are still leafless, and the sun is higher in the spring compared with mid-winter and enough radiation comes in, though less than in the open, to warm up the forest litter, which has a low specific heat by volume and a low heat conductivity as compared with a mineral soil outside the forest. Moreover, the forest has as yet no ground herb layer which would transpire. Finally, the warmed-up soil will not cool off so rapidly by convection because there is less wind than out in the open. An example: on April 5, with a wind speed of 5 m/s and an air temperature of 14 °C a litter temperature of 40 °C was measured in an oak coppice. Of course on dry litter in an open place surrounded by vegetation, even higher temperatures can be measured and inside the forest warm and sunny patches occur side by side with shady patches and this pattern shifts as the sun moves through the sky.

This extra heat may be important for the early development of many spring herbs in the forest (Firbas 1927, cf. Geiger 1961), especially those with food reserves in rhizomes, tubers or bulbs (geophytes), for instance *Adoxa moschatellina*, *Gagea sylvatica*, *Ornithogalum umbellatum*, *Ranunculus ficaria*, *Galanthus nivalis*, *Anemone nemorosa*. This time is favourable for these species because the forest floor receives very little photosynthetically active light later in spring once the trees are in leaf. The above-ground parts of such vernal species die off by May or June.

Evidently the time of leafing is very important for the development of a specific microclimate and the development of an understorey. Oak and ash woods (*Quercus*, *Fraxinus*) are more favourable for vernal species than beech and hornbeam woods (*Fagus*, *Carpinus*). Birch (*Betula*) also leafs relatively early but the leaves are small and transmit more light.

During the day, the vapour pressure is higher, and the saturation deficit is lower than outside the forest. In humid forests there is a clear decrease of the saturation deficit *SD* with height, whereas the relative humidity *RH* shows an increase. In relation to this we often observe a zonation of epiphytic mosses and lichens in a forest, while in tropical forests it also includes epiphytic ferns and members of the *Bromeliaceae* and *Orchidaceae*.

The higher the overall relative humidity in a forest, the higher up certain epiphytic communities may be found on the trees and the more humidity-requiring communities are found, the most sensitive ones at lower levels. Consequently the vertical zonation of epiphytes is more diversified (Barkman 1949, 1958). Many epiphytes seem to be favoured by higher daytime humidity in the forest, in spite of the fact that at night the forest may be drier than the open field with respect to *SD*, *RH* and dew fall.

3.4.2 Precipitation

> *"And Noah he often said to his wife when he sat down to dine: "I don't care where the water goes if it doesn't get into the wine"*
> *G. K. Chesterton*

A forest canopy intercepts both rain and snow. The fate of the rain drops is decided by the crown density and especially the density of the leaves. On large leaves rain drops merge towards the leaf tip. On each leaf tip only one rain drop may remain for some time. Coniferous forests with their narrow needles keep the rain in the canopy longer than broad-leaved forests. The percentage of intercepted rain is also dependent on the amount of rain per shower: with a very light shower, all the rain is intercepted; the more rain that falls, the more rain passes through immediately, especially if the wind is strong. The interception decreases asymptotically to a lower value which is characteristic for the type(s) of leaf(ves) and varies between 20 and 70 %. For a precipitation of 10 mm the following percent interception occurs:

Pinus cembra	74%
Juniperus communis	70%
Picea abies	45%
Larix	37%
Fagus silvatica	37%
Quercus	32%

(Geiger 1961; Mitscherlich 1971; Larcher 1973; Barkman, Masselink & de Vries 1977). In tropical rain forest 100 mm precipitation per day may be needed before substantial stem flow occurs (Landsberg 1984).

Of the rain passing through the crown leaves, some falls directly onto the ground; this is called the 'throughfall'. Some runs partly down the branches and finally down the tree stems; this is called the 'stem flow'. In relation to throughfall and stem flow we distinguish:

Table 20. Distribution of precipitation (in %) in two types of forest. Int = interception; Flow = stem flow; Fall = throughfall.

Species	Season	Int	Flow	Fall
Picea abies	summer	32.4	0.7	66.9
	winter	26.0	0.7	73.3
Fagus sylvatica	summer	16.4	16.6	67.0
	winter	10.4	16.6	73.0

- trees with sloping branches: the centripetal type, examples *Fagus*, *Populus*, some *Salix* species, *Cupressus*, *Juniperus*;
- trees with bent branches, the arcuate type, examples *Sambucus*, *Crataegus*, *Ulmus*, *Tilia*;
- trees with horizontal branches, examples *Cedrus*, *Larix*, *Pinus sylvestris*, *Quercus*;
- trees with hanging branches, the centrifugal type, examples *Picea abies*, *Pseudotsuga menziesii* (Linskens in Geiger 1961).

Only in centripetal and arcuate types is rainwater collected on the stem; this is especially so with the arcuate type. In some trees of centripetal and arcuate types most of the water is kept in the canopy, e.g. *Juniperus*. In the case of stems with bark fissures, water can be kept there. The largest stem flow values are found with trees of the centripetal type which have broad leaves, smooth stems and big crowns. *Fagus* is a good example of this type and *Ulmus* another. As a contrast *Picea abies* retains more rainwater, but has about the same throughfall (Table 20 after Geiger 1961). Differences between summer and winter are found in interception (also for the evergreen *Picea*) but not in stem flow. The two broad-leaved species *Acer platanoides* and *Quercus robur* have stem flow values of 5.9 and 5.7 and are intermediate between spruce and beech.

Rainwater always flows to the lowest possible place. Since no tree trunk is exactly vertical, and water is guided by large branches there is always a side of the tree that the water moves to and then descends along special tracks. After heavy rains the water can run with some force along such tracks. Here moss species such as *Orthotrichum affine*, *Dicranoweissia cirrhata*, *Bryum capillare*, *Ceratodon purpureus* and filamentous algae (*Prasiola crispa*, *Hormidium flaccidum*) are found, the latter with their threads in the direction of the stream. During a heavy thunderstorm as much as 16 l water can move down one single beech tree. Where the water meets the soil, erosion occurs and bark material with minerals is deposited. Again, particular communities of mosses and algae occur here, adapted to mineral soil and periodic inundation.

The distribution of rain over the forest floor is irregular. In a natural forest there are always gaps. In a mixed forest patterns occur because different trees have different crown densities. Extremely dense crowns are found in *Picea*, *Pinus cembra*, *Castanea sativa*, *Juniperus* and *Crataegus*; moderately dense crowns are found in *Pinus sylvestris*, *Quercus* and *Alnus*, and open crowns in *Larix*, *Betula*, *Salix* and *Robinia*. Further crown density variation is found within each species in relation to habitat and age. Apart from stemflow we distinguish three routes for rain water:

a. Dripping from the periphery of the crown; this is an important route for the arcuate and especially the centripetal type, e.g. *Picea abies*.
b. Falling through gaps.
c. Falling through the crowns; this is a smaller amount.

The relation between a, b and c is roughly 6 : 5 : 3 for a wood of Norway spruce . Component c can be divided further into precipitation halfway between the trunk and the crown periphery, c_1, and near the stem, c_2. This relation is 1.7 : 1 (calculated after Kern 1966, see Mitscherlich 1971). Very little rain falls near the base of the trunk of spruce trees, which can be experienced when looking for shelter from rain in a spruce forest. A similar pattern occurs with snow. In order to obtain a representative picture of precipitation in a forest we need 30 - 40 of rain gauges.

The unequal distribution of rain within the forest may be accompanied by special micro-patterns in the distribution of species. In wet spruce forests there is a mosaic of *Polytrichum commune*, *Sphagnum girgensohnii* and *Equisetum sylvaticum* in open places and *Ptilium crista-castrensis* and *Vaccinium uliginosum* around the base of trees. In moist spruce forests the latter species combination is found in the gaps, while different species including *Hylocomium splendens* and *Vaccinium myrtillus* appear around the tree stems. In still drier spruce forests the latter combination is found in gaps and *Pleurozium schreberi* and *Vaccinium vitis-idaea* near the trees.

Juniper scrub transmits very little precipitation. One fungus, *Geastrum floriforme* occurs underneath juniper scrub in the Netherlands; this is the only Dutch habitat where this xerophytic species (common in dry grasslands of continental Europe) is found. In the more atlantic, moister Netherlands the species is limited to relatively rain-free microsites, especially the 'dense needle' microsite type, where only 24% of the free precipitation is transmitted (Ch. 3.6, Table 23).

3.4.3 Light climate

As sunlight passes through vegetation it is not only gradually weakened by absorption, but its spectral composition changes as well. Anderson (1966) remarked that as openings in the canopy occur especially near the zenith, a larger fraction of the diffuse light is transmitted than direct radiation. Chlorophyll has a very specific absorption spectrum with a strong absorption of wave-lengths < 700 nm and a weak absorption in the near infra-red (>700 nm). Radiation transmitted through the leaves is rich in dark red and infra-red, and relatively rich in green light (Fig. 36). This effect becomes stronger as it passes through more leaves. One leaf transmits 50% of the infra-red and 10% of the short wave-lengths but for two leaves these figures drop to 25% and 1% respectively.

In Fig. 38 examples are presented of transmission depending on wave-length for two types of forest, based on measurements under a clear sky and by comparing them with the radiation intensity outside the forest. For the darkest patches in both the beech forest and the mixed forest the relative intensity of radiation in wave-lengths < 700 nm could hardly be measured by the instrument used, i.e. ca. 0.1%. The relative intensities are similar to those given by Zavitkowski (1983) for dense young plantations of *Populus*. There is a slight increase, though, for the green light between 500 and 600 nm. For this measurement the darkest places in the forests have been chosen, and the measurements do not indicate the average relative radiation intensity. In as much openings in the canopy do occur and consequently sun spots on the forest floor, light intensity is less there than one would assume. The smallest, entirely circular spots are weak pictures of the solar disc formed by small openings in the canopy, according to the principle of the pinhole camera. In tropical mountain forest the relative intensity - in the wave-lengths 400 - 700 nm in such sun spots is only 1%, which is only twice as much as in the shade (Stoutjesdijk 1972b). Only in larger, irregularly formed sunny spots may the light intensity be much higher.

If the canopy opening is large enough to allow observation of the entire solar disc, the

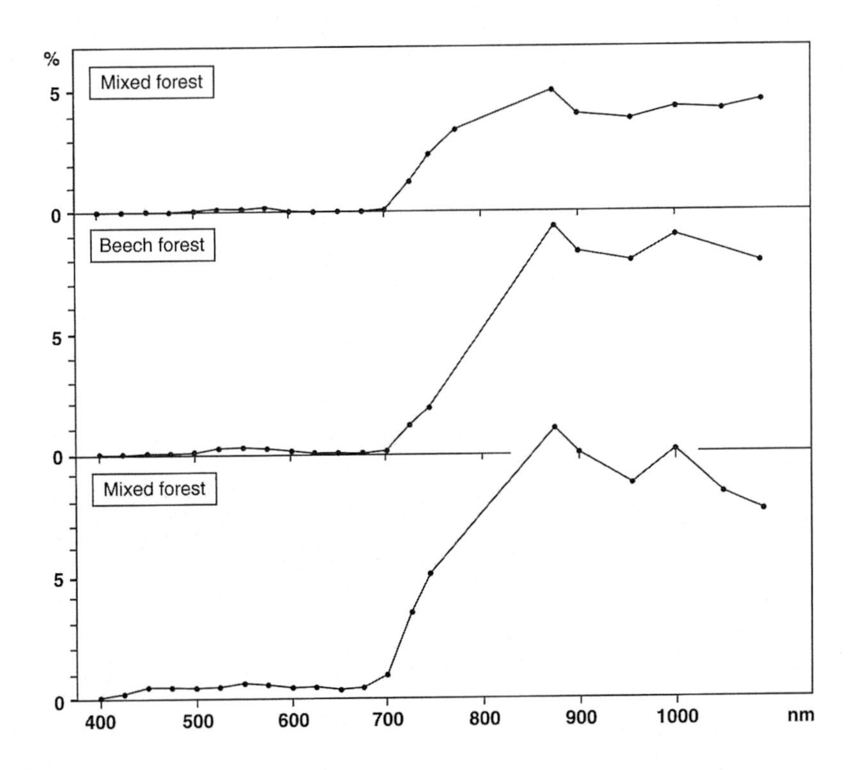

Fig. 38. Relative transmission for beech wood and mixed broadleaved wood in July around midday After Stoutjesdijk (1972a).

radiation intensity on the forest floor is of course much higher, at least equal to that of direct radiation. Dirmhirn (1964) found that in sun spots with a diameter of between 5 and 30 cm the light intensity increased regularly with the diameter, from 20% to 90% of the intensity in the open. This was in 10 m tall mixed wood.

Of course, sun flecks move over the forest floor with the changing position of the sun. For flower mites which live in the otherwise uniform microclimate of the tropical rain forest, the passing of a sun fleck is a stressfull event (Dobkin 1985). Forest plants react immediately to a sun fleck by elongation and increased photosynthesis (Morgan & Smith 1978; Knapp, Smith & Young 1989). At changing light intensities the photosynthetic efficiency increases with a higher rate of fluctuation, as long as the period does not exceed 100 seconds (Pollard 1970). This could explain Horn's (1971) remark that if the average leaf size is smaller, the forest will develop more layers (Barkman 1979). With smaller leaves sun flecks will be smaller as well, and because of the changing position of the sun and the changes caused by wind, the pattern of sun and shade spots on the forest floor as well as the lower leaf layers will change more rapidly with smaller leaves than with larger ones.

Maybe we should take the total Leaf Area Index, LAI, leaf area in m^2 per m^2 surface area, as a measure of comparison instead of the number of layers. The only figures available refer to the tree layer only. Still, they give an indication and indeed LAI values can be very high: maximally 28 for coniferous forests (two-sided leaf area) and 22.2 for an *Alnus* wood (Larcher 1973). Walter (1977; see also Walter & Breckle 1985) presents somewhat lower figures: 22 for boreal spruce forest, 18 - 20 for North-European pine forest, 16 for moist

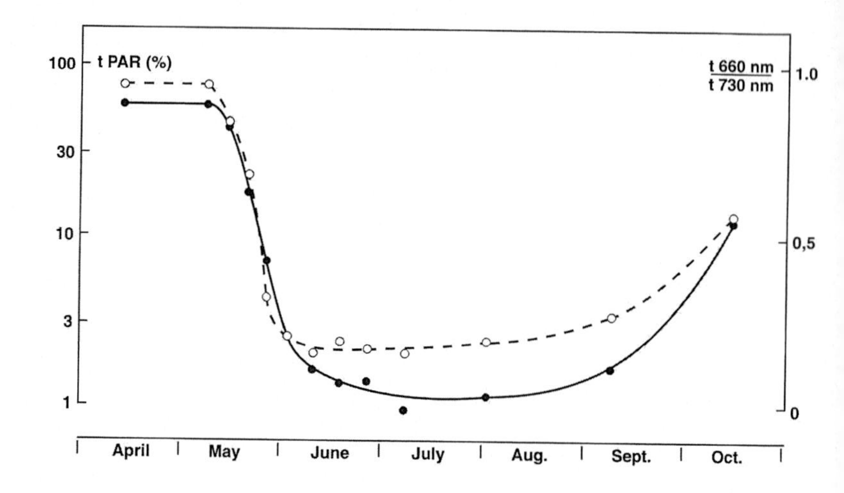

Fig. 39. Transmission of PAR (●) and the ratio between transmission at 660 nm and 730 nm (○). Ash coppice, growing season 1981. After Pons (1983).

mixed deciduous forest (including the shrub layer), and 10 - 12 for oak forest ('in moist years higher').

In order to obtain an idea of the relative radiation intensity in vegetation from one single measurement, this should be done under cloudy conditions; the intensity of transmitted radiation varies more when the sun shines. The ratio between transmission of radiation < 700 nm and > 700 nm is higher the more light is transmitted, but the differences are not as great as the LAI and the above theoretical example suggests where the transmission of one and two leaves was compared. The probable reason is that part of the light passing through the canopy is either filtered only once or not filtered at all. For the germination of seeds the relation between near red (ca. 660 nm) and far red (ca. 730 nm) is important. In full sunlight the ratio is 1.2 : 1. Fig. 39 shows how in a 7-yr old ash coppice the ratio of transmitted near red and far red first decreases in spring and increases again during summer. As can be expected, this ratio is dependent on the transmission of the wave-length interval 400-700

Table 21. Transmission in % of the intensity outside the forest, for a beech forest (B) and a spruce forest (S) in three spectral intervals. Measurements under a clear sky. For the beech forest, measurements in a leafless situation (26 November) are also presented. After Brasseur & De Sloover (1976).

Date	Wave-length interval nm					
	400-500		600-700		700-800	
	B	S	B	S	B	S
6 August	1.5	3.3	1.9	3.0	5.8	4.0
26 August	2.0	2.4	3.5	2.9	9.1	3.5
23 September	1.7	1.6	4.1	2.1	12.6	3.5
16 October	6.1	4.0	6.4	2.4	29.5	6.1
Average	3.4	2.6	4.6	2.4	17.8	4.2
26 November	33.3		51.3		54.5	

nm, the photosynthetically active radiation (PAR). Under a dense canopy the spectral composition of light can be unfavourable for the germination of several species (Stoutjesdijk 1972a; Pons 1983). The germination of both light- and dark germinators can be completely or partly suppressed under a canopy.

Brasseur & De Sloover (1976) measured the transmission for three wave-length intervals in old beech forest and spruce forest (Table 21). The canopy was not entirely closed: 70% for beech and 80% for spruce. The transmission values are averages over a longer period and approach the average in such forests better than those in Fig. 38. For the beech forest, values for the leafless situation are also given. The transmission ratio near red/far red is ca. 1: 3 in high summer, increases when the leaves become coloured, and drops to 1 : 1 after the leaves have fallen. The transmission in the blue part of the spectrum is somewhat less than that of other parts, probably because the skylight is intercepted somewhat more than the direct sunlight.

In the spruce forest the difference in transmission between near red and far red is much smaller. Here the non-selective absorption by many thin twigs is more important, and also absorption by needles is probably so strong that a greater proportion of the light transmitted consists of unfiltered light, passing through the small openings between the needles.

Long-term measurements in a German beech wood showed that in August the transmission reached a minimum, i.e. 4.4% for global radiation and 1.6% for PAR. When the wood was leafless (April) the transmissions were 29.9% and 30.7% respectively. For a *Pinus sylvestris* wood the (June) values were 14.2% and 13.2% respectively (van Eimern 1986).

Fig. 39 shows an annual curve representing the wave-length interval 400 - 700 nm for ash coppice (Pons 1983). Transmission in winter is 66%, slightly resp. much higher than in the beech forests mentioned above. During the summer the transmission decreases to something over 1%. These measurements were made under a cloudy sky; in the shade the transmission values are always higher than when the sky is clear. Comprehensive descriptions of the light climate in many types of forest and also in other vegetation types are presented by Tranquillini (1960) and Walter (1985).

3.5 The microclimate of heathland

On bare inland dune sand the first colonizers are the grass *Corynephorus canescens* and the moss *Polytrichum piliferum*, followed by lichens (Ellenberg 1988). The albedo thus decreases, although the vegetation is still open; sunlight penetrates down to the soil surface. There is no humus, so heat conduction to and from deeper soil layers is relatively easy. In this plant community, the *Spergulo-Corynephoretum*, *Calluna vulgaris* can establish and heathland of the type *Genisto-Callunetum* develops with *Calluna* becoming dominant. A humus layer is now formed, and the microclimate under the heather is shady and thus moderate.

This picture can change drastically if the heather dies of old age (usually after 15 yr) or from an attack by the heather beetle *Lochmaea suturalis* (Berdowski & Zeilinga 1983). The eggs of this beetle need a moist microclimate for their development. This is only found in a closed heathland of more than 3 yr old. If the heather does die off, *Festuca ovina* and *Carex pilulifera* and especially lichens of the subgenus *Cladina* establish.

This community, the *Cladonietum mitis* is low and open and the sun penetrates down to the dark humus. Thus, extreme temperatures can be expected and indeed the most extreme temperatures of all plant community types have been found here. Only on vegetation-free spots with an organic soil, can still higher temperatures be found (see Fig. 2), for instance

Table 22. Temperature extremes (°C) in a heathland complex. Measurements at 0 cm, but at 10 cm in the grassland.

Vegetation	Max	Min	Diff
Grassland	34	–14	48
Empetrum	21	–8	29
Calluna	42	–17	59
Spergulo-Corynephoretum	59		
Cladonietum mitis	65	–19	84

ant hills, or the bare south side of juniper scrub or bare dried up peat (cf. Fig. 1).

Establishment of young heather plants in the *Cladonietum mitis* is difficult, compared with the *Spergulo-Corynephoretum*; many seedlings die after a short time. Re-establishment of *Calluna* dominance takes a long time. If this does happen the so-called 'little heather cycle' is closed. The *Cladonietum* phase can easily be damaged by trampling by man or sheep, because the lichens do not have roots and they are very vulnerable, especially in dry weather. If the humus layer is also blown away we are back to the original pioneer situation, which is succeeded by the stage with *Corynephorus*, etc. This is the so-called 'big heather cycle' (see Gimingham 1975, 1988).

In the northern Netherlands *Empetrum nigrum* establishes occasionally in the *Spergulo-Corynephoretum. Empetrum* may also replace *Calluna. Empetrum* has a dense regular canopy as opposed to a looser and irregular canopy of *Calluna.* Therefore moderate temperatures occur under an *Empetrum* canopy (see Ch. 2.11). As an illustration, an example from a heathland complex in Drenthe, the Netherlands, is shown where temperatures were measured on an extremely hot summer day and a very cold winter night (Table 22).

Such diverging temperatures can be measured within only a few m of each other. Clearly, the *Cladonietum* is hottest in summer. Still, the soil does not feel hot, while the sandy soil of the *Corynephoretum* does. Earlier measurements were therefore probably made where the highest temperatures were expected: on bare dry sand facing south and not on the humus surface of the *Cladonietum.* That the highest temperatures occur in the latter community can be easily explained. Humus has a low specific heat per volume, but our hand, with which we try to estimate temperatures, has a very high specific heat per volume because of the high water content (85%). At first contact between hand and humus an equilibrium temperature may be established not far from the body temperature, i.e. not far from 37 °C, say 42 °C. The heat needed to warm up our hand from 37 to 42 °C is taken from the humus as a result of which its temperature rapidly decreases from 65 to 42 °C. Subsequent transfer of heat from deeper humus layers can be neglected because of the low k. The sand on the other hand, has a larger ρC and a larger k, as a result of which our hand becomes much warmer (one can make a drastic comparison by touching stone or iron of 100 °C and wood or turf of that temperature!). It is quite illustrative to walk barefoot on such a heathland mosaic of vegetation types with such extremely different surface temperatures. Differences between the different patches in this heathland mosaic can also be illustrated with weekly minimum temperatures, which were recorded over a number of years. In some years night frost occurred under *Empetrum* during only four months of the year, under *Calluna* during eight months, but in the *Cladonietum* all year round. Such variation results in differences in the species composition of mosses, lichens, fungi and beetles - the

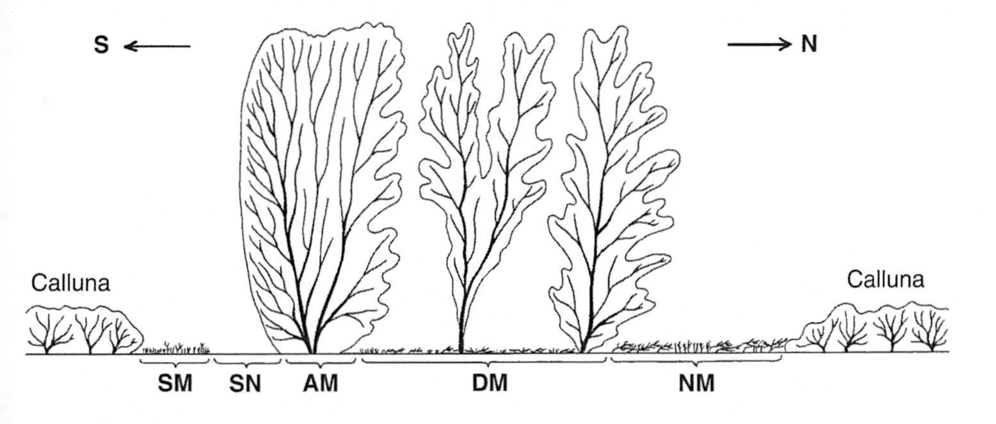

Fig. 40. Micro-communities in a juniper scrub (idealized north - south cross section), after Barkman, Masselink & de Vries (1977). SM = South, moss-rich; SN = South, needle-rich; AM = Absence of mosses; DM = Dense moss cover; NM = North, moss-rich.

organisms living at or just below the surface (den Boer 1967).

Snowfall provides another phenomenon. The irregular *Calluna* canopy gives a broken snow cover. The very cold air formed at night above this canopy sinks onto the snow-free patches in this mosaic, which is impossible with a close *Empetrum* canopy. On the soil underneath the *Empetrum* canopy, i.e. under a cover of 20 cm air and 21 cm snow, a temperature of only −0.6 °C was measured against an air temperature of – 10 °C and we must expect that the temperature on the snow-free patches was at least as low. The difference in winter minima between *Calluna* and *Empetrum* will be still greater under a snow cover, which implies that in the boreal region of Scandinavia, where the same mosaic is found, the long-lasting snow cover will cause even more pronounced temperature differences and consequently a stronger impact on the microdistribution of organisms.

3.6 The microclimate of juniper scrub

3.6.1 Vegetation

The microclimate of juniper scrub will be treated in more detail because it is found throughout the northern hemisphere; it is the only type of scrub which has been thoroughly investigated (e.g. Ellenberg 1988) and finally because it shows some interesting features and strong spatial differentiation. The following information refers to juniper scrub in Drenthe, the Netherlands, belonging to the association *Dicrano-Juniperetum* (Barkman 1985) which occurs in NW Europe on dry, poor, acid soil, mostly sand.

Juniper scrub may form an open to closed mosaic in heathland. In the case where the heathland dominants *Empetrum nigrum* and *Deschampsia flexuosa* are found these species border the juniper scrub and *Empetrum* may even behave as a climber here, reaching 1.8 m height among the junipers. *Vaccinium myrtillus* and *V. vitis-idaea*, which also belong to this type of heathland may penetrate to the centre of juniper shrubs. In this case the differentiation into micro-communities is small. On the other hand, if the surrounding heathland is dominated by *Calluna vulgaris* a transition zone is formed around the juniper shrubs, 1 - 1.5 m wide at the north side and 0.5 - 1 m at the south side, where neither heather nor juniper

Table 23. The microclimate in juniper scrub. For abbreviations see Fig. 40.

Climate factor	Microbiotopes				
	SM	SN	AM	DM	NM
Mean weekly t_{max} at 0 cm (°C)	28.6	28.3	19.3	16.6	18.7
Absolute t_{max} at 0 cm	75.0	72.0	36.0	35.5	35.0
Mean weekly t_{min} at 0 cm	0.2	–0.2	1.0	1.2	0.8
Absolute t_{min} at 0 cm	–14.6	–13.8	–12.2	–11.6	–14.9
Mean temperature amplitude	28.4	28.5	18.3	15.7	17.9
Absolute temperature amplitude	89.6	85.8	48.2	47.1	49.9
Mean relative light intensity (%)	41	51	6	7	18
Precipitation (throughfall) *PT* (mm/yr)	508	123	203	334	472
Potential evaporation *E* (mm/yr)	702	677	405	353	300
Precipitation surplus *PT* - *E* (mm/yr)	–194	–554	–202	–19	172
Mean *SD*[1] at 0 cm (mbar)	29.1	49.1	9.2	8.0	7.4
Max. SD[1] at 0 cm	212.7	172.2	27.3	18.9	20.7
Mean *SD*[1] at 10 cm	10.2	15.8	10.0	9.6	10.0
Mean soil moisture (Aucon[2] values)	21.8	2.7	26.3	63.1	80.0
Litter fall (dry weight $gm^{-2}yr^{-1}$)	107	365	529	290	145

[1] Mean day values (no measurements at night) in the summer. [2] Aucon values are a relative measure of soil humidity, max 100 (see Ch. 5.2.3).

dominates (Fig. 40).

Several characteristic microsites in and around juniper scrub can be distinguished. The north-facing transition zone bears a moss-rich grassy vegetation characterized by big pleurocarpous mosses, big liverworts and species of grassy heaths (phytosociological alliance *Violion caninae*) such as *Sieglingia decumbens*, *Anthoxanthum odoratum*, *Viola canina* and *Veronica officinalis*. This microsite is indicated NM or 'North, moss-rich' (Fig. 40). The east and west sides form a transition to the south-facing dry border with mosses (*Polytrichum piliferum* and *Ceratodon purpureus*) and lichens (*Cladonia*) occurring but hardly any vascular plants. This is SM or 'South, moss-rich'. A narrow zone, 20-40 cm wide, directly bordering the juniper shrubs, is free of vegetation; only juniper needles are found here: SN or 'South, needle-rich'.

In the juniper scrub there is a pattern of moss-rich sites: DM or 'Dense moss cover' and moss-free sites: AM or 'Absence of mosses'. The moss species in DM differ partly from those in NM and SM, while the vascular species in DM also differ, with ferns such as *Dryopteris carthusiana*, *D. dilatata* and *Polypodium vulgare*, the grass *Deschampsia flexuosa*, *Hieracium lachenalii* and *H. laevigatum* and *Corydalis claviculata* present. Nitrophilous species can also be found in DM, but this group is more common in type AM, which does contain vascular plants as opposed to type SN. Characteristic species are *Epilobium angustifolium*, *Rumex acetosella*, *Galeopsis tetrahit* and *Senecio sylvaticus*, all constant in the relevés of this type, as well as *Sambucus nigra*, *Solanum dulcamara*, *Stellaria media* and *Urtica dioica*.

3.6.2 Temperature and radiation

Where the soil type is the same for all microsites, the differences between the sites must be related to differences in microclimate and in litter deposition. This is amply indicated by

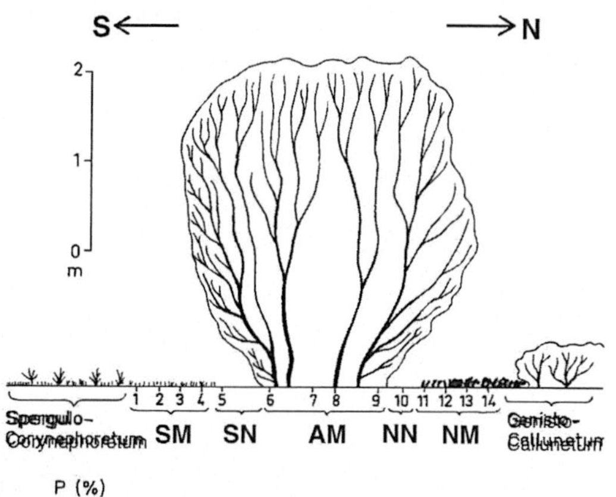

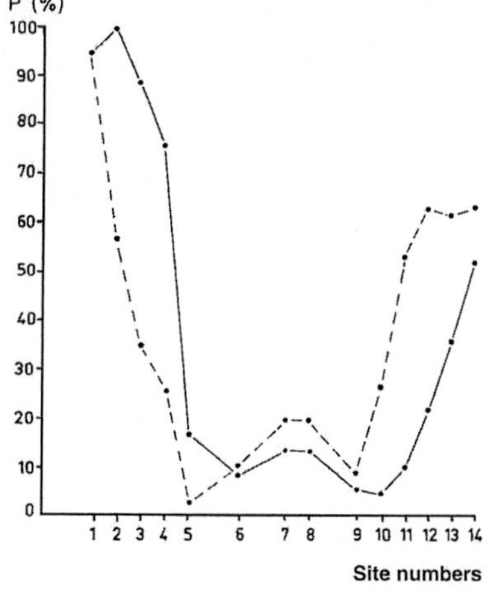

Fig. 41. Distribution of rain within and around a juniper shrub as measured by 14 rain gauges along a north-south transect. Solid line: precipitation with southerly wind; interrupted line: with northerly wind. Mantinger Zand, NE Netherlands, 1972-04-04 to 1972-06-11. From Barkman, Masselink & de Vries (1977). For abbreviations see Fig. 40.

the results of many years of measurements as summarized in Table 23.

It appears that SM and SN are characterized by very high maximum temperatures (the values, surpassing 28 °C, are averages over the entire year. The absolute maxima are still higher than in the *Cladonietum mitis* in the heathland and are probably only exceeded by the temperature of the southern slopes of ant hills. Not only have juniper scrub and the needles underneath a very low *k* and *ρC*, but reflection of solar radiation to the soil also occurs from the dense, vertical or overhanging side of the shrub. The long-wave radiation from the overhanging crown at the same time prevents marked night cooling the soil surface. Nevertheless, the yearly average of the weekly minimum temperature of SN is below zero. This average is highest in the shrubs themselves.

The lowest minima, occurring with frost but no snow, are found in SM and NM. With snow the relations are different, however. We can take the night of 12/13 February 1969 as an example. Both minimum temperatures and an indication of snow cover are given:

NM	DM	AM	SN	SM
–7.0 °C	–9.4 °C	–11.2 °C	–16.5 °C	not measured
much snow	less snow	very little snow	no snow	no snow

Clearly, the minima are also determined by the snow cover. A further difference between AM and SN was that AM was much less exposed to the sun and the open sky than SN.

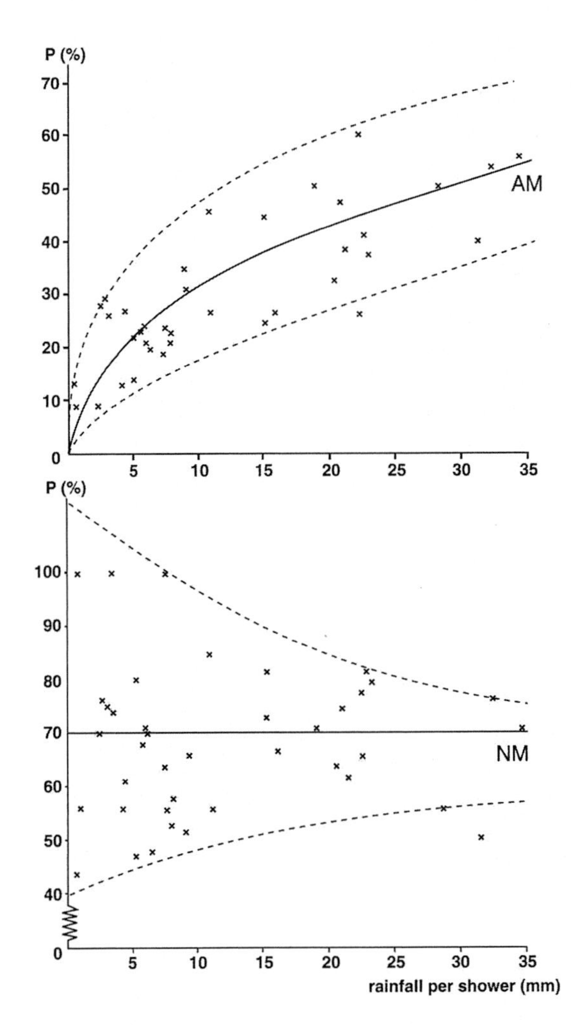

Fig. 42. Percentage rain passing through the microsites AM (absence of mosses) and NM (northern moss) in juniper scrub in relation to the size of rain showers (from Barkman, Masselink & de Vries 1977). Solid line: regression line (no equation provided); the interrupted lines indicate the extreme values.

It also follows from Table 24 that the south side receives more light than the north side and little light penetrates through the bushes. SN receives more light than SM; this can be explained by the reflection by the dense shrubs towards SN and also by the partial and temporary shading by the neighbouring *Calluna* shrubs towards the south.

3.6.3 Precipitation

The distribution of precipitation is curious (Table 23). Most precipitation falls outside the bushes (SM and SN), though less than in the open field (856 mm), this because the shrubs catch the rain when this comes with wind. In DM there is more precipitation than in AM, because the crown cover above it is less. But the low precipitation in SN was not expected. The explanation is found in Fig. 40: the shrubs are much denser here, probably because the needles receive more light, even with a dense crown cover, through a favourable exposure towards the sun. Because most shrubs become wider towards the top and hang in all

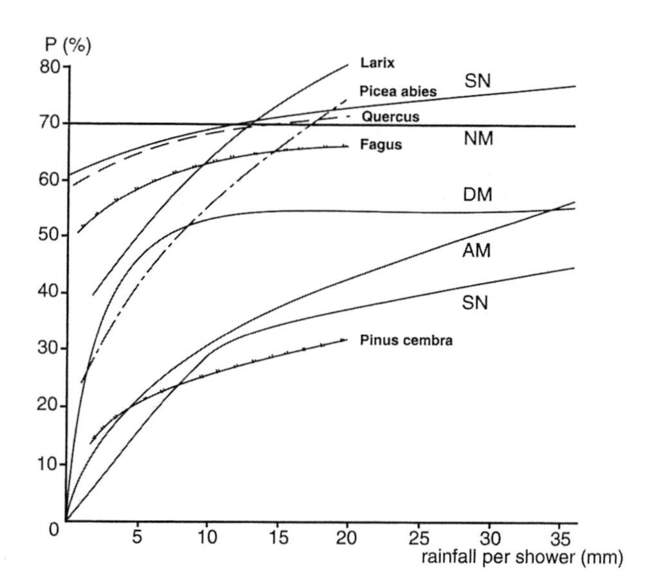

Fig. 43. Rain transmitted as a percentage of the precipitation in the open field for various woodlands and the microtopes in juniper scrub. From Barkman, Masselink & de Vries (1977). For abbreviations see Fig. 40.

directions, SN lies just under the dense crown. Although wind can carry some rain here- with a south wind to the south side, with a north wind to the north side (Fig. 41) - the precipitation transmitted is always low in SN, even with southerly winds.

If the amount of precipitation is compared with the part of the sky covered by the juniper crowns above the rain gauges, seen in different directions around the zenith, the highest correlation is found between precipitation and coverage within 20 ° from the zenith. Thus, by far the most rain falls at angles between 70 and 90 °, i.e. straight downward.

According to the relation between amount of precipitation per shower and transmission percentage (Figs. 42 and 43), all microtopes behave 'normally', with the exception of NM, where the percentage is constant at 70% disregarding the size of the shower.

It may be interesting to compare the various microtopes with the situation in woodlands. It then appears from Fig. 43 that the transmission curves vary strongly and also that SM behaves like an oak wood, SN like a *Pinus cembra* wood, while the curves of DM, AM and NM cannot be compared with any other woody vegetation.

The number of rain gauges that can be used in such an analysis is always limited and it is uncertain whether the microsites richest and poorest in rain have really been included. On the other hand, snow is visible precipitation and one would like to map the snow depth in order to obtain the complete precipitation picture. Unfortunately this is not possible, at least not generally, for the following reasons:

1. In some cases, for example with an *Empetrum nigrum* cover discussed above, there may occur a layer of air between snow and soil surface. As far as we know this complication does not occur under *Juniperus*.
2. Snow is much more wind-sensitive than rain and thus falls more obliquely, which may lead to special patterns of snow depth under vegetation; snow accumulates in the lee of the wind.
3. Relatively more snow than rain remains in the canopy.
4. The weight of a thick snow cover causes branches and even stems to bend outward,

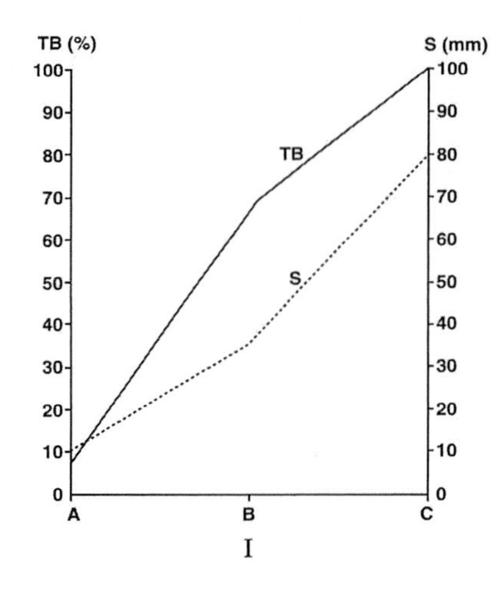

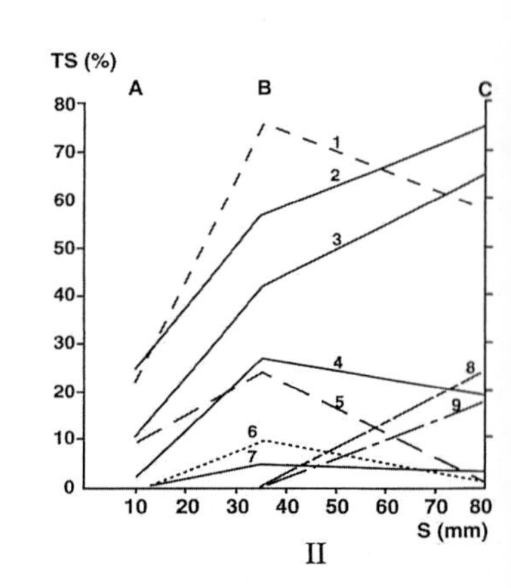

Fig. 44. Average snow cover and average cover of moss layer and nine moss species in juniper scrub. From Barkman, Masselink & de Vries (1977). I. Total moss cover TB (%) against snow depth. II. Cover (%, estimated) of individual moss species against snow depth. A = microsites with little snow; B = microsites with moderate snow depth; C = microsites with deep snow. 1 = *Hypnum jutlandicum*; 2 = *Dicranum scoparium*; 3 = *Pleurozium schreberi*; 4 = *Barbilophozia barbata*; 5 = *Brachythecium rutabulum*; 6 = *Pseudoscleropodium purum*; 7 = *Lophocolea bidentata*; 8 = *Cladonia portentosa*; 9 = *Dicranum polysetum*.

making room for more snow to fall into the centre of a shrub.
5. Fresh snow on non-frozen soil will melt away partly, but not evenly, because the soil temperature may show a pattern in relation to varying conductivity - also in relation to the thickness of the litter cover.
6. Snow can be removed in a capricious way by the wind; drifting may occur.

Measurements of snow depth variation as an estimation of the rainfall pattern would only make sense if:
- the soil was frozen before the snow came;
- the snow was dry, and during and after the snowfall there was little wind;
- not too much snow had fallen;
- the temperature remained below zero before the measurement.

All these conditions were satisfied on a few occasions during the observation period from 1965 to 1975. On such days a total of 2000 snow depth measurements were taken; the measurement points were marked, and, the following summer, the vegetation composition and structure were described. The average snow depth, expressed as a % of the average snow cover in the open field, varied greatly with microsite type:

SM	SN	AM	DM	NM
25%	10%	18%	31%	66%

The general conclusion is that with the exception of NM little snow reached the soil surface. There was a large difference between the northern and southern edges of the juniper scrub. There was much more snow in the northern moss microsite, which is related to the prevailing wind (if only weak) during snowfall, i.e. NW to NE. The difference in rainfall is the opposite, which is explained by the fact that a great deal of rain comes with the prevailing wind during rain showers, i.e. SW.

It follows from Fig. 44 that the total cover of the moss layer is related to snow depth, and that some mosses (no. 2, 3, 8 and 9 in Fig. 44II), all occurring optimally in microsite NM, attain their greatest cover where snow cover is most extensive. The mosses 1, 4, 5, 6 and 7 show an optimum with moderate snow cover, the latter coinciding with microsite DM; indeed, mosses 1, 4 and 5 are preferential for microsite DM, but species 7 is indifferent and 6 prefers NM.

3.6.4 Evaporation, air and soil moisture

The potential evaporation, *E*, has been measured on many places with Piche-evaporimeters (see Ch. 5.2.7), at 3 cm above the soil surface. Despite some problems discussed in that section the results can be used for a general comparison of the microsites. Evaporation is highest in microsites SM and SN, which is related to the high temperatures there. In NM, however, evaporation is smaller than in DM, although the maximum temperature is higher. This is explained as follows: NM gets more rain and keeps it better through the thick moss cover, so that the air remains humid for a longer time after a shower.

Of course, one cannot simply subtract the Piche-evaporimeter value *E* from the throughfall rain *PT* when calculating the water budget of the soil. On dry well-drained sandy soils the real evaporation stops soon and also with a wet soil the evaporation differs from that of the standard wet filter paper used. However, $(PT - E)$ is an indication of the water stress to which plants are subject as well as the length of time of such stress. Table 23 shows the desert microclimate of microsite SN, where $(PT - E) - 554$ mm/yr.

The latter fact is also clearly expressed in the high average and maximal saturation deficit values for SN, which can be as high as 213 mbar. The saturation deficit at 0 cm is meant here as the difference between e_{max} at the surface temperature and the vapour pressure of the air immediately above it. It gives an impression of the speed with which mosses and lichens on the surface are dried out. For a good understanding of the vegetation as a whole one should realize that the microclimate for mosses and herbs is completely different, and that the differences between the microsites even can change sign. It follows from Table 23 that the saturation deficit by daytime in SM and SN is much higher at 0 cm (mosses) than at 10 cm (herbs), but in the other microsites lower. This is due to the large temperature differences on the south side, where the temperature at 0 cm is often 22 - 25 °C higher than at 10 cm and thus much drier. In other (shaded) microsites this difference is much smaller, so that the higher vapour pressure becomes decisive. In NM the temperature near the surface during the day in summer can be even lower than at 10 cm. An example: 22 June 1973; 13.13 h, air temperature at 2 m 25.1 °C, in NM at 10 cm 24.1 °C, at 0 cm 19.0 °C. Often there was a negative radiation balance.

From the data presented on the various parameters it may be deduced that the average soil humidity will increase in the order SN, SM, AM, DM, NM. This sequence has been found, with values from 2.7 increasing to 80 (Aucon values, see Ch. 5.2.3). These are summer values. They also show that the only microsite without vegetation, SN, has the lowest soil humidity. There is a good correlation between the Aucon-soil humidity and the difference $(PT - E)$.

A final remark on litter moisture (*SH*). This factor has been measured as volume %, all year round, by drying and weighing. Litter moisture was always lower in DM + AM (no distinction has yet been made between the microsites, dense moss and dense needle) as compared with NM, and lowest in July - September, although most of the rain falls then.

Only after the temperature starts to drop and evaporation decreases, does *SH* increase.

3.6.5 *Periodicity of fungi*

Weekly quantitative investigations of fruit-bodies of macrofungi (removed after counting) showed that the total number is positively correlated with litter humidity (Kendall's $\tau = 0.78$; $p = 0.002$). Similarly high correlations were found with precipitation and especially with number of rainy days the week before the count, and a significant negative correlation appeared between fruit-body number and potential evaporation at 3 cm. Furthermore, the species appeared only after the litter humidity had exceeded a certain threshold value, for most species between 17 and 26%.

Many species, 31 in total, occurred both in NM as in DM+AM and 23 of them usually appeared 6 - 8 weeks earlier in NM. This time difference is in accordance with the observation that the threshold value for the litter moisture in NM was reached 8 weeks earlier than in DM+AM. In the autumn of 1971, the species *Clitocybe vibecina* reached its greatest abundance in NM on 25 October, in DM on 2 November and in AM as late as 16 December.

Later in the autumn fluctuations in fruit-body numbers are not synchronous with litter moisture, but show a delay of one week, probably because the development slows down with decreasing temperature.

As to effects of temperature, abundance of *Marasmius androsaceus* in June showed a significantly negative correlation with the maximum temperature at the soil surface. This is probably an indirect effect working through a more vigourous soil evaporation and lower soil moisture with higher soil surface temperatures. If the correlation was corrected for this indirect influence via litter moisture, the number of *Marasmius* fruit-bodies was positively correlated with maximum temperature. This positive influence of surface temperature on growth stops at 30 °C, and for other species at 20 °C, and is often overruled by desiccation of litter and soil surface.

Late in the autumn night frost is another decisive negative factor in fruit-body development; fluctuations in numbers are then correlated with minimum temperatures.

In conclusion, these results demonstrate the considerable influence of the microclimate on both the periodicity of fungi and the microdistribution of mosses.

3.6.6 *Litter fall*

Litter fall, though not a direct microclimatological factor, is important in that it contributes to the formation of micropatterns in the juniper scrub. The differences between the microsites are large (Table 23) with a maximum of over 500 g/m^2 in the microsite named after this needle litter concentration. On sites with more than 300 g/m^2 on average no mosses were found, but there were (in AM) some vascular plants. Clearly, mosses become suffocated more easily because they are smaller. In the critical range, between 250 and 350 g/m^2 litter, other factors influence the appearance (or disappearance) of mosses, particularly trampling by man and animals.

In view of the variation in microclimatological factors as described above, the (non-) occurrence of vascular plants, mosses, lichens and fungi at the different microsites within the juniper scrub is determined by a combination of microclimate and soil surface factors.

3.6.7 Microclimatological experiments

In order to disentangle this complex of factors, a number of field experiments, lasting three years, were carried out in the juniper scrub where either one of the factors, rainfall and litter fall, was varied at the time. Circular plexiglass discs of 30 cm diameter were placed on 10 cm tall metal legs in order to intercept rain. The litter collected on the discs was removed regularly and spread underneath; the further effects of this installation on the microclimate: decreasing wind speed, absorption of UV, greenhouse effect, were diminished by removing the discs during periods of little rain. Enough rain was intercepted, as followed from a comparison with rain gauges in the open field.

In parallel experiments plots of a similar size to the discs were watered with collected rainwater, in microsite AM 5 mm/week and in microsite SN 10 mm/week. Litter was intercepted with square frames 20 cm × 20 cm, fitted to legs and covered with nylon netting, which allowed transmission of rain. The interception of light was small. All litter was intercepted on microsites SN and AM, and spread out weekly over parallel plots in SM and DM.

According to Table 23, SM and SN differ particularly in precipitation and litter fall. The effects of the experiments were directed so as to simulate changes of one microsite type into another. Since similar differences were found between DM and AM the same treatments were carried out, as follows:

Less rain	in	SM ≈ SN	in	DM ≈ AM
Extra rain	in	SN ≈ SM	in	AM ≈ DM
Extra litter	in	SM ≈ SN	in	DM ≈ AM
No litter	in	SN ≈ SM	in	AM ≈ DM

Some results of these experiments are presented in Tables 24 and 25. Neither all experimental plots nor all species are represented, but what is shown is certainly quite surprising. Almost all species behave differently and some species behaved in an irregular way. To mention some striking examples: the moss *Polytrichum piliferum* and the grass *Festuca ovina* declined strongly under extra litter (L+ in Table 25) on microsite SM, but did not establish in SN when litter was intercepted there. Other cryptogams were not influenced negatively by extra litter, as one might expect. For instance, growth in *Dicranum scoparium* and the lichen *Cladonia portentosa* was even increased. As regards less litter (L–, Table 25, plot 15a) two major species here behave differently: *Agrostis vinealis* increases and *Rumex acetosella* decreases.

Less rain (P–, Table 24) in SM caused the decline of many species, but not of *Cladonia portentosa* and *Agrostis vinealis*. The latter species was found germinating occasionally on SN with P+. Its counterpart *Rumex acetosella* strongly increased in one plot (5) but strongly decreased in the other (15). Apparently other factors than the experimental ones play a part here.

Comparison of DM and AM was also confusing. Here variation in litter fall was more important, whereas variation in rainfall was more important in SM and SN. This is probably because rainfall is the limiting factor in SN (where it is lowest) and litter fall in AM (where it is highest). *Dicranum scoparium* shows a typically contrary behaviour: in SM it declines in response to less rain and not to more litter, in AM it declines under more litter but not under less rain.

More rain in AM led to the establishment of some mosses: *Lophocolea bidentata*, *Pleurozium schreberi* and *Plagiothecium denticulatum*, and the herb *Stellaria media*; less rain caused the decline of some species such as *Hypnum jutlandicum*, *Atrichum undulatum* and *Polytrichum formosum*, but also the increase, for instance of *Dicranoweisia cirrhata*,

Table 24. Rainfall experiments on microsites in juniper scrub and their effect on the cover percentage of mosses and liverworts (mo), lichens (li), grasses (gr) and herbs (he) in plots of 700 cm². (.+ = present; .r = sporadic). Extra rain is indicated by P+, less rain by P–; microsite SM = 'South, moss rich'; SN = 'South, needle rich'; DM = 'Dense moss cover'; AM = 'Absence of mosses'.

Exp.	Site	Species		Mar '73	Jul '73	Feb '74	Aug '74	Mar '75
SM P–	13a	*Polytrichum piliferum*	mo	95	75	40	20	10
		Agrostis vinealis	gr	5	3	1	10	10
	6a	*Cladonia portentosa*	li	2	5	5	8	10
		Cladonia arbuscula	li	6	4	2	.r	.r
		Agrostis vinealis	gr	15	3	.+	.r	.r
		Corynephorus canescens	gr	8	20	3	1	.r
		Rumex acetosella	he	4	.+	.+	—	—
	14a	*Dicranum scoparium*	mo	40	30	25	20	15
		Cladonia portentosa	li	15	18	15	10	10
		Agrostis vinealis	gr	3	.+	.+	.+	2
		Rumex acetosella	he	5	3	5	3	3
	9a	*Dicranum scoparium*	mo	85	55	35	25	20
		Polytrichum piliferum	mo	8	4	2	.+	.r
		Cladonia arbuscula	li	.+	.r	—	—	—
		Rumex acetosella	he	—	—	4	10	2
SN P+	8, 12			—	—	—	—	—
	5	*Agrostis vinealis*	gr	—	—	.r	.r	2
		Rumex acetosella	he	—	.r	20	55	70
	15	*Agrostis vinealis*	gr	1	1	5	30	35
		Rumex acetosella	he	30	30	.+	.r	.r
DM P–	4a	*Polytrichum formosum*	mo	70	50	70	30	
		Pleurozium schreberi	mo	10	2	2	2	
	10a	*Dicranum scoparium*	mo	35	30	40	20	20
		Hypnum jutlandicum	mo	25	30	10	7	15
		Pleurozium schreberi	mo	15	12	20	35	20
		Lophocolea bidentata	mo	10	6	3	3	12
		Pohlia nutans	mo	—	.2m	.1	.1	2
		Dicranoweisia cirrhata	mo	—	.1	—	2	5
	16a	*Brachythecium rutabulum*	mo	40	60	70	75	90
		Pseudoscleropodium purum	mo	15	—	2	2	2
		Lophocolea bidentata	mo	—	—	1	—	5
		Deschampsia flexuosa	gr	4	10	6	20	10
		Agrostis tenuis	gr	2	.+	.+	2	2
	18a	*Hylocomium splendens*	mo	30	10	8	10	20
		Pseudoscleropodium purum	mo	30	25	20	12	35
		Lophocolea bidentata	mo	20	10	1	2	10
		Deschampsia flexuosa	gr	20	15	15	25	15
		Dryopteris carthusiana	fern	—	2	2	4	8
		Pleurozium schreberi	mo	—	2	5	15	8
AM P+	7, 19			—	—	—	—	—
	2	*Pleurozium schreberi*	mo	—	1	10	2	2
		Lophocolea bidentata	mo	—	.+	—	.r	—
	11	*Plagiothecium denticulatum*	mo	—	2	.+	2	2
	17	*Lophocolea bidentata*	mo	—	.r	—	.1	8

Table 25. Litter fall experiments in microsites in juniper scrub and their effect on the cover percentage of mosses (mo), lichens (li), grasses(gr) and herbs (he) in plots of 400 cm² (.+ = present; .r = sporadic). Extra litter is indicated by L+, no litter by L–; microsite SM = 'South, moss rich'; SN = 'South, needle rich'; DM = 'Dense moss cover'; AM = 'Absence of mosses'.

Exp.	Site	Species		Mar '73	Jul '73	Feb '74	Aug '74	Mar '75
SM L+	13	*Polytrichum piliferum*	mo	85	75	70	30	20
		Dicranum scoparium	mo	5	3	8	4	5
		Agrostis vinealis	gr	.1	2	2	5	5
	6	*Cladonia portentosa*	li	8	10	10	15	25
		Cladonia arbuscula	li	12	5	2	—	.+
		Cladonia gracilis	li	—	—	.2m	.2m	3
		Festuca ovina	gr	20	10	1	2	2
		Rumex acetosella	he	6	2	4	.r	.r
		Corynephorus canescens	gr	2	2	1	2	2
		Agrostis vinealis	gr	4	15	15	5	5
		Spergula morisonii	he	—	.+	6	—	10
	14	*Dicranum scoparium*	mo	65	80	80	80	90
		Cladonia portentosa	li	15	15	15	15	15
		Cornicularia aculeata	li	3	3	1	3	.1
		Agrostis vinealis	gr	10	6	10	2	2
		Rumex acetosella	he	2	4	10	—	4
	9	*Dicranum scoparium*	mo	60	40	65	50	80
		Polytrichum piliferum	mo	.1	2	1	.1	.1
		Rumex acetosella	he	—	2	6	12	8
		Rumex acetosella (number of plants)	he	—	2	10	13	22
SN L–	5a, 8a, 12a			—	—	—	—	—
	15a	*Cladonia portentosa*	li	.r	.+	.+	.+	.+
		Agrostis vinealis	gr	5	8	3	15	25
		Rumex acetosella	he	15	3	.r	10	.+
DM L+	4	*Polytrichum formosum*	mo	50	70	40	8	
		Pleurozium schreberi	mo	25	2	5	6	
		Lophocolea bidentata	mo	3	.+	—	.1	
	10	*Dicranum scoparium*	mo	25	20	20	12	10
		Hypnum jutlandicum	mo	40	35	40	25	15
		Pleurozium schreberi	mo	15	15	6	10	7
		Lophocolea bidentata	mo	8	—	.r	.r	—
		Pohlia nutans	mo	2	1	2	—	1
		Ptilidium ciliare	mo	1	4	.1	.1	3
	16	*Brachythecium rutabulum*	mo	35	25	5	10	15
		Pseudoscleropodium purum	mo	25	15	4	3	6
		Lophocolea bidentata	mo	.+	—	.+	.+	5
		Pleurozium schreberi	mo	—	—	2	.1	10
		Deschampsia flexuosa	gr	10	7	1	5	10
		Agrostis tenuis	gr	2	7	2	3	5
	18	*Dicranum scoparium*	mo	20	20	5	3	8
		Pseudoscleropodium purum	mo	20	15	12	1	10
		Lophocolea bidentata	mo	15	8	.+	.+	.+
		Hypnum jutlandicum	mo	10	.+	—	—	—
		Hylocomium splendens	mo	5	4	1	1	2
		Deschampsia flexuosa	gr	10	40	6	10	5

Table 25 (continued).

Exp.	Site	Species		Mar '73	Jul '73	Feb '74	Aug '74	Mar '75
AM L–	2a, 7a, 19a			—	—	—	—	—
	3a	*Hypnum jutlandicum*	mo	—	.+	3	6	25
		Lophocolea bidentata	mo	—	.+	2	6	15
		Dicranum scoparium	mo	—	.+	.+	.+	2
		Pleurozium schreberi	mo	—	.r	.+	.+	2
	11a	*Lophocolea bidentata*	mo	—	.1	3	10	25
		Hypnum jutlandicum	mo	—	.r	.r	.r	.r
		Pohlia nutans	mo	—	—	.r	—	.+
		Campylopus flexuosus	mo	—	—	—	.r	4
	17a	*Lophocolea bidentata*	mo	—	—	.1	.2m	35
		Pohlia nutans	mo	—	—	—	—	.+

Pohlia nutans and *Brachythecium rutabulum*. The behaviour of other bryophytes: *Lophocolea bidentata*, *Pseudoscleropodium purum* and *Pleurozium schreberi* was irregular: in some plots with less rain they decreased, in others they increased and in some they remained constant.

More litter in DM exerted a negative influence on most species, whereas less litter in AM had a very positive effect. Most species were thus missing in AM because of the litter accumulation, some others because of the lack of precipitation and some because of both factors. Finally, some species appeared to be promoted by both the adding of litter and interception of rain. Maybe, this is a form of competition, other species suffering still more from the conditions imposed.

3.7 The microclimate of grassland

> *"Soon we shall have equally anxious discussions on the conservation of microclimates"*
> *B. Kullenberg (1962)*

3.7.1 Microclimatological grassland types

Grassland covers large areas in temperate regions both as natural or as anthropogenic vegetation. The following information stems largely from some grassland types in the central Netherlands and some coastal dune grasslands in the southwestern Netherlands. Large differences in structure, height and density occur in grasslands. Hence the fate of incoming radiation varies greatly (Figs. 45 and 46). In the first case we are dealing with dense, tall grassland with dominating *Phleum pratense* of 40 cm height. Only ca. 3% of the incoming solar radiation reaches the soil surface and almost 80% is already absorbed in the upper 10 cm of the sward. The second case concerns a dry grassland with *Holcus lanatus* and *Anthoxanthum odoratum* with a height of only 19 cm. Here 42% of the solar radiation reaches the surface. In the *Phleum* grassland the highest temperature is found halfway down

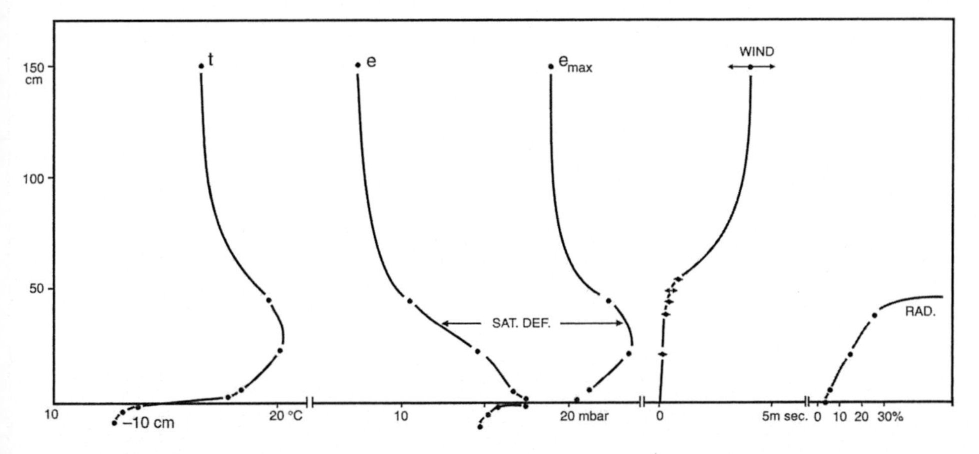

Fig. 45. Profiles of temperature, vapour pressure, wind and radiation in tall dense grassland. 13 May, 1980. Global radiation 836 W/m^2.

the vegetation, in the low open grassland near the soil. In both grasslands the vapour pressure is highest near the ground, but as compared to the free air, the saturation deficit is markedly decreased in the tall grassland, and markedly increased in the dry grassland, because here, as a result of the high temperature, the maximum vapour pressure increases more than the real vapour pressure.

If we want to characterize many grasslands in a short time, it seems most effective to measure temperature and saturation deficit near the ground. Here, differences between grassland types are greatest. We use measurements here at 1 cm with a miniature psychrometer (cf. Ch. 5.2.3). Of course, the height of 1 cm is not exact because of small irregularities in moss and litter layer. In order to obtain an overview we have indicated in Fig. 47 how temperatures at 1 cm differ from those at 1 m. These are measurements at midday in sunny weather and a radiation intensity between 600 and 900 W/m^2, carried out

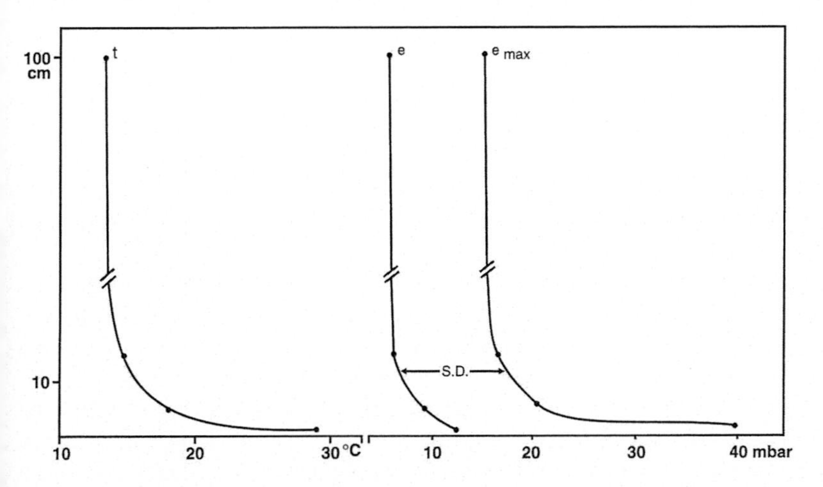

Fig. 46. Temperature and vapour pressure profiles in low open grassland. 22 May, 1980. Global radiation 838 W/m^2.

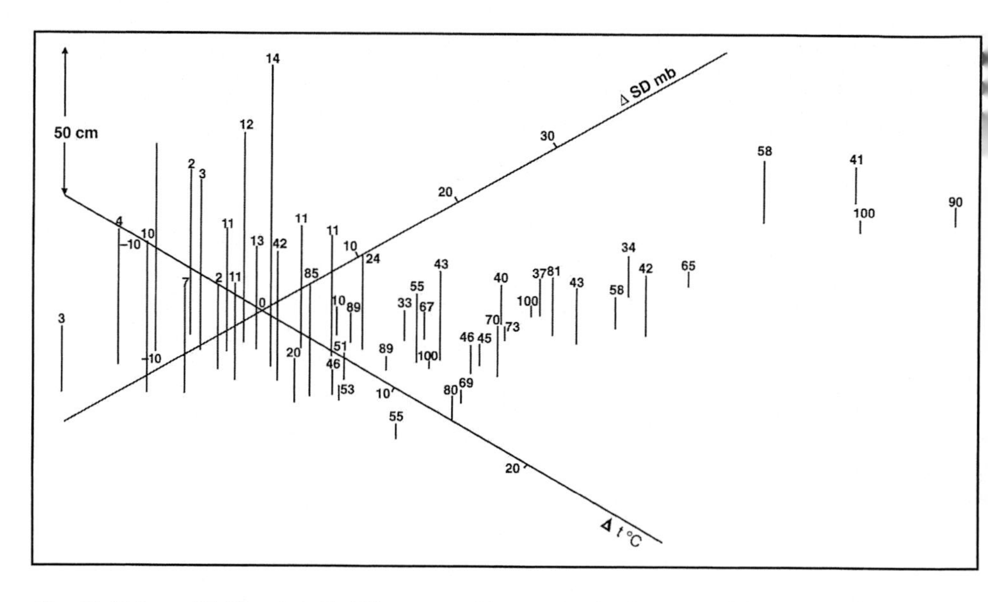

Fig. 47. Values of ΔSD and Δt (in °C) measured in grassland. The bases of the vertical lines indicate the measured values, their lengths indicate the height of the vegetation and the figures on top the transmission of sunlight. After Stoutjesdijk (1981).

at least 24 h after the last shower and not after a period of extreme drought.

The diagram can be divided into four sectors: cool - dry, cool - moist, warm - dry and warm - moist. Cool means that the temperature in the vegetation is lower than that of the surroundings, warm means the opposite; Δt is negative or positive. Moist means that the saturation deficit is lower than outside the vegetation: ΔSD is negative. The combination cool - dry does not occur; in that case the vapour pressure in the vegetation would have to be lower than in the air above, which does not happen with a strongly positive radiation balance. In other words: because the grass and the soil release water vapour, the vapour pressure in the vegetation is higher than outside the vegetation and in that case the saturation deficit can only be higher if the temperature is higher than in the open as well.

The combinations of Δt and ΔSD are found in a long, narrow sector, meaning that these variables are correlated. Another clear relation is found between the density and the height of the vegetation. In closed, tall grassland the saturation deficit is always small. The temperature can be somewhat lower than outside, but only in grassland with a transmission of sunlight which is less than 3%. This grassland is usually heavily fertilized. In less heavily fertilized hay meadows with a higher transmission the temperature is always some degrees above the air temperature.

As could be expected the oligotrophic dry grasslands are found in the sector warm - dry. Here, the vapour pressure is always higher than in the open, but the effect of the higher temperature is stronger. This is a grassland with species such as *Festuca ovina*, *F. rubra*, *Anthoxanthum odoratum*, *Luzula campestris*, belonging to the subassociation *Lolio-Cynosuretum luzuletosum campestris*. More important than the composition is the height of the grassland, which is low, and its density, which is also low. The soil surface is dry, even if the top few cm often have a high moisture content. The highest values of ΔSD, > 40 mbar with Δt > 20 °C, are found on dry sandy soils with a low humus content and a low water capacity (phytosociologically such grasslands are assigned to the orders *Corynephoretalia* or *Festuco-Sedetalia*). These dry grasslands may be warm - moist after rain, but usually warm - dry again after one day of warm sunny weather. With dry grassland on peaty soil the drying up can take longer. After one dry day Δt was 13 °C at a ΔSD of –4.7 mbar.

The oligotrophic wet hay meadows are situated in the sector warm - moist, at least as long as the surface is moist. The temperatures are higher here (Δt = 5 - 8 °C) than in the closed grassland, which could be expected in view of the more open structure. The saturation deficit is reduced, but not as strongly as in the closed grassland (with a ΔSD of –3.6 to –4.7 mbar). These are measurements in June, canopy height was 40 cm then and transmission 42 - 88%. Only later in the summer does the vegetation close and then it becomes comparable to a closed grassland.

It is remarkable that, as long as the structure is open, the oligotrophic wet grassland is similar to the dry grassland as regards the microclimate. This is the case as soon as the moss layer (with *Hypnum cupressiforme* or *Sphagnum*) dries up superficially. Temperature and moisture near the ground are determined by the vegetation structure and the situation at the surface rather than by the moisture storage in the soil.

As far as the saturation deficit of the free air is concerned, the measurements in Fig. 47 were made under strongly varying circumstances. They show how much the saturation deficit in the vegetation had been increased or decreased, but not what the actual value was. Of course, the absolute value of *SD* in free air also determines the possible value of ΔSD in the vegetation.

Fig. 48 provides us with more information. Because we are dealing with simultaneous measurements here, we also indicate the real value of the saturation deficit *SD* as well as the increase in vapour pressure in the vegetation as compared with that in free air (Δe). Further, the height of the vegetation and the transmission of solar radiation is shown. The origin of

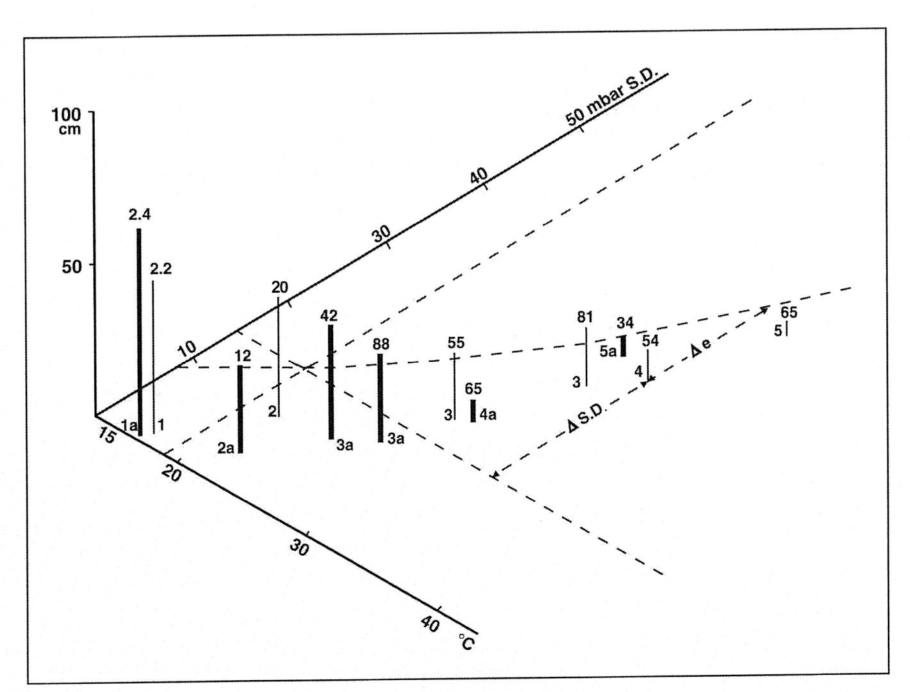

Fig. 48. Temperature and saturation deficit in different types of Dutch grassland. Measurements at about the same temperature and saturation deficit above the vegetation. After Stoutjesdijk (1983).

the axes Δt and ΔSD indicates the situation of the air above the vegetation. The temperature there is 18.7 °C, the vapour pressure 7.1 mbar, e_{max} = 21.6 mbar, and SD = 14.5 mbar. The stippled curve indicates how the saturation deficit of this air would change if only the temperature changes and the vapour pressure remained the same. If, for example the air, with a vapour pressure of 7.1 mbar, warmed up to 33 °C, then e_{max} = 50.3 mbar and SD = 43.2 mbar. The distance from the stippled curve indicates the value of Δe, for instance at point 4 this would be 12.7 mbar. The area above the stippled curve is empty, meaning that with a strongly positive radiation balance, the vapour pressure near the ground cannot be lower than at 1 m in height. The thin vertical bars represent the situation in different types of grassland after a dry period of one week.

In dense tall grassland (1) on wet soil, 20 - 30 cm above the phreatic level, the saturation deficit is only 1.2 mbar, while the vapour pressure is 11.1 mbar higher than in the free atmosphere. The two measurements in wet oligotrophic grassland show a rather different result. Here the values of Δt are 9.2 and 12.3 °C and of the saturation deficit 17.8 and 27.1 mbar. The big differences in saturation deficit between the open and the closed grassland are certainly due to the big difference in temperature; the values of Δe and hence of the vapour pressure do not differ that much.

The tall moist grassland (1) is situated alongside the wet oligotrophic grassland (3), it has in fact been derived from this natural grassland through heavy fertilization and intensive use. In both fields the phreatic level is 20 - 30 cm below the surface and the soil is almost saturated with water. The open oligotrophic grassland has a moss layer which is dried up after one week without rain. In the dense grassland (1) it is the surface of the crop which transpires heavily. This water is taken from the root zone in the soil and the soil surface is protected from evaporation.

Unfertilized hay meadows (2) on a moderately moist light clay soil (25 vol. % water) are intermediate between the oligotrophic (3) and the fertilized wet grassland, regarding height and density as well as temperature and humidity.

A low grassland with *Festuca ovina* (4) on sandy soil, under the influence of capillary water (40 vol. % water) is warmer and drier than the oligotrophic grassland. Still warmer and drier is the dry grassland (5) on dry sandy soil (7 vol. %).

The measurements show once more how the microclimate is determined by the height, density and structure of the vegetation rather than by the water content of the soil. For comparison measurements have been included (thick bars in the figure) which were carried out 36 h after the last rain, under almost equal conditions to the earlier measurements. In the tall moist grassland (1a) there are small changes as compared with the dry situation; the soil was already relatively moist and at both times the vapour pressure was near the saturation point. In the more open hay meadow (2a) the saturation deficit is less now. The soil surface was almost dry and is now moist. The wet oligotrophic grassland (3a) now within the warm - moist sector with temperatures about 8 °C above air temperature. The saturation deficit has diminished to about half the earlier value, but is still ten times higher than in the dense grassland. Both the grassland with *Festuca ovina* on moist sandy soil (4a) and the dry grassland (5a) dry up soon after rain. Now they are less warm and the saturation deficit is lower than after the dry period, but both are back in the warm - dry sector of the diagram. We may expect them to have been in the warm humid sector a few hours after the rain until the vegetation dried up.

The measurements in grasslands provide a good general impression of the different combinations of temperature and moisture which may occur in low vegetation, near to the ground. Very low values of the saturation deficit only occur in dense vegetation on moist soil, 'filled' with leaves and stems. If the soil is not under the influence of capillary water,

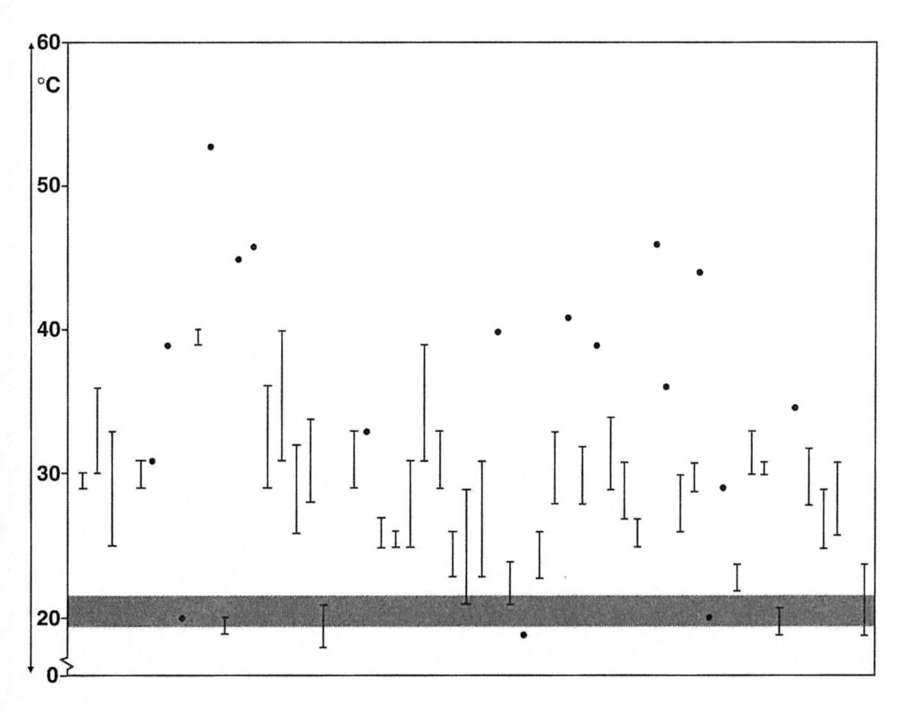

Fig. 49. Surface temperatures and their variation in a semi-natural grassland, indicated by the length of vertical lines measured by radiaton thermometer. Hatched: range of surface temperature in a neigbouring cultivated grassland. Air temperature was 20 °C.

open vegetation can maintain a moist microclimate only for a short time; even a thin moss layer, if dry, can form an important barrier for the transport of water vapour from the moist soil below. The question is: how long can dense vegetation maintain a moist microclimate? Here, we have data available on grassland after a period of more than one month of largely dry weather.

The measurements were taken on 3 August, 1982 after a very dry month in July, in dense grasslands varying in height between 25 and 70 cm on soils without any influence from the capillary zone. Between 4 and 7% of the solar radiation reached the soil. The saturation deficit was greatly reduced and attained values of between 33 and 66% of the value above the vegetation, which was high, i.e. up to 27 mbar. Thus, such dense vegetation can maintain a moist microclimate during several weeks, as long the plant roots can take up water.

3.7.2 Vegetation mosaic and temperature mosaic in grasslands

Only very homogeneous vegetation such as anthropogenic grassland can be properly characterized by profiles of temperature, vapour pressure, wind and radiation, i.e. using vertical differences.

A more natural grassland shows various forms of horizontal heterogeneity. Many types of semi-natural grassland, i.e. grassland with a spontaneously developing flora but a structure influenced by man, through mowing or grazing or both, have a tussock structure.

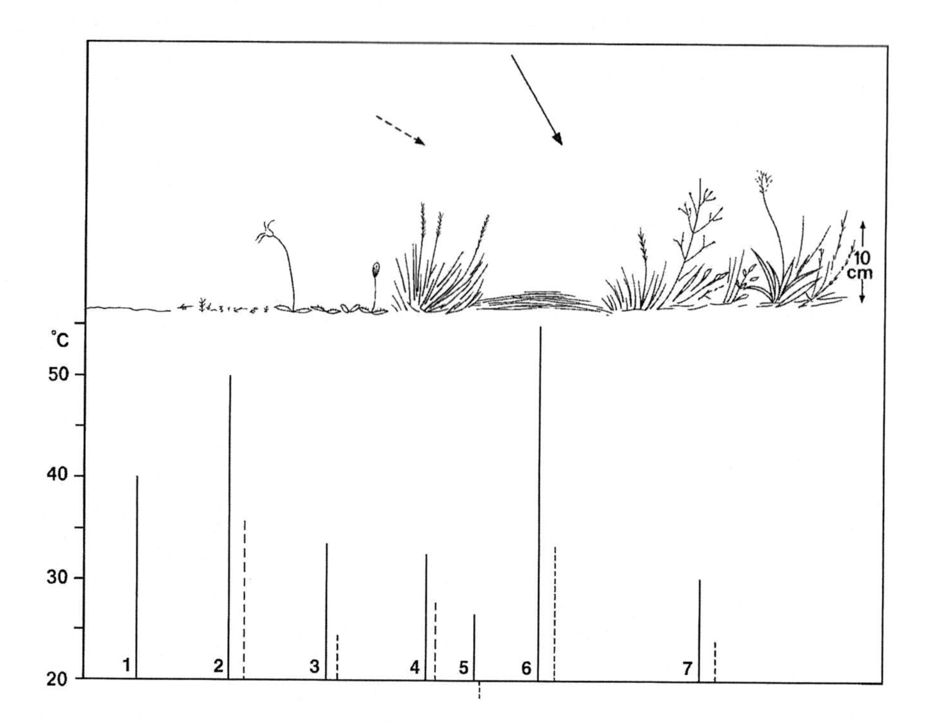

Fig. 50. Surface temperatures of a dry dune grassland. The base line indicates the air temperature, the arrows the direction of the solar radiation and the radiation intensity. Solid line: measurements at high elevation of the sun, global radiation 839 W/m^2. Interrupted line: measurements at low elevation of the sun, global radiation 419 W/m^2. After Stoutjesdijk (1966).

This implies differences in aspect and strong temperature differences between the north and south side of a tussock. Other features are spots with dead material, often recolonized by mosses or lichens, and developing a warmer and drier microclimate. Such heterogeneity will develop in production grassland as soon as it is neglected for some years.

We need a simple measurement to be carried out rapidly, scores of times, to describe small-scale patterns of variation. Such an easy measurement is the effective radiation temperature of soil and vegetation. Because heat and water balance are linked, such a measurement also gives an indication of the humidity in the vegetation. Fig. 49 shows a transect in a grassland reserve in the polder Oostelijk Flevoland, where the only management device since establishment after reclamation has been mowing in late summer; in this way the grassland undergoes an 'oligotrophication', i.e. it slowly changes to a lower nutrient status (cf. Oomes 1990). Here, surface temperatures of the vegetation were measured every tenth step with an instrument having a visual field of 10 cm in diameter. The variation in temperature within a circle of 1 m diameter is indicated by the length of the vertical bars. Sites with strongly deviating temperatures are indicated by points.

With the grasses *Holcus lanatus*, *Phalaris arundinacea* and *Phragmites australis* the surface temperatures do not differ much from the air temperature, being only 1 - 2 °C lower or higher. *Anthoxanthum odoratum* and *Festuca rubra* are warmer, up to 10 °C and 18 °C above the air temperature, respectively. Where dead grass or moss covers the soil, surface temperatures may be up to 30 °C higher. Even under a cloudy sky, surface temperatures are

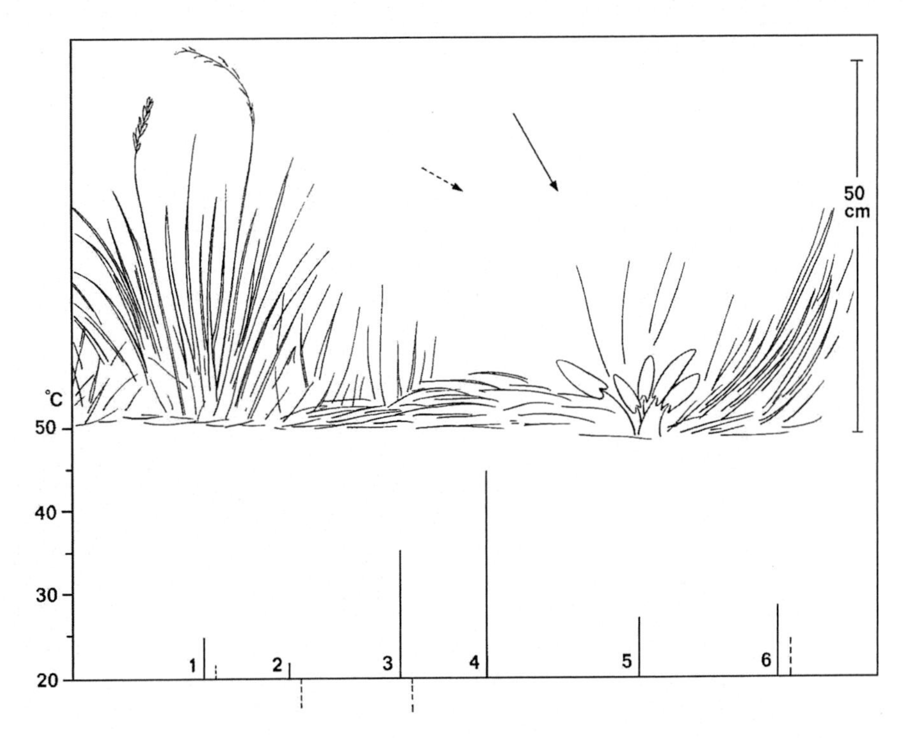

Fig. 51. As Fig. 50 for a tall *Calamagrostis epigejos* vegetation in a dune hollow. After Stoutjesdijk (1966).

locally much higher than the air temperature. As an example, under a homogeneous cumulostratus cover with a solar radiation intensity not higher than 315 W/m^2, and at an air temperature of 15.7 °C surface temperatures between 14 and 30 °C were measured.

It is interesting to compare this grassland subject to 'oligotrophication' with the heavily fertilized productive grassland on the neighbouring field outside the reserve. In this grassland temperatures do not deviate more than from 2 °C above to 1 °C below the air temperature. As a result of agricultural use grassland has a different microclimate, with lower temperatures and a loss of microclimatic diversity. This has immediate ecological consequences; e.g. van Wingerden & Kreveld (1989) found that in these grasslands the eggs of most grasshopper species cannot develop because the temperatures are too low.

Figs. 50 and 51 present more in detail surface temperature values for a dry grassland in the coastal dunes of Voorne, the Netherlands and for an adjoining *Calamagrostis epigejos* grassland in a dune hollow. From left to right we see how Δt (the difference between surface and air temperature) reaches 20 °C at a spot of bare sand, and even over 30 °C on dry moss (*Polytrichum piliferum*). A patch of *Hieracium pilosella* and a tussock of *Festuca tenuifolia* both have a Δt of 15 °C. A spot with half-decayed grass remains has a surface temperature of 55 °C, but the tuft of *Anthoxanthum odoratum* only 30 °C. A tussock of *Calamagrostis* is only 4 °C warmer than the air. The dead grass (4, Fig. 51) is not as warm as compact dead material (6, Fig. 50) because of its looser structure. At a low elevation of the sun and radiation intensity (in September) the differences remain clear, but now the shady spots near the tall grass tussocks are clearly cooler than the air (see Ch. 3.8.2).

3.8 The microclimate of vegetation edges and gaps

3.8.1 Introduction

Small-scale variation in vegetation structure and in microclimate occur especially in natural and semi-natural vegetation. As we said before, horizontal variation is one of the differences between natural vegetation and stands of an agricultural crop. An important aspect of horizontal variation not yet discussed sufficiently is the effect of sharp boundaries, both natural and artificial, between adjoining tall and low vegetation. This will now be discussed (see also Mattson 1970, 1979).

3.8.2 The open shade

"The cypresses cast a good shadow, cold as well water"
G. Durrell: My family and other animals

Inside the shadow cast by an obstacle, a hedge or a rock, or a north-facing impermeable wall, when the sun is in the south, only diffuse solar radiation from the northern sky is received. This is only roughly 25% of the total diffuse solar radiation from a clear sky, and only 5% of the global radiation received by a free surface (Table 1, p. 16). We have the curious situation here that the net radiation loss in the low wave-length intervals is about equal to the short-wave radiation received; in other words: the radiation balance is zero or even somewhat negative.

At some distance from the wall, but still inside the shadow cast, somewhat more diffuse light is received, but less long-wave radiation: there is more 'free sky', but less radiation from the wall, which is warmer than the sky. Near a forest edge or a hedge sunlight filtered by the leaves is also received.

A concrete example is given in Table 26. The net radiation is practically 0, while no less than 62.9 W/m^2 in short wave-lengths is received. The heat flow into or from the soil is less than the sensitivity of the instrument (2.5 W/m^2). The wet surface uses heat for evaporation, which has to be taken from the air, because there is virtually no energy available from radiation. As we can see, the temperature is close to the wet-bulb temperature. The measurements are not accurate enough to explain this small difference. Net radiation can also be weakly negative in the middle of the day (Stoutjesdijk 1974b, 1977a). The soil

Table 26. Energy balance (W/m^2) in the open shade on 3 September, 1980 at 12.00 h under a clear sky. Radiation from 400 - 700 nm was 10.3% of that in the open field. Surface temperature 12.1 °C, air temperature 17.6 °C, wet-bulb temperature 13.0 °C.

In		Out	
Diffuse solar radiation	62.9	Reflected solar radiation	8.9
Long-wave radiation from atmosphere and surroundings	322.3	Long-wave radiation from the surface	374.6
		Evaporation minus heat uptake from the air	1.7
		Heat flux to the soil	0
Total	385.2	Total	385.2

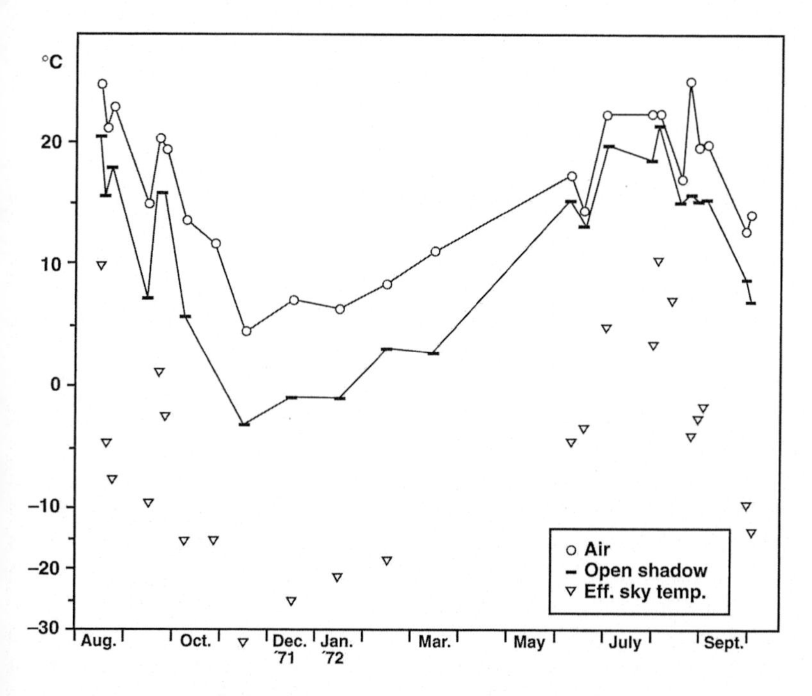

Fig. 52. Surface temperatures in the open shade compared to the air temperature at 1.50 m and the effective radiation temperature of the sky. After Stoutjesdijk (1974b).

surface is colder than the air, even with strong global radiation, in contrast to most other environments.

An interesting situation arises when clouds appear during sunny weather. Whereas in the open and at the south side of a wall, hedge or forest edge the temperature of soil (and air) decreases, it rises on the north side, this because the diffuse radiation and the atmospheric long-wave radiation increase (the clouds have a much higher temperature than the free sky dome). The radiation balance in the open shade may then move from negative to positive. It follows that the typical open shade conditions rarely occur in cloudy regions such as western Scotland and Norway. The effect is strong in the high mountains of continental areas. Measurements in the Netherlands show that the open shade effect can occur all year round (Fig. 50). At the time of these measurements, at midday, there was always dew.

Air temperature differences may be up to 10 °C. Such effects may be compared with those in the open during a clear night when similar temperature differences occur, though the net radiation loss is much greater. An important difference is that at night a great deal of energy is released by dew formation, while during the day energy is taken from the surface for evaporation. When such situations are compared it is not only a matter of radiation balance but also of the total energy balance.

In the open shade situation at night the net radiation is strongly negative. Even when, during the day, the net radiation is weakly positive, the overall 24-h radiation balance under a clear sky is negative. This can also be expressed as follows: in addition to the particular radiation situation during the day the cold reservoir built up at night contributes to the cool and moist microclimate in the daytime.

At first sight it is hard to understand how the net radiation can be constantly negative, but the total thermal budget must be considered. The radiation loss can be compensated for by heat taken from the air, where wind is important, by the heat flow from the soil, and during cloudy and rainy days, when there is a positive radiation balance.

The biological significance of the open shade situation is found in the longevity of dew and hoarfrost rather than in low temperatures. Especially in long periods of dry weather in late summer and autumn with clear skies and little wind, the moss and litter layer in the open shade remain moist with dew whereas they dry up completely in woods and shrublands.

As may be expected the open shade situation is favourable for many mosses, e.g. the thin-leaved *Mnium undulatum* and *M. cuspidatum* and some species appear to occur almost exclusively there, for example in the Netherlands *Hylocomium splendens*. According to D. Vogelpoel (pers. comm.) fructification of the moss *Dicranum scoparium* is largely restricted to north slopes and typical open shade environments. In dry autumn periods mushroom bodies are only found in the open shade situation in a park landscape. In juniper scrub they are then only found at the north side of dense shrubs.

Another example of an open shade species is the boreal species *Cornus suecica*. It occurs only on one site in the N Netherlands, where it is confined to northwestern, north and northeastern edges of a woodland and, at the same time, southern edges of open areas covered by *Calluna vulgaris*, and only occurs abundantly in a narrow zone, halfway from the woodland edge to the open area, where the high June sun casts a shadow (Fig. 53).

The north side does not only receive less light and less heat radiation, but the light is also of a different spectral composition. Short-wave radiation is dispersed to a greater extent than long-wave; in a clear sky the light is bluer. Seybold (1936) spoke of 'Blauschatten',

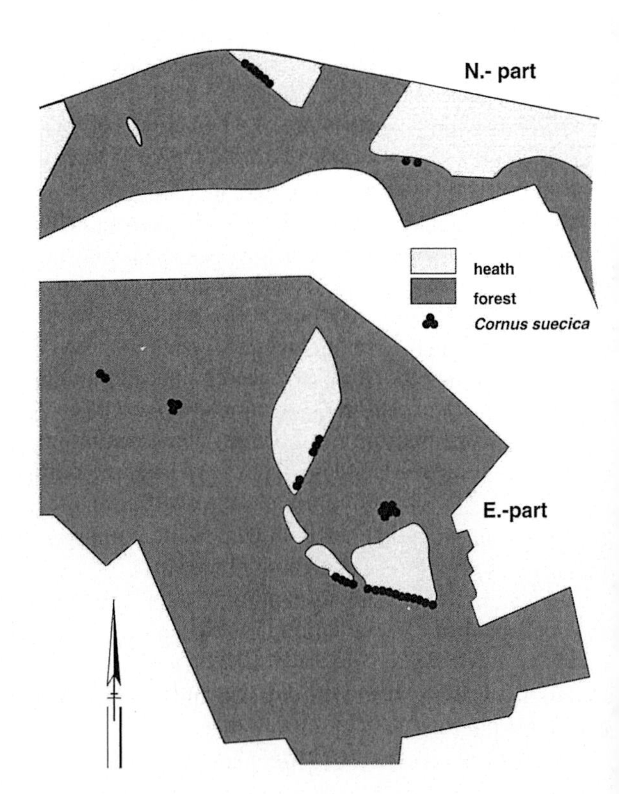

Fig. 53. Occurrence of *Cornus suecica* in the woodland complex Zeijer Strubben in Drenthe, the Netherlands. After E. Stapelveld (unpubl.).

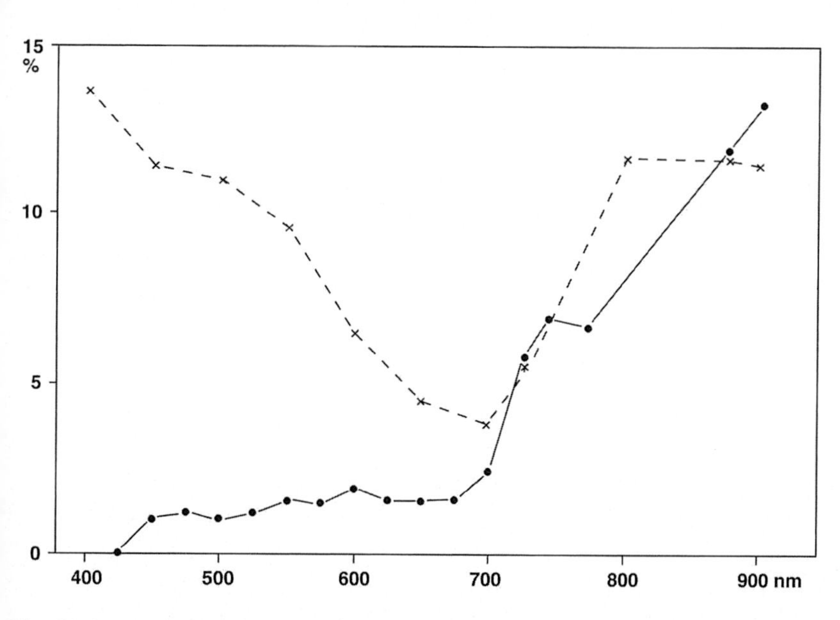

Fig. 54. Comparison of the spectral composition of radiation in open shade (interrupted line) with that under *Crataegus monogyna* scrub (solid line); both are given as a percentage of free sunlight. After Stoutjesdijk (1974b).

'blue shade'. This was long recognised by painters and other careful observers. The fact that diffuse skylight is blue can also be seen in the shadow on a snow cover in a clear sky. On a northern forest edge, green-infra-red light filtered by the forest canopy also arrives. In that situation we speak of blue-green shade. The most spectacular example of a blue-shade plant is the bright yellow foliose lichen *Letharia vulpina* occurring on *Pinus sylvestris*, *P. cembra* and *Larix europaea* in the driest and most sunny mountains, e.g. in Scandinavia only in E Norway, in the Alps only in the Central Alps (Wallis, Engadin, Aosta valley) and there only between 1700 and 2300 m. It is especially found there on north slopes and on northern sides of trees in open woodland or bordering open spots and only above the snow level. This species never receives direct sunlight but only diffuse light, rich in blue and ultraviolet (few clouds, high altitude, dry and clear atmosphere). The radiation balance is often negative and the temperatures are low. Interestingly, the yellow lichen has a surface colour which is complementary to that of the prevailing light. On the other hand, in lowland open shade environments plants are often blue-green, for instance the herbs *Glechoma hederacea* and *Vinca minor* and the liverwort *Calypogeia*.

Fig. 54 shows how the composition of the solar radiation in an open shade situation is compared with that in the open and under the canopy of a *Crataegus* scrub: it has shifted definitely towards the blue. Infra-red has also increased, no doubt because the open shade was caused by the *Crataegus* canopy which transmits much of the infra-red (see Ch. 3.4.4). In the case where the shade is caused by a wall or rock this transmitted infra-red does not occur. Another difference is that regarding radiation intensity. In the scrub the relative intensity of light between 400 and 700 nm is 0 - 2 %, in the open shade 4 - 14 %. In other words, radiation important for photosynthesis in the open shade is about 7 × as high as in the scrub (see also Ch. 2.2). In cloudy weather the difference is much higher.

Open shade may exhibit small-scale pattern especially at low sun elevations, as soon as areas with low vegetation are found in taller structures. Open gaps in between grass tussocks remain cool and covered with dew (cf. Fig. 51).

In winter the open shade effect is more pronounced and, because of the low elevations of the sun, more extended as well. The open areas may regularly have temperatures of 10 °C below the air temperature (Table 27). In February or March, under a clear sky puddles may freeze (again) and remain frozen the whole day even if the air temperature rises to +5 °C. This happens to the top soil as well in such places (cf. Fig. 1).

3.8.3 Warm fringes, south sides

The counterpart of the open shade situation is found along the southern edges of woods, hedges and rock walls. As regards long-wave radiation there is a clear 'gain' with clear weather: long-wave radiation of the northern half of the cold sky dome is replaced by the radiation from the vegetation or rock to the north, having a temperature near to the air temperature. This is especially important at night, when the minimum temperatures are higher than in the open.

As to the short-wave radiation the situation is more complicated. The wood intercepts the diffuse radiation from the northern sky but reflects some direct solar radiation. This is usually much less, however, than the intercepted radiation. The effect is that the total amount of radiation received, hardly differs from that of the open field. There are no extensive measurements available, but as an indication we mention that at the south side of a pine wood in the Netherlands the global radiation was 40.5 W/m^2 less than in the open, while the heat radiation received was 52 W/m^2 more. At the foot of closed juniper or spruce vegetation, however, the net radiation gain can be more than 100 W/m^2. Of course, this all holds true only for the situation at midday. In the temperate and boreal summer there is a loss of solar radiation, as compared with the situation in the open, in the early morning and the late evening. At night there is a gain in the energy budget, or rather, the loss is less. Over the 24-h day as a whole, the loss in solar radiation is compensated for by the gain in long-wave radiation. In winter, the south side is not shaded in the morning and evening and when

Table 27. Surface temperature t_s (°C) along a transect across a clear-cutting in a spruce woodland in the central Netherlands, compared with air temperature t_{air}. Measurements in the middle of the day, clear sky; * = snow cover; wo = in wood; os = open shade; su = in sun; oe =outer edge; ie = inner edge.

Date	t_{air}	t_s				
		wo	os	su	oe	ie
1976-12-10	5.5	0.0	–3.5	0.5	2.0	9.0
1976-12-14	2.0	–2.0	–3.0	2.0	4.5	8.0
1976-12-23	3.7	–1.5	–3.5	–1.0	4.5	5.5
1976-01-10	–5.0	-	–15.0 *	–10.0 *	–8.0	–5.0
1976-01-30	–6.0	–9.0 *	–12.0 *	–9.5	–5.0	–5.0
1977-01-31	2.0	–1.0	–2.0	3.0	7.0	7.0
1976-02-02	–1.6	–5.7	– 11.5 *	–5.5 *[1]	10.0	10.0
1976-02-20	14.1	6.5	4.0	13.0	17.0	20.0
1976-03-03	6.9	1.5	–1.0	9.3	17.0	20.0
1976-03-22	3.2	2.0	0.5	13.0	16.0	25.0
1976-04-09	15.0	8.0	8.3	34.0	39.5	39.5

[1] - A patch without snow was 3.0 °C.

total radiation is small, nightly differences in long-wave radiation mean a lot for the 24-h radiation balance; the south side then has a more positive balance than the open field. Even with an equal radiation balance the south side can become warmer than the open field, because the wind is less. Moreover, soil and vegetation are often different (cf. Ch 3.6).

Examples from the literature showing differences in microclimate between wood edge and open field suffer from these complications. Table 27 presents an example of measurements in a situation where differences in soil and vegetation were eliminated. Here, the edge of a spruce wood was compared with a recent clear-cutting - surrounded by wood - where the soil was covered with the same layer of needle litter. Temperature differences of 12 °C (always measured on a horizontal surface) were found between the clear-cutting and the edge. There were also clear differences in snow cover. A large contrast is found with the south side of the clear-cutting, which has an open shade situation due to the wood to the south. Two situations were possible: the inside edge of the wood where overhanging branches cover more than half of the sky dome, and the outside edge where exactly half the sky dome was screened. Temperature differences are greater than can be explained from the differences in energy balance at the time of the measurements, which were always made after clear nights with strong hoarfrost formation in the open field but little at the edge of the wood. Along the edge of the wood cooling off at night is less and this is found again in daytime temperatures. Thus, the energy balance must be studied over an entire 24-h period.

Apart from fundamental differences in radiation balance, which appear mainly in the winter, there are some additional effects. For example, wind speed is often lower than in open fields, especially with northerly winds. In the example given however, this cannot have been important in view of the weather conditions. Further, the wood edge is drier than the open field because the crowns of the trees effectively intercept the rain, especially light showers. As a result, the vegetation is often open, and dry litter covers the bare areas. So, (apart from radiation effects) the woodland edge often has already a warmer microclimate than the adjoining vegetation. At the northernmost limit of its distribution (60 °N in Sweden) the sand lizard *Lacerta agilis* is found in south-exposed forest edges, both in natural situations, at the edges of small lakes, and in artificial situations (Berglind 1988).

So far we have discussed forest edges with a sharp, usually anthropogenic transition to lower vegetation. In cases where there is a more gradual transition between forest and grassland with shrubs and forbs ('fringe communities') the microclimate is transitional as well (e.g. Dierschke 1974).

Near the forest edge the gain in long-wave radiation in the daytime is largely lost by the loss in diffuse solar radiation. We have a different situation, however, with a closed vertical surface warmed by the sun and reflecting solar radiation, for example a rock wall. Considerable gains in both long-wave and short-wave radiation can be observed at the foot of such vertical objects, especially when the sun is low and the radiation comes in almost perpendicular to the surface. This will also happen on a smaller scale, for instance near the foot of a tree trunk. In early March the extra radiation gain for a horizontal surface at the foot of a vertical wooden wall with a low sun, was 239 W/m^2. The greater the albedo of the reflecting surface, the more reflected solar radiation is received by the surroundings, and the less additional long-wave radiation, because the temperature of the reflecting surface rises less. Which of the two counteracting effects is stronger, depends, of course, on the albedo of the receiving surface as well as on the wind velocity and the thermal properties of the wall.

Mosses at the south side of an obstacle dry out much faster than at the north side. J. J. Barkman (unpubl.) developed the hypothesis that mosses receive more photosynthetically active radiation over the entire day on north sides than on south sides, in comparison with vascular plants. Experiments by J. Bouterse (unpubl.) confirm this. A tuft of *Dicranum scoparium* was divided into four parts and one piece placed in the soil at the north, east, south and west sides respectively of a dense, isolated juniper bush. On a sunny day the maximum light intensity on the north side was only 66% of that on the south side, but nevertheless the moss patch on the north side received more PAR which could actually be used for assimilation. The moss patch on the north side did not really dry out, whereas the patch on the south did regularly. For instance, on June 14, 1974 the moss patch on the east side dried up completely at 7.21 h, the patch on the south side followed 7.55 h and the west patch at 12.18 h. However, the north patch remained moist. The 'effective' light sums (intensity × time) received on E, S, W and N were in the proportion of 288 : 253 : 281 : 836. In the period until when the south moss patch dried up the north patch received only 33% of the light at the south side, but by the end of the day 330% of what the south side had received before 7.55 h, i.e. the moment photosynthesis stopped due to desiccation. Similar relations are found on steep north slopes as opposed to south ones. Mosses and lichens on a north slope are in a sense more 'light plants' than those on a south slope.

3.8.4 The microclimate of gaps

In natural forests open areas are an essential part of the vegetation structure. They have their own microclimate. This is important for forestry and nature management because many trees regenerate in gaps and many plants and animals are bound to such open places.

In the case of large open places such as forest meadows, large clear-cuttings and storm-fellings, a typical warm microclimate arises at the northern edge of the open place and a typical cool moist open shade microclimate on the south side. In regions with prevailing westerly winds the west side of gaps will be sheltered from wind and will receive sun only in the morning; while the east side receives afternoon sun. At all sides of the gap the nightly cooling off is stronger than in the forest but less than in the open field (because of the radiation from the forest). In the centres of large gaps the microclimate will resemble that of the open field, although in the first few years, in the absence of herbs and shrubs, more extreme temperatures will occur on the bare humus as compared with an open field. On the other hand, the wind strength is always less. Near the centre of a clearing (diameter of 500 m) in a dry tropical forest in Thailand, where the trees were 20 - 35 m tall, the wind velocities at 2m height were ca. 40% from what could be expected in open terrain (Pinker 1980).

Very small open areas show a microclimate that deviates little from that in the forest, mainly in that there is somewhat more diffuse light from the sky. On the other hand, the zenith is open and the sky dome is coldest there. In temperate regions areas with a diameter of at most 70% of the height of the surrounding trees will not receive any direct solar radiation and areas with a d/h ratio of 1 still receive very little. As a result day and night temperatures do not differ from those in the forest. Krečmer (1966, 1967, 1968) found in an open area in a pine forest ($d/h = 0.66$) no difference in maximum temperature, and minimum temperatures only 0.3 °C lower (on a clear spring night). Open areas with a diameter of at least the tree height do have somewhat more rain and dew, however. Gaps arising from tree fall, as occurring in natural forests will, at least in one direction, have a dimension roughly equal to tree height. With uprooted trees another differentiation arises: the steep upper side of the clod receives less rain and moreover the rain water runs off faster. The overhanging lower side receives no rain at all and little light.

The amounts of incoming radiation on level ground and the relative amount of effective radiation lost to the free sky as compared to an open situation for different values of the ratio d/h are as follows: (Lauscher in Geiger 1961):

d/h	1	1.5	2
incoming radiation	35	46	64%
relative radiation loss	24	42	55%

The long-wave radiation values of the surrounding trees are in the proportion 76 : 58 : 45.

Gaps with a diameter/height ratio of 1.5 to 2 are microclimatically the most deviating ones. Geiger (1941) created seven circular gaps in a mixed forest with d/h ratio varying from 0.5 to 3.4. The highest maximum temperatures were measured in gaps with ratios 1.5 and 1.8, more than 5 °C higher than in the forest. These values were even higher than in the open field. This is because of the incoming radiation, but also reflection by the forest edge and little wind: wind speed at $d/h = 1.5$ was only 20 - 28% of the open field. These values are summer values. In winter in open places (d/h from 0.7 to 2.0) day and night temperatures were not more than 1 °C lower than in the forest; because of the low elevation of the sun there is no direct incoming radiation (Krečmer 1966, 1967, 1968). At night an open place functions somewhat like a frost pocket (German 'Frostloch') with no wind to carry the cold air away. Where the gaps are small this frost pocket effect is compensated for by radiation from the trees and by run-off of cold air to the warmer interior of the forest. Geiger (1961) stated (p. 368) that enlargement of clear-felled areas increases the risk of night frosts late in the spring. Only if the gap becomes very large, can wind come in and blow away the cold air so that the risk of night frost diminishes again. Still, measured on still nights, he actually found continuously decreasing minimum temperatures with an increase in gap size, apparently because the corresponding open sky increased as well. Probably gaps with a d/h ratio of 2 are coldest during clear and windy nights (see also Mattson 1970). Smaller gaps are warmer because of the long-wave radiation, larger gaps because of more wind.

In gaps there is not only more assimilable light, but also the ratio red / far red is more like that under normal daylight conditions, as compared with the light climate under the forest canopy. We present some data from Chazdone & Fetcher (1984) on photosynthetically active radiation in % of that on a large clearing (PAR) and red / far red ratio (R/FR) on different sites in a Costa Rican rain forest tropical rain forest:

Site	PAR	R/FR
Clearing, 5000 m^2	100	1.23
Gap, 400 m^2	25	0.86
Gap, 200 m^2	9	-
Understorey	1	0.42
Sun flecks	-	0.99

The essential role of gaps in seed germination and development of trees in the rain forest was studied extensively by Bongers & Popma (1988). Oblong gaps can be expected to transmit more direct solar radiation with a high rather than with a low elevation of the sun, so their orientation NS or EW is of importance. In this respect it is interesting that Faliński (1978, 1986) found that in a natural spruce-oak forest in Poland 75% of the tree-fall caused by wind were found in the sectors N to W and S to E.

Because rain often comes in at an oblique angle, tree crowns catch part of the rain from a gap and in small gaps ($d/h < 1$) precipitation is less than in the open field. But in larger gaps more rain, up to 105 %, can reach the ground because of whirlwinds. Geiger (1941) found the following relation:

d/h	0.46	1.47	3.36
precipitation	87	105	102%

This is still more pronounced for snow. In gaps, snow cover can be 167% of that in the open

Table 28. Temperature and saturation deficit at midday for different sites in a tropical rain forest. After Fetcher et al. (1985).

	Understorey	Gap	Clearing	Canopy	Pasture
Temp. °C	25	27	30	27	(28)
S.D. mbar	2.2	5	11	9	(10)

field. Moreover, snow remains for a longer period of time (Krečmer 1966, 1967, 1968). Slavík, Slavíková & Jeník (1957, see Geiger 1961) found large differences in precipitation between the east and west side of a gap ($d/h = 1$). The wind-exposed east side had more rain than the open field.

For the potential evaporation (measured with the Piche-evaporimeter; see Ch. 5.2.7) at 20 cm these authors found a south-north gradient of 80 to 120% of the value in the forest. Krečmer found a higher potential evaporation in summer in gaps with $d/h = 2$ than in the open field, probably because of the high air temperature and thus the higher saturation deficit. Absolute humidity in a gap is higher than in the open field because of the transpiration from the surrounding trees and, because of lower wind speed, the potential evaporation is on average 90% of the open field value, and on the sunny north side where air temperatures are higher than in the open field, up to 110% (Mitscherlich 1971). At night dew is formed in the gaps, though less than in the open, at the edges less than in the centre, and in large gaps more than in smaller ones (Krečmer 1966, 1967, 1968). In the middle of the smallest gap and on the south side of the largest, dew remains for a longer period than in the open field.

Gaps with a d/h of 1.5 - 2 have a more continental character than the open field as far as temperature is concerned, but a more oceanic character regarding precipitation, air humidity and evaporation.

Epiphytic lichens are generally favoured by light and humidity. However, the combination of sun (light) and humidity is contradictory. Epiphytic lichens especially dry up rapidly in the sun. Small to middle-sized gaps are amongst the few environments where the combination of sun and moisture occurs regularly. Trees in the sun near waterfalls are also an ideal habitat for epiphytic lichens. From the Netherlands to northern France some epiphytic lichens produce soredia only in forest gaps: *Hypogymnia physodes*, *Parmelia subrudecta*, *Ramalina farinacea* and *Cetraria glauca*. For *Chaenotheca melanophaea* forest gaps are the optimal habitat in the Netherlands; for *Graphis elegans* and *Graphina platycarpa* in southwestern France (Barkman 1958).

A special case is formed by small patches of deciduous wood in dense coniferous forest. We may expect that these act as deciduous wood in the summer, i.e. cool and moist and with largely green and infra-red light, while in winter they act as gaps, cold and with mainly blue light. As far as we know this situation has not yet been investigated. According to observations made of such deciduous islands in the Netherlands some epiphytic mosses occur higher up the tree stem than anywhere else.

Fetcher, Oberbauer & Strain (1985) took measurements in a natural gap of 400 m^2 caused by the falling of a tree and in a clearing of 0.5 ha in a Costa Rican rain forest 30 m high. Their results, presented as averages over several months of readings at midday, are summarized in Table 28. The measurements were made at 70 cm height, the differences between understorey, gap and clearing stand out clearly. The data for the pasture are placed in brackets as they were gathered in a different period from those of the understorey.

During the night the temperature (20 °C) and the S.D. (2 mbar) differed insignificantly between the sites.

4. The biological significance of the microclimate for plants and animals

4.1 The significance of temperature and air humidity for living organisms

4.1.1 Introduction

> *"There is a well developed and readable scripture in this open book of nature but we are not yet able to read it fluently"*
> *H. Boyko (1962)*

The heat and water budgets of living organisms are closely linked through the energy balance. On body temperature we divide organisms into ectotherms and endotherms. The ectotherms need heat from the environment to keep their body temperature because they produce too little metabolic heat.

There are exceptions. Some bumble bees and large moths produce enough heat during their flight to keep their body temperature 20 - 30 °C above the air temperature, and they literally warm up before a flight. In the spadix of some *Araceae* the temperature can be 12 °C above the air temperature. Colony insects such as bees and termites produce enough heat collectively to keep their nests much warmer than the air around. Moreover such animals can regulate the temperature through ventilation and transpiration. Except for such special cases ectotherms change their body temperature with the temperature and radiation conditions of the environment. So, while their body temperature, in the sun, may be over 40 °C, higher than that of many endotherms, ectotherms are far from being cold-blooded.

Homoiotherms (endotherms) can keep their body temperature constant by regulating the intensity of their metabolism and the dissipation of heat, the latter through perspiration through the skin (some mammals), through panting (mammals and birds) or through secretion of saliva on the fur (mammals). Some endotherms hibernate and allow the body temperature to drop while the metabolism is reduced to a very low level.

A division similar to that made between ectotherms and endotherms can also be made with moisture. Ectohydrics (poikilohydrics) cannot regulate their body water content, endohydrics (homoiohydrics) have regulatory mechanisms. Ectohydrics dry up rapidly as the air gets drier until equilibrium with air humidity is reached. In this case the equilibrium depends upon relative humidity. When the air becomes moister they can take up water vapour, but this process is slow. For organisms which are never in contact with fluid water the relative humidity *RH* is the decisive environmental factor, i.e. active life is possible only above a certain *RH*. The speed of water loss after a rain shower is still more important and this speed is determined entirely by the saturation deficit of the air (Ch. 2.6). Homoiohydrics rarely dry up completely. For them the factors controlling the speed of transpiration are decisive, that is the difference between e_{max} at body temperature (constant for homoiotherms, variable for poikilotherms) and the vapour pressure of the air.

Poikilohydric organisms include mosses, lichens, aerial algae, some vascular desert plants (e.g. some *Selaginella* species), part of the *Rotatoria*, *Tardigrada* and *Nemathelminthes*.

In other animal groups cysts, larvae and eggs can be poikilohydric. Their moisture content changes rapidly with the micro-weather. The moisture content can drop to very low values without lethal effect, e.g. for lichens down to 5% of the dry weight (Lange 1969), or even 2% (Neubauer 1938). In poikilohydric organisms metabolism decreases and some sort of 'drought sleep' is adopted. Some lichens, e.g. *Ramalina maciformis* in the Negev desert, remain active (with a positive biomass balance) at a moisture content of 20% and a *RH* of 80% (Lange 1969).

Poikilohydric organisms can take up water vapour from relatively dry air; *Ramalina*, for instance, from air with a temperature of 10 °C with *RH* = 80% and a saturation deficit of 2.4 mbar and a suction pressure of 280 bar (Lange 1969). Some lichens never touch rain or phreatic water and obtain all their water as vapour from the air (Barkman 1958). Willis (1964) showed how rapidly shoots of the moss *Tortula ruraliformis* dried out down to 5 % water content (dry weight), took up water from air of different relative humidity values. At *RH* = 80 the water content of the moss is 20 %, a level reached after a few hours. Also various invertebrates may absorb water vapour from an unsaturated atmosphere, e.g. larvae of the grasshopper *Chortophaga* at *RH* = 82% and prepupae of the flea *Xenopsylla brasiliensis* even at *RH* = 45% (see Wigglesworth 1953; Edney 1957; Solomon 1966).

Most organisms, all vertebrates and almost all vascular plants are homoiohydric. They have a constant body water content, except for extremely dry situations, which are usually lethal. Strong desiccation is irreversible. The protoplasm of all plants and animals consists of water up to 70 - 80%; the total water content of insects varies from 50 - 90%, of terrestrial vascular plants, if not withered, from 60 - 95%. Homoiohydric organisms replenish their water supply regularly, plants mainly through water uptake by the roots, animals by drinking and eating, or they have a large store and limit transpiration (cacti, camels), or they stop transpiration completely by encapsulating themselves like amphibians, fish and molluscs. Kangaroo rats (*Dipodomys spectabilis*) restrict their water excretion and produce concentrated urine (Schmidt Nielsen 1972).

If the water supply stagnates for a longer period of time homoiohydric organisms die, even if the relative humidity of the air is high. The snail *Vitrina pellucida*, for instance, can loose 60% of its body fluid after three days at 97 % *RH* and it is killed after five days (van der Maarel 1965). The wilting point, i.e. the moisture state whereby most vascular plants cannot take up water from the soil any more is at about 16 bar, even when the air in the soil still has a relative humidity of almost 99% (cf. Ch. 2.5). Below this level it is impossible for them to take up water from the atmosphere. Thus, homoiohydric organisms cannot take up water vapour from the air, in contrast to lichens. In vascular plants water uptake must compensate for loss through evaporation, at a speed dependent on the saturation deficit.

4.1.2 Evaporation and transpiration

In Ch. 2.6 we have seen how evaporation of a wet surface follows from equations of the type 2.19 and 2.23. The diffusion coefficient *D* is inversely proportional to atmospheric pressure. The thickness of the boundary layer *d* is inversely proportional to the square root of the atmospheric pressure and with the square root of the wind speed. We have also seen that for most vascular plants and animals transpiration occurs over only a small part of the organism's surface. Diffusion resistance is often great; in stomata of leaves and in tracheas of insects the boundary layer is inside the organism, so wind plays a minor part here. In other words, the diffusion resistance within the organism is equivalent to a laminar boundary layer which is thick relative to the boundary layer on the outside. The thickness *d* of the 'internal boundary layer' is independent of wind speed. For a given value of $(c_s - c)$ or $(e_s - e)$ transpiration is only proportional with D, thus inversely proportional to atmospheric pressure. On a wet surface both D and d are important and evaporation is inversely proportional to the square root of the atmospheric pressure (Ch. 2.6).

Evaporation of a superficially dried-up soil, just like transpiration of vascular plants and animals, is determined by the resistance of the internal boundary layer, in this case the upper soil layer.

Transpiration of animals and higher plants, homoiohydrics, is partly a biological process which can be regulated by the organism through changes in the osmotic value (suction tension) of the cellular liquid, and by opening and closing the stomata, tracheas and sweat glands. Mosses, lichens and other poikilohydric organisms do not have such regulation mechanisms. They evaporate over the entire surface. Mosses store water 'externally' in capillary spaces between the leaves and the stems. Plants and animals living in a constantly moist environment, in water, in the spray from waterfalls, in humid caves, have a constant water content, but cannot really regulate their hydric condition and thus are poikilohydrics.

With homoiohydric organisms transpiration is always less than that of a water surface of the same size, because they have impermeable shields, a horny skin, cuticle, or a dense pelage reducing wind speed and thus transpiration. With snails, transpiration through the skin is equal to that of a free water surface but transpiration can be reduced by withdrawal into the shell and closure of the opening by an operculum. In this state they can survive months of drought (van der Maarel 1965). In the case of a forest the total transpiration may be larger than the evaporation of a water surface of equal size, but forest is of course multi-layered.

Because of their different physical constitution organisms differ in how they experience air humidity and effects on transpiration from radiation, air temperature and wind. As to evaporation, this is proportional to vapour pressure on the evaporating surface, e_s minus the vapour pressure in the air, e_{air}. Here, e_s is only dependent on the temperature of the surface. Hence, evaporation is greater in parallel with the higher temperature t_s of the evaporating surface and lower vapour pressure e_{air}. Here we distinguish three different cases.

1. t_s and t_{air} are about equal. This is the case with poikilothermic animals and poikilothermic (all) plants, if not subjected to direct solar radiation. (This is not entirely correct because t_s may be somewhat lower than t_{air} because of transpiration.) Now, $e_s = e_{max-s} = e_{max-air}$, and hence $e_s - e_{air} = SD_{air}$. Thus, the evaporation is only dependent on the saturation deficit of the air. The slight difference between t_{air} and t_s is dependent on the saturation deficit as well.
2. t_s is higher than t_{air} as a result of radiation. This situation occurs with poikilotherms in the sun, especially basking reptiles and dark-coloured insects such as ants and some beetles, as well as for dark green to almost black mosses and lichens, particularly those found in high mountains. Evaporation in these cases is mainly determined by the surface temperature of the organism (especially with a high transpiration resistance, see Ch. 4.1.3), because $e_s - e_{air}$ is hardly influenced for high values of e_s. This holds for low air temperatures in particular, because in this case e_{air} can vary only a little.
3. t_s is constant and high: homoiothermic animals. Here e_s is constant as well and, as far as the microclimate is concerned, the transpiration rate is dependent only on vapour pressure e_{air}. Because of this variation plants, animals and men experience some climates in completely different ways, as was explained long ago in a series of little known papers by Szymkiewicz (1923 - 1930).

An example may clarify this. The climate in the wet tropics is known as sultry, i.e. hot and humid, while boreal winter weather as dry. But for plants it is the reverse. The tropics are not at all humid for plants. They may show xeromorphic characters and the tallest trees in the rain forest, exposed to the open air may have thick leathery leaves. Only within the forest is there a wealth of hygrophytic thin-leaved mosses, bushes and lianas, but this is because of the special microclimate inside the forest.

The explanation for this is as follows. Suppose the temperature in the wet tropics is 28 °C and *RH* = 85% and on a boreal winter day 5 °C and *RH* = 95 %. At 28 °C e_{max} = 37.8 mbar, thus e_{air} = 0.85 × 37.8 = 32.1 mbar. For a plant we have e_s = 37.8 mbar, thus $e_s - e_{air}$ = 5.7 mbar. The human skin has a temperature of 33 °C, with e_{max} = 50.1 mbar and $e_s - e_{air}$ = 18.0 mbar. In the boreal situation e_{max} = 8.6 mbar, e_{air} = 0.95 × 8.6 = 8.2 mbar. For the plant $e_s - e_{air}$ = 0.4 mbar, for the human skin $e_s - e_{air}$ = 50.1 – 8.2 = 41.9 mbar. Thus, in the boreal situation a human being is subject to a desiccation which is almost 2.5 × faster than in the tropics, but for the plant the desiccation process is 14 × slower than in the tropics! These figures refer of course to a situation in the shade or under a cloudy sky.

Let us return to the poikilo- and homoiohydric organisms. Poikilohydrics are only active if they contain enough water. As far as water in their microclimate is concerned, the frequency and duration of rain and mist periods are the most important aspects. The total time after a shower during which they remain moist is also important. This depends on the speed of transpiration, thus on the saturation deficit of the air. For some organisms periods with a high *RH* (above 80%) are important as well, because they can take up water vapour from the air, e.g. lichens. Finally, for some hygrophytic mosses, periods with a low *RH* are of importance because they can be damaged, even in a state of latent dryness, by a low water content, the latter being determined by the relative humidity.

For homoiohydric organisms the saturation deficit is only decisive for survival if the loss of water proceeds so rapidly that water uptake cannot make up this loss (irreversible wilting of plants).

4.1.3 Heat and water budget of vascular plants

The living organism is influenced by the same processes regarding its heat and water budget as discussed for a soil surface. We know, relatively, a good deal about these processes occurring on green leaves, so we will concentrate on this situation. The situation for a leaf is less simple than for the earth's surface as a whole. The main differences are that leaves dissipate water vapour from both surfaces of the leaves, as well as releasing and taking up heat. The two surfaces differ in radiation and diffusion resistance for water vapour. Moreover, the leaf transmits a considerable proportion of solar radiation. Lee (1978) found, for leaves of 15 American tree species, that on average 22.2% of the total solar radiation was transmitted, 29.6% reflected and 48.2% absorbed. As an approximation we may say that about half the solar radiation received by a mesophyllic green leaf is absorbed.

Both the reflection and the transmission of radiation are strongly selective regarding wave length. There is a weak transmission and reflection between 400 and 700 nm and thus a strong absorption. For wave lengths > 700 nm, which are unimportant for photosynthesis - both reflection and transmission are considerable.

Lee (1978) also demonstrated that large differences occur between species. We give some examples below. We found considerable differences for western-European plants as well. Fig. 55 shows the situation for three coastal dune plants. Marram (*Ammophila arenaria*), a pioneer of the foredunes, has rather a high reflection both for visible and near infra-red radiation. A similar spectrum is found for the companion species *Elymus arenarius* (the major pioneer of northern European foredunes) and the mediterranean-atlantic *Eryngium maritimum* and *Euphorbia paralias*. A relatively low reflection is found

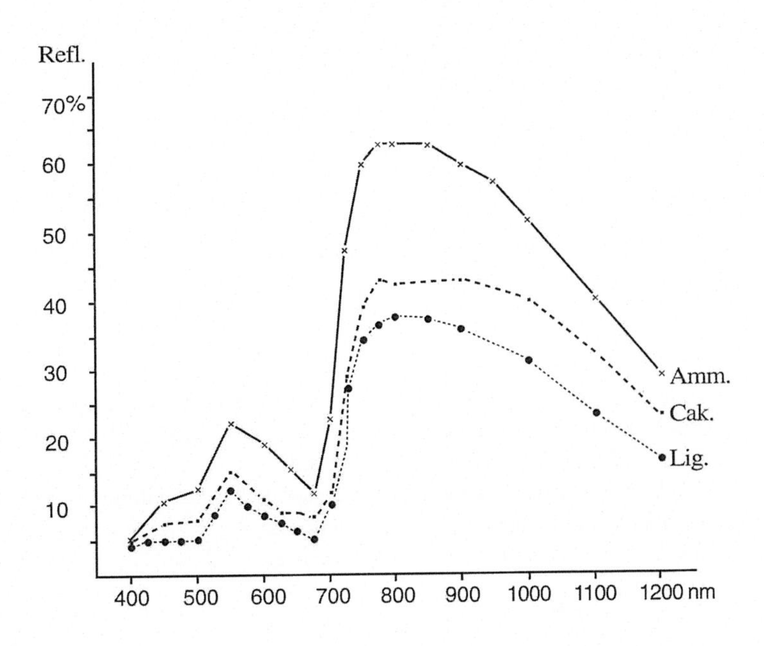

Fig. 55. Spectral reflection curves for leaves of *Ammophila arenaria*, *Cakile maritima* and *Ligustrum vulgare*. After Stoutjesdijk (1972c).

for the shrub *Ligustrum vulgare*, occurring behind the foredunes. The strandline colonizer *Cakile maritima* has an intermediate position as far as reflection is concerned. We mention that a high reflection in the near infra-red is common for organic material e.g. bird eggs (Dirmhirn 1964; Bakken et al. 1975). Still higher reflection values are reached by desert plants (Billings & Morris 1951). *Agave americana* reflects as much as 80% of the wave-lengths around 800 nm (Gates 1980).

Strong reflection occurs especially from thick sclerophyllous leaves. These have a lower transmission, so that the amount of absorbed radiation need not be less than with a thin leaf which is reflecting less strongly. Reflection was high over a broad spectrum on leaves of *Dudleya brittonii*, covered with a whitish wax layer. On the same plant leaves without the typical wax layer occur. These leaves have a much lower reflection.

Wave-lengths nm	290 - 400	400 - 700	700 - 1000
Leaves + wax	70 - 80%	60 -70%	60 - 75%
Leaves –wax	5 - 7%	7 - 30%	20 - 55%

In both cases the transmission of these thick succulent leaves is very small and thus absorption was almost the complement of reflection (Mulroy 1979).

We now elaborate on the heat and water budget of a leaf and choose *Plantago lanceolata* to do so. Leaf temperature and other parameters were measured in the middle of a 4 cm wide leaf and compared with values for a leaf covered with vaseline to prevent transpiration. Solar radiation received was 866 W/m^2, wind speed 15 cm/s. At an air temperature of 24.0 °C and a vapour pressure of 14.1 mbar the temperature at the surface of the transpiring leaf was 25.5 °C and of the non-transpiring leaf 34.2 °C.

Table 29. Energy balance (W/m^2) for a *Plantago lanceolata* leaf. Air temperature 24.0 °C, wet bulb temperature 16.7 °C, Δt transpiring leaf (tr) 1.5 °C, Δt of not-transpiring leaf (–tr) 10.2 °C. Values calculated per unit of leaf surface (two-sided).

In		Out	–tr	tr
Solar radiation	454	Long-wave radiation	1015	905
Long-wave radiation, sky	338	Convection + evaporation	245	355
Long-wave radiation, environment	468			
Total	1260	Total	1260	1260

The transpiring leaf is much cooler and apparently uses a good deal of energy. Table 29 presents some data on the energy balance. On the basis of Lee's (1978) data it is assumed that 50% of the solar radiation is absorbed.

The net absorbed radiation R_{net} is taken as the balancing item - here all errors are combined with the measurements. In this case $R_{net} = 245$ W/m^2, which amount is dissipated to the air as sensible heat. The heat transfer coefficient for the leaf can be calculated from

$$R_{net} = 2\alpha \cdot \Delta t \tag{4.1}$$

as in the equilibrium situation the total energy received is R_{net} which is completely given to the air. The factor 2 (compare Eq. 2.10) refers to the dissipation of heat on two surfaces of the leaf. For $\Delta t = 10.2$ °C and $R_{net} = 245$ W/m^2 we get $\alpha = 12$ $Wm^{-2}K^{-1}$. This value is much higher than that which can be expected from the technical literature (Eq. 2.10). For $l = 2$ cm (see Eq. 2.10) - this because we measure at the midpoint of the leaf - we calculate $\alpha = 5.9$ Wm^{-2} K^{-1}. It is a general phenomenon for leaves that higher values of α are found than those predicted by theory (Grace 1981). This might be explained by the turbulence of natural air currents and also by the fact that for small objects the decrease of α is stronger with decreasing l than the relation with $\sqrt{u/l}$ would suggest (Gates & Papian 1971; Sinclair 1970; Clark & Wigley 1975; Eq. 2.11 and 2.14).

It is also remarkable that the value of Δt is much lower than that is found at a soil surface under similar conditions. Here the explanation is that a smaller fraction of the radiation is absorbed. Also, the leaf's lower surface dissipates heat through convection more easily than the soil surface does through conduction.

If we assume that the intact leaf absorbs as much solar energy as the non-transpiring one we can also now write down the energy balance for the transpiring leaf. The situation for a transpiring leaf regarding the incoming energy is the same as for the vaseline-covered leaf. As regards the outgoing energy fluxes, the loss through long-wave radiation is much smaller because of the lower temperature, and therefore the net radiation energy received is much greater than with the vaseline-covered leaf. From the heat transfer coefficient we can calculate that 30.9 W/m^2 of the available energy is dissipated as heat to the air and the remaining 91.3% is used for transpiration.

Transpiration reduces the temperature of the leaf. The maximum reduction is provided in the situation with a wet surface, for example a piece of wet filter paper. If there is no net absorption of radiation its temperature is equal to that of the wet bulb of an ideal psychrometer (cf. Ch. 5.2.3). If a considerable amount of energy is absorbed, the temperature will be higher. It will be clear that this temperature rise on a wet or at least evaporating surface will be smaller than on a dry surface under the same conditions, this because the item of the balance concerned is used partly in the temperature rise and partly in evaporation. In Ch. 2.7

Table 30. Leaf temperatures and n_r values of some plants from tropical mountains and lowlands, some shrubs from a Dutch dune area, some coastal marsh species (Stoutjesdijk 1970a), and a desert species (Lange & Lange 1963). Δt tr = Δt transpiring leaf; Δt –tr = Δt non-transpiring leaf; t_{air} = Air temperature; t_w = Wet-bulb temperature; Loc. = Locality.

Plant species	t_{air}	t_w	Δt tr	Δt –tr	n_r	Loc.
Musa acuminata	19.6	17.3	11.6	21.4	5.0	Java, 1500 m
Musa acuminata	22.5	18.8	13.1	15.9	25.0	Ibid.
Rhododendron javanicum	21.8	18.2	9.1	17.2	5.6	Ibid.
Polypodium feei	22.7	20.3	10.8	11.1	182.0	Ibid.
Saccharum officinarum	31.3	26.2	4.6	12.5	5.5	Java, lowland
Saccharum officinarum	31.7	25.4	1.6	9.8	5.2	Ibid.
Calotropis gigantea	30.8	25.1	2.2	10.7	3.7	Ibid.
Calotropis gigantea	33.6	21.9	– 1.0	8.1	5.9	Ibid.
Rhamnus cathartica	20.4	16.4	8.0	16.3	4.2	the Netherlands
Salix cinerea	20.4	16.4	5.7	13.2	2.6	Ibid.
Verbascum nigrum	23.9	19.8	5.8	17.0	2.8	Ibid.
Cynoglossum officinale	21.0	16.2	5.7	11.1	5.7	Ibid.
Eryngium maritimum	18.0	14.4	5.9	13.6	3.2	Ibid.
Aster tripolium	18.0	14.5	10.1	18.4	4.5	Ibid.
Phragmites australis	23.0	16.9	1.0	7.6	3.5	Ibid.
Crataegus monogyna	24.3	15.7	8.5	9.0	128.0	Ibid.
Ligustrum vulgare	24.3	15.7	9.5	10.0	138.0	Ibid.
Hippophaë rhamnoides	22.1	17.2	7.7	11.0	12.8	Ibid.
Convolvulus arvensis	18.1	14.6	14.2	14.6	255.0	Ibid.
Citrullus colocynthis	27.5	20.2	7.5	15.5	7.3	S Spain
Citrullus colocynthis	50.5	24.0	– 13.0	10.0	3.5	Sahara

we explained how the temperature of a wet surface in the sun can be calculated. The highest temperature to be attained by a leaf under these circumstances follows from the temperature of a non-transpiring leaf.

Of course, a leaf loses less water than a free water surface of the same temperature and under the same conditions. However, early this century Brown & Escombe (1900) pointed out that the transpiration of a leaf is only a few times less than the evaporation of a free water surface under the same conditions, though stomata, which are responsible for the major part of the water loss, cover only 1% of the leaf surface.

This is the well-known edge effect of the stomata (and of small openings in a closed surface generally). If, in a concrete case the reduction factor is called n_r (Raschke 1956; Stoutjesdijk 1966,1970a), the energy balance of a unit surface of leaf can be written as

$$(1/n_r)\,(e_{leaf} - e_{air})\,\frac{2\alpha}{\gamma} + \Delta t_{leaf} \cdot 2\alpha - R_{netl} = 0 \qquad (4.2)$$

The first term of the equation is the amount of energy (per unit of time) used for evaporation (see also Ch. 2.8); the second term can be negative. For the maximum vapour pressure at the temperature of the transpiring leaf the symbol e_{leaf} can be used, for the vapour pressure of the surrounding air e, then the term Δt_{leaf} is valid for a transpiring leaf.The factor 2 is due to the leaf having two surfaces. The factor $1/n_r$ is an average value for the upper- and under-side of the leaf. From the above equation we calculate:

$$n_r = \frac{e_{leaf} - e_{air}}{(R_{netl}/2\alpha - \Delta t_{leaf})\gamma} \qquad (4.3)$$

$R_{netl} / 2\alpha$ has the dimension °C. If we substitute R_{netl} = 355W/m^2 (Table 29) and α = 12 Wm^{-2}K^{-1}, then $R_{netl} / 2\alpha$ = 14.8 °C. Thus, $n_r = (32.6 - 14.1) / (14.8 - 1.5) \cdot \gamma = 3.0$. Now we have calculated n_r exactly. This is possible because we have a complete energy balance available. However, we can obtain a reasonable estimation if we only measure the temperature of a transpiring and a non-transpiring leaf (Stoutjesdijk 1970a).

In Table 30 we have collected a number of leaf temperatures from transpiring and non-transpiring leaves. The values of n_r are usually included. For non-transpiring large leaves Δt values of over 20 °C are found, e.g. *Musa acuminata* (the banana) in a gap in the tropical mountain forest of Java. Here the absence of wind, the strong solar radiation and the large dimensions of the leaf together lead to a maximum Δt. *Agave* leaf temperatures are up to 26 °C above the air temperature. In these succulent leaves a large temperature difference can occur between the upper and lower surface of a leaf, because the heat transport through the leaf is small. In a thin leaf this temperature difference can be ignored (Pieters 1972). In a thick leaf the upper surface becomes warmer than the lower surface because less heat is dissipated downward. Upper and lower surfaces do not only differ in temperature because of the difference in net radiation but also possibly because the lower surface transpires more because it has more stomata.

The significance of leaf size for leaf temperature can be understood from the relation between surface size and heat transfer coefficient. In still weather α would depend only a little on size. However, this is only true for larger objects. Sinclair (1970) found that for very small or narrow leaves α increases strongly with decreasing size. Small size can be very favourable in cases where overheating might occur as a result of high radiation intensity, high air temperature and water shortage (Gauslaa 1984). The frayed leaves of the banana may have a function here (Taylor & Sexton 1972); on the other hand in the natural habitat of *Musa*, sheltered gaps, the leaves remain undamaged and here the risk for overheating is greatest.

Not only the size of a leaf but also the surface structure influences the heat transfer. Hairs on a leaf create a stagnant air layer at the leaf's surface, which does not always prevent radiation from reaching the epidermis and being absorbed. Krog (1955) found that willow catkins under subarctic conditions could be heated 15 - 25 °C above the air temperature (0 °C). When the hairs were removed from the catkin the value decreased to half the Δt value of the normal catkin. When the catkin was covered with soot (in order to maximize absorption) the surface temperature was in between the hairy and the hair-free catkin. The absorption of radiation through the soot was made maximal, but heat transfer was moved to the outside of the catkin. Similar results were obtained with hairy caterpillars (Kevan et al. 1982). An effect similar to that of the hairs on the willow catkin is exerted by the transparent bracts of *Eriophorum vaginatum*. In both cases one could speak of a 'greenhouse effect'. Wuenscher (1970) found that hairs on the leaves of *Verbascum thapsus* reduce heat transfer and evaporation. As a result high leaf temperatures may occur. Our measurements on this species on the dunes of Voorne, the Netherlands, showed a surface temperature on the leaves of 43 °C against an air temperature of 18 °C. The influence of a hair layer on the absorption of solar radiation is apparently less important than the influence on heat transfer and evaporation.

Hairs can also have an entirely different effect. In the dry summer the American semi-desert plant *Encelia farinosa* forms leaves with a thick layer of partly white and partly air-filled hairs. But in the wet season the leaves are only a little hairy. The absorption of light between 400 and 700 nm by slightly hairy leaves is about 81% against only 29% for strongly hairy leaves; for total solar radiation these percentages are 46% and 16% (Ehleringer & Björkman 1978). As we saw already, small leaves are less warmed up in the sun than are larger leaves and because of this they transpire less at the same level of internal resistance (Stoutjesdijk 1970a; Givnish & Vermey 1976).

While looking at the mechanism of transpiration we can also estimate the influence of wind on the transpiration and temperature of a transpiring leaf. The factor n can be regarded as a resistance to water loss, where the resistance of a free water surface under the same conditions will have the value 1. For a leaf the resistance is partly in the leaf, between the mesophyll cells and in the opening of the stomata, and partly in the boundary layer around the leaf. The latter layer controls the transport from the opening of the stomata to the free air. This resistance is not simply identical to that of a free water surface as is sometimes suggested. It is rather the resistance of a thin membrane with pores separating a water surface from an air surface, and this resistance is a factor 2 to 3 greater than that of a free water surface.

If a leaf has a high n_r value it will have only weak transpiration and when in the sun it will attain a temperature close to that of a non-transpiring leaf. If the wind speed increases and with it α, the leaf temperature may drop a great deal, but the resistance to water loss will not change much, because this is largely found in the interior of the leaf. For this very reason the already low transpiration will decrease further, this because of the lowered temperature, $e_{leaf} - e_{air}$ has dropped much more than the total resistance (cf. Eq. 4.2). At a low value of n_r transpiration may increase with wind speed or it may decrease. The cooling effect of wind and the decrease of transpiration will be large for large leaves and small for small leaves which have low Δt anyway. For a low value of n_r a general statement for the effect of wind cannot be given. It might, however, be calculated for individual cases.

The entire complex of interactions between air temperature, leaf temperature, wind, vapour pressure, and absorbed radiation was analyzed for a large number of cases by Gates & Papian (1971). Not only are the size of the leaf, position towards the sun, albedo and n_r important, but weather conditions such as air humidity and the presence or absence of sunlight, will also determine the effect wind will have on transpiration.

Of course, air humidity can have a big influence on the intensity of transpiration and with it also on leaf temperatures. In dry air and with a low transpiration resistance the leaf temperature can be considerably below the air temperature. Lange (1959) found that leaves of desert plants can attain temperatures in the sun which are 13 °C below air temperature and this level of cooling through transpiration is essential to avoid lethal temperatures.

The temperature in flowers may differ considerably from the air temperature. On the arctic Ellesmere Island (81 °N) the flowers of *Dryas integrifolia* are always directed towards the sun. In the centre of the flower temperatures of 8 °C above the air temperature were measured and insects living in the interior of the flower reached temperatures 15 °C above the air temperature (Kevan 1975). Half this effect disappeared if the petals were removed. Kevin considers the parabolic form of the flower crown very important because it focusses the sunbeam. We assume that the parabolic form is not perfect enough really to give a focussing effect; reflection is too diffuse. Still, some concentration in the centre of the flower is possible; moreover the crown has the effect of reducing wind speed.

4.1.4. Heat and water budget of poikilothermic animals

We can look at the heat and water budget of many organisms in the same way as we did for plants. Regarding the necessity of water loss there is a big variation indeed. A green plant must withdraw CO_2 from the air where there is a concentration of 0.03 %. CO_2 follows the same route as water vapour, but in the reverse direction. The intensity of the CO_2

Table 31. Values for the transpiration reduction factor n_r for various arthropods.

Species	n_r
Acarus siro (*Acaridae*)	1175
Tenebrio molitor, larva (*Coleoptera*)	333
Calliphora sp. (*Diptera*)	317
Glossina sp. (*Diptera*)	999
Periplaneta americana (*Orthoptera*)	74
Scolopendra sp. (*Centipeda*)	35
Porcellio scaber (*Isopoda*)	29
Glomeris sp. (*Millipeda*)	12.5
Philoscia sp. (*Isopoda*)	7.7

assimilation is therefore strongly linked to the intensity of transpiration. Limiting water loss leads, of necessity, to a reduction in photosynthesis.

If the main physiological process is respiration (animals, plants at night), the uptake of oxygen is usually connected with some loss of water. The difference with CO_2 uptake under photosynthesis, though, is that the atmosphere has 20% O_2 against only 0.03% CO_2. So the intensity of air transport may be much less, and thus the loss of water for respiration may be much less than for photosynthesis. Moreover, animals with a low transpiration resistance (and fungi as well) can live in an environment with a low radiation intensity which usually has high air humidity.

Machin (1964a,b) found that water on the soft parts of snails evaporates in a similar way to a free water surface. Amphibians have a moist skin and consequently strong transpiration. A bullfrog maintains a temperature only 3 °C above the air temperature and 10 °C above the wet bulb temperature in the sun (700 W/m^2) and a wind speed of 30 cm/s (Tracy 1975). This shows that a considerable part of the energy taken up disappears again via evaporation. These animals compensated for the water loss through the uptake of water from the moist soil. According to Wygoda (1984) the resistance to water loss in frogs (with the exception of tree-frogs) is almost as low as that of a free water surface.

For a number of arthropods transpiration intensity as a function of surface size and saturation deficit is known from the literature. We have calculated n_r values (cf. Ch. 4.1.3) using these figures (Table 31). This can be done approximately if we can realistically estimate the heat transfer coefficient. In many cases the transpiration is so small that it can be neglected in the energy budget. Many insects behave thermically like dry objects, but with strongly transpiring arthropods such as the shore slater *Ligia oceanica* (*Isopoda*) the transpiration will doubtless influence the body temperature (Edney 1957). Barton-Browne (1964) pointed out that the transpiration of insects via their stigmata is somewhat similar to the transpiration of leaves through the stomata. Colour and hairiness may vary strongly and have a larger influence on the energy budget than for a green leaf.

Willmer & Unwin (1981) measured temperatures of insects in the sun (840 - 900 W/m^2). They found Δt values of ca. 10 °C for bigger insects (body weight 90-100 mg) and of ca. 5 °C for smaller insects (26 - 27 mg). For the smallest insect (3 mg) Δt was almost zero, so there was hardly any difference between body temperature and air temperature. So the size of the insects is very important, which agrees with the findings of Sinclair (1970) for leaves of different size (and Eq. 2.11). Wilmer & Unwin (l.c.) compared insects of different size which were placed freely in the air. We can also consider another aspect of size, the relation between size and the possibility of exploitation of small-scale differences in the microclimate.

As regards temperature we know that there can be big differences between the surface temperature and the free air and how strong the temperature changes are in the first few cm above the surface. An organism of a mm size living on a surface is virtually part of the surface and has the same temperature, which can be 40 °C higher than the air temperature. A small mobile organism can move to that part of the temperature gradient where it is best adapted. We found flies (*Musca autumnalis*) sunbathing in the cracks of the bark of pine trees where the surface temperature was 25 °C, 5 °C higher than elsewhere on the bark. A larger organism cannot choose the right spot quite so exactly. On the other hand, because of its larger size it has a much smaller heat transfer value and is thermally more independent.

We can demonstrate this with the following experiment: Under a strong artificial radiation source the following temperatures were measured:

Surface of dark humus	69.7 °C
Small brass cylinder (3 mm diameter) on surface	59.8 °C
Cylinder at 1 cm above the surface	38.3 °C
Cylinder in free air	18.1 °C
Air temperature	9.8 °C

A fine wire in free air does not have a much higher temperature than the air, but placed on the surface its temperature approaches the surface temperature very closely.

A larger organism can also move faster and escape unfavourable conditions in a boundary layer. And, of course further away from the surface conditions are less extreme. On the other hand, a larger size also means more air movement around the organism and bigger thermal inertia e.g. a slower warming up. Large reptiles can only reach the required body temperature in warm regions.

Regarding moisture, similar gradients exist from a surface into the surrounding air. But with water vapour transfer out of the organism the internal resistance is often larger and not related to body size (see Chs. 2.4 and 2.6). High degrees of humidity are often found in dense vegetation, where smaller animals can move more easily than larger ones. Finally, a larger size gives a ratio between surface area and volume which is more favourable, as far as moisture is concerned. For a small organism the thermal conditions of a warm but dry boundary layer may be favourable, but the moisture conditions may be stressful there.

4.1.5 Heat and water budget of homoiothermic animals

"You know what a pig is, don't you? If not, there's a lot to explain"
P. G. Wodehouse

Birds and mammals are homoiotherms (endotherms). These warm-blooded animals can maintain a constant high body temperature thanks to considerable metabolic heat production and an isolating layer of feathers or hair. Heat production can be adapted to local environmental conditions, but it cannot be increased indefinitely; neither can it sink below a minimum for a resting healthy animal. This minimum is called the basal metabolic rate (BMR). This BMR is proportional to the 3/4 power of weight. This relation holds over a large range of weights, from 'mouse to elephant' (Schmidt-Nielsen 1972) though with a finer analysis there are differences between taxonomic groups and between times of the day or the year. A well-known exception is the birds of the order *Passeriformes* which have a

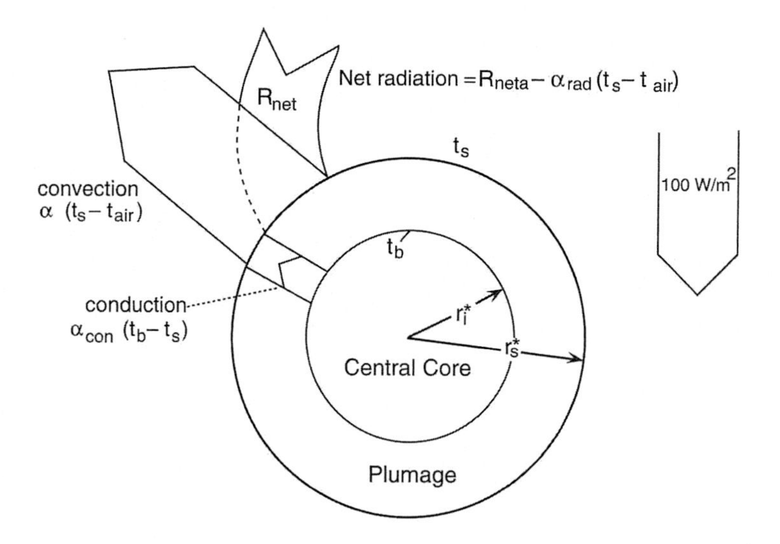

Fig. 56. Energy exchange at the surface of a bird or mammal, by radiation, conduction and convection. r^*_i = inner radius; r^*_s = outer radius; t_b = body temperature; t_s = surface temperature; t_{air} = air temperature.

roughly 30% higher BMR than would be expected. So this obligatory heat production alone already makes it comprehensible that an endothermic animal may suffer from heat stress under conditions which are optimal for an ectotherm.

The thermal essence of an endothermic animal is a core of living tissue kept at a constant temperature surrounded by insulating plumage or pelt. If the temperature of the body t_b is higher than the surface temperature t_s of the hair or feathers, then there is heat flow to the outside. What concerns us here is the influence of the surroundings on this flux.

The energy balance for a piece of the surface is as follows (Fig. 56):

$$R_{net} + \frac{k}{d}(t_b - t_s) = \alpha\,(t_s - t_{air}) \tag{4.4}$$

Here R_{net} is the net radiation absorbed by the surface. The second term is the unknown heat flow H_{con} we wish to know, where k is the heat conduction coefficient of the pelage or plumage and d is the thickness. The term at the right of the equality sign is the heat transfer to the air with t_{air} as the air temperature. R_{net} is the sum of all long-wave and short-wave radiation received from the sun, the sky and the surroundings, minus the reflected short-wave radiation and the emitted long-wave radiation ($\varepsilon\sigma T_s^4$). For any set of parameters in Eq. 4.4, except T_s we can, by iteration, determine the value of T_s that satifies the equations and so for a great number of combinations of radiation fluxes received, wind, plumage thickness etc. find the heat loss or gain. Several other approaches are possible (Gates & Porter 1969; Monteith 1973; Bakken & Gates 1975; Gates 1980). It is, however, possible to show the relation between the environmental parameters, the plumage or pelage properties and the heat flow by means of an explicit equation.

When we write $(t_s - t_{air}) = \Delta t$ we can replace Eq. 4.4 by

$$R_{neta} - \alpha_{rad} \cdot \Delta t + \alpha_{con}\,(t_b - t_{air} - \Delta t) = \alpha \cdot \Delta t \tag{4.5}$$

Table 32. Some values of α_{con} for various birds and mammals in relation to body weight. Resistance units are obtained by rewriting $r = \rho C_p / \alpha_{con}$ with $\rho C_p = 1210$ J/m^3.

Species	BW	α_{con}
Birds		
Verdin[1]	6.5 g	8
Black-capped chickadee	11 g	2.5
Little penguin[2]	990 g	2.9
Sharp-tailed grouse	1000 g	
Breast		1.62
Neck		9.4
Snowy owl	2000 g	0.6
Mammals		
Antelope ground squirrel	100 g	20
Arctic ground squirrel	1000 g	10
Arctic fox[3]	6 kg	0.8
Reindeer[3]	120 kg	1.0

[1]Webster & Weathers (1990); [2]Stahel et al. (1987); [3]Scholander et al. (1950), other references in text.

Here R_{neta} is the net radiation if the surface has the same temperature as the air and α_{rad} indicates the increase in long-wave radiation for each °C temperature increase of the surface relative to the air (see Ch. 2.6). Because we are dealing with a parameter with the same dimension as the heat transfer coefficient (Wm^{-2}K^{-1}) we write α_{rad}. Likewise we write k/d as α_{con}, the conductance (see Fig. 56).

Then we resolve Δt from Eq. 4.5:

$$\Delta t = \frac{R_{neta} + \alpha_{con}(t_b - t_{air})}{\Sigma\alpha} \tag{4.6}$$

Here $\Sigma\alpha = \alpha + \alpha_{rad} + \alpha_{con}$.

It is interesting to compare Eq. 4.6 with the measurements of the surface temperature of the sharp-tailed grouse (*Pedioecetes phasianellus*) by Evans & Moen (1975). These measurements were made with a radiation thermometer, in still air and with a negligible value of R_{neta}. Under these conditions we have:

$t_s = t_{air} + (t_b - t_{air})\alpha_{con} / \Sigma\alpha$, or rearranged:

$$t_s = t_{air}\ (1 - \alpha_{con} / \Sigma\alpha) + t_b \cdot \alpha_{con} / \Sigma\alpha \tag{4.7}$$

Evans & Moen found the regression equation $t_s = 0.87 t_{air} + 5.25$. Obviously this is a special case of Eq. 4.7; with $t_b = 41$ °C there is an excellent fit with $\alpha_{con} / \Sigma\alpha = 5.25/41$. With $\alpha =$ 6 Wm^{-2}K^{-1}, for still air and $\alpha_{rad} = 5$, we find $\alpha_{con} = 1.62$ Wm^{-2}K^{-1}. What was said above also applies to the measurements of Hill, Beaver & Veghte (1980) on the Black-capped chickadee (*Parus atricapillus*).

We return to the heat flow by conduction H_{con} for which we have (see Eqs. 4.4 and 4.5 and Fig. 56):

Table 33. Heat loss by conduction in W/m^2 (after Eq. 4.9) for a model bird or mammal with two values of α for still air and a wind of 1 m/sec, two values of R_{neta} representing shady and sunny conditions, and two values of α_{con} representing thin and thick plumage. The air temperature t_{air} = 10 °C, the body temperature t_b = 40 °C; α_{rad} = 5 W/m^2 °C.

R_{neta}		0	200
α	α_{con}		
6	10	157.1	61.9
	3	70.7	27.8
20	10	214.3	157.1
	3	80.3	58.9

$$H_{con} = \alpha_{con} \ (t_b - t_{air} - \Delta t) \tag{4.8}$$

If we substitute the value of Δt from Eq. 4.6 in Eq. 4.8 we obtain after some rearranging

$$H_{con} = \frac{\alpha_{con}}{\Sigma\alpha}\{(\alpha + \alpha_{rad})\ (t_b - t_{air}) - R_{neta}\} \tag{4.9}$$

Eq. 4.9 can represent local values both of H_{con} and of the parameters on the right side of the equals sign. R_{neta} especially can show considerable local differences e.g. 800 W/m^2 on the sunny side and 0 W/m^2 on the shade side of the animal. It is also possible to let Eq. 4.8 represent average values for the whole animal as we shall do when not otherwise stated.

To fix our thoughts it is worthwhile giving some plausible numerical values to the parameters in Eq. 4.9. We assume as a first approximation that α_{con} is independent of wind velocity, or rather that the plumage or pelage is impermeable to wind. Some typical values of α_{con} are given in Table 32. As shown in Table 33, wind has a relatively small effect on the heat loss when R_{neta}= 0, but quite a strong effect when R_{neta} = 200. With α = 6 and R_{neta} = 200, the heat loss H_{con} is reduced to 39% of the value at R_{neta} = 0. This relative effect of radiation remains the same when the value of α_{con} is changed.

A higher α_{con} means a stronger effect of wind on heat loss, even if α_{con} is independent of wind velocity. To what extent is it realistic to assume that α_{con} does not increase with wind velocity? Bakken (1990) has given an overview of the effects of wind on heat loss by birds. When the heat loss (H_{con}) in still air is 1, the effect of wind velocity u (m/sec) is given by $H_{con} = 1 + c\sqrt{u}$; The factor c usually has values between 0.2 and 0.4 and occasionally as low as 0.1 or as high as 0.7. When these data are compared with those of Table 33 it can be concluded that quite often the effect of wind on H_{con} can be understood mainly through its effect on α alone. This is the more so as the highest values of c were found for birds which had the head freely exposed, and the lowest for resting ruffed grouse (*Bonasa umbellus*) with the head tucked under the feathers. The thinly covered head has a high α_{con} and thus the heat loss via the head will be more sensitive to wind than that of the rest of the body.

Grubb (1975, 1977) has shown through many observation series how foraging woodland birds, especially Carolina chickadee (*Parus carolinensis*) and tufted tit (*P. bicolor*) avoid windy situations in winter and favour sheltered places, even though wind velocity did not exceed 3 m/sec. This preference was quite clear with temperatures below 0 °C and occluded sun. When the sun was shining, avoidance of windy sites was not clear; maybe because no sunny sheltered sites could be found. A calculation as presented above (Table

33) makes clear that a higher R_{neta} can compensate for a windy site, and surely energy aspects are not the only ones determining the behaviour of a bird. Jackdaws often roost in town centres which are warmer than the surrounding countryside - by several °C - but the energy gain often does not compensate for the energy expenditure in flying the often long distance to the roost site (Gyllin, Källander & Sylvén 1977). The same applies to wintering starlings in a natural landscape in Israel (Yom-Tov, Imber & Otterman 1977).

Chappell (1981), in his study on the thermoregulation of the arctic ground squirrel (*Spermophilus undulatus*) found that up to a wind velocity of 1.5 m/sec α_{con} (10 W/m² °C) was not affected by wind speed but then it increased rather strongly to 30 W/m² °C at 9 m/sec. We may assume that a rather small animal (1 kg) in a vegetation of willow scrub (60% cover with lower vegetation intermingled), can usually avoid high wind velocities - the average at 20 cm height was 1.5 m/sec.

When discussing the energy balance of a leaf it was useful to compare the temperature of the actual leaf with that of a non-transpiring leaf. Similarly we can compare the real animal with a model that produces no metabolic heat. A life-size model of a bird or mammal with the same reflection for short wave and long-wave radiation which is placed in the same position as the actual animal will have the same R_{neta}, α and α_{rad}. When it has a high thermal conductivity and thus a rather uniform temperature, its average equilibrium temperature t_e will be such that $R_{net} = (t_e - t_{air})\alpha$ or: $R_{neta} = (t_e - t_{air})(\alpha + \alpha_{rad})$. A good approximation of t_e is obtained when the temperature in the centre of a model of thin copper or aluminium foil is measured. When we replace R_{neta} in Eq. 4.9 with $(t_e - t_{air})(\alpha + \alpha_{rad})$ we get the simple expression:

$$H_{con} = \frac{\alpha_{con}}{\Sigma\alpha} (t_b - t_e) \quad (\alpha + \alpha_{rad}) \qquad (4.10)$$

The temperature t_e is identical to the operative temperature (Gonzalez, Nishi & Gagge 1974; Bakken 1980).

By means of this t_e we have a simple method for characterizing environmental conditions. In t_e the effects of air temperature, wind (α) and R_{neta} are integrated. In the numerical example $t_e = t_{air} + 200/(6+5) = 28.2$ °C for $\alpha = 6$ W/m² °C and 18 °C for $\alpha = 20$ W/m² °C (cf. Table 33).

To make field measurements comparable mutually and with those made in metabolic chambers one has defined the standard operative temperature (Bakken 1980). In a metabolic chamber α is usually lower (less wind) than in the field situation, also α_{con} may be somewhat wind-dependent, i.e. lower in still air. In Eq. 4.10 a change of α, say a reduction, also means a reduction of H_{con} and the same applies to α_{con}. When we change t_e in such a way that H_{con} remains the same, the corrected t_e is the standard operative temperature.

Taken as an average over an approximately spherical shape R_{neta} can reach a value of 300 W/m² in summer for a black animal and even more on a bright winter day over a snow surface. The operative temperature is often 15 -20 °C and occasionally up to 30 °C above the ambient temperature. When t_e is t_b, say about 41°C, it is clear that a bird or mammal cannot dissipate any heat by conduction via the plumage or pelage. Respiration becomes the main pathway for heat dissipation by evaporation and as sensible heat for birds and non-sweating mammals.

As an illustration of Eq. 4.9 we will now calculate the necessary metabolic activity of a great tit (*Parus major*) with a body temperature of 42 °C, spending the night in a field or a dense spruce wood. We assume that the tit puts its head into its plumage and in this way turns into a little sphere of living tissue with a diameter of 3 cm, surrounded by feathers, so

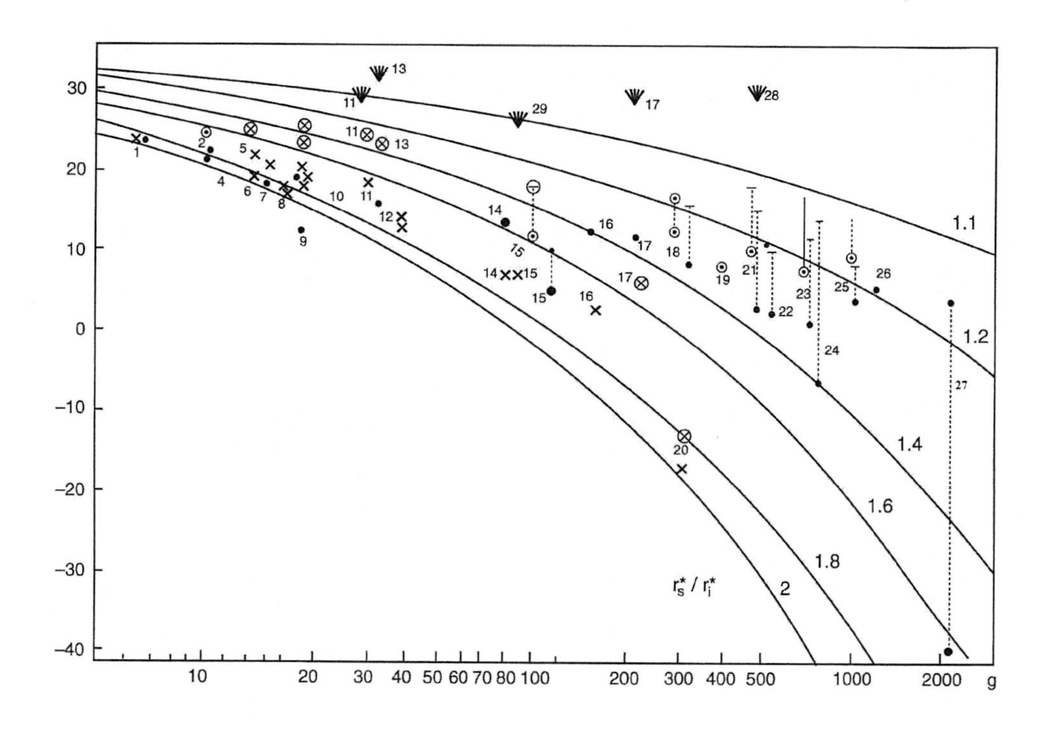

Fig. 57. Relation between weight and lower critical temperature of spherical model birds (*Passeriformes*) with various insulation values (r_s^*/r_i^*; see text).

- • = The insulation value of the plumage of real birds expressed as the r_s^*/r_i^* of the model bird with which it is equivalent. For passerine birds the approximate value of the lower critical temperature can be read from the temperature scale.
- ⫪ = Measured lower critical temperature for non-passerine birds. (• and ⫪ after data from Kendeigh, Dolnik & Gavrilov 1977);
- × = Insulation value calculated for birds of which the plumage volume was determined photographically;
- ѱ = Temperature above which heat stress or avoidance was observed;

winter values: × • ⫪

summer values: ⊗ ⊙ ⊙̣

1. Goldcrest *Regulus regulus*
2. Wren *Troglodytes troglodytes*
3. Long-tailed tit *Aegithalos caudatus*
4. Blue tit *Parus caeruleus*
5. Marsh tit *Parus palustris*
6. Willow tit *Parus montanus*
7. Redpoll *Acanthis flammea*
8. Siberian tit *Parus cinctus*
9. Great tit *Parus major*
10. Nuthatch *Sitta europaea*
11. Yellowhammer *Emberiza citrinella*
12. Bullfinch *Pyrrhula pyrrhula*
13. Red-backed shrike *Lanius collurio*
14. Blackbird *Turdus merula*
15. Great spotted woodpecker *Dendrocopus major*
16. Jay *Garrulus glandarius*
17. Jackdaw *Corvus monedula*
18. Black-headed gull *Larus ridibundus*
19. Rook *Corvus frugilegus*
20. Hawk owl *Surnia ulula*
21. Partridge *Perdrix perdrix*
22. Willow grouse *Lagopus lagopus*
23. Wigeon *Anas penelope*
24. Coot *Fulica atra*
25. Mallard *Anas platyrhynchos*
26. Raven *Corvus corax*
27. Snowy owl *Nyctea scandiaca*
28. Hooded crow *Corvus corone*
29. Starling *Sturnus vulgaris*

that the outer diameter becomes 5.5 cm. The surface of the sphere is 95 cm^2. We assume the heat conductivity k to be 0.04 $Wm^{-1}K^{-1}$ (Evans & Moen 1975). See Table 33. The value of α_{con} cannot simply be put equal to k/d. This would only hold for a flat surface and would only be approximately true if the thickness of the insulating layer were small relative to the diameter of the sphere. According to the technical literature:

$$\alpha_{con} = k / (r_s^* - r_i^*) (r_s^* / r_i^*) \qquad (4.11)$$

The radius of the outer sphere $r_s^* = 2.75$ cm and of the inner sphere $r_i^* = 1.5$ cm. On the basis of the above data α_{con} will be 1.75 $Wm^{-2}K^{-1}$.

It is interesting to compare the calculated figures for a 'schematisized' bird with the measurements of metabolic activity of the actual animal. Mertens & Gavriloff in Kendeigh, Dolnik & Gavrilov (1977), give a metabolic rate of 0.50 - 0.71 W for a resting tit in a neutral radiation environment i.e. with $R_{neta} \approx 0$ with an ambient temperature of –2°C. We calculate for our model tit, with the same environmental parameters, $H_{con} = 0.64$ W.

It is also instructive to vary the parameters of Eq. 4.8 and to analyze the resulting variation of H_{con}. For example, if we reduce the thickness of the plumage from 1.25 cm to only 1 cm, $r_s^* = 2.5$ cm; this will have only a small influence on the total heat dissipation of the tit model, **α increases and the total surface decreases**, the latter from 95 to 78.5 cm^2 (Eq. 4.10). We can also vary the total size of the animal, assuming that the thickness of the plumage is proportional to body diameter. Also, we can vary α with varying values of R_{neta}, both positive and negative ones. In summary, it is both easy and instructive to play with this bird model.

We shall now elaborate the idea of the spherical model of a bird and its thermal relations with the environment somewhat further and compare the results with data from real birds. We assume that the weight (P, in g) of the bird is equal to $4/3\ \pi \cdot r_i^{*3}$ (r^* in cm).

In Fig. 57 we have tried to evaluate the effect of different amounts of plumage on the heat loss by conduction when there is no excess heat to get rid of. In the first place the concentric model was used with values of r_s^*/r_i^* of 2, 1.8, 1.6, 1.4, 1.2 and 1.1. The heat loss by conduction was calculated for 'birds' of different weights in W/°C using Eq. 4.9 or 4.10 and 4.11. The heat loss was calculated for still air (low α) as a connection was sought with the laboratory studies on heat loss found in the literature, which were collected by Kendeigh, Dolnik & Gavrilov (1977). The lines in the diagram are drawn in such a way that they indicate the temperature t_e, or t_{air}, when R_{neta} is negligible, at which a bird can dissipate in still air an amount of heat equal to its basal metabolic rate. BMR, in W, is related to weight (P, in g) as follows: BMR = 0.0544 $P^{0.66}$ which is the standard value for *Passeriformes*. It was assumed that the specific conductivity (k) of the plumage was 0.04 W/m °C, this is equal to the value of the best insulating materials given in the technical literature and is also realistic when compared to the values for plumage and pelage.

To compare the model birds with real ones, we used, among other things, actual data on the lower critical temperature, which for many birds were collected by Kendeigh, Dolnik & Gavrilov (1977). A bird which is completely at rest has the BMR down to a certain temperature at which the feathers are fluffed completely, i.e. the insulating properties of the plumage are maximum. Below this 'critical temperature', LCT, the metabolic rate must be increased above the BMR to keep the bird warm.

Data on lower critical temperatures and metabolic rates were used to estimate the insulating properties of the plumage expressed as r_s^*/r_i^*. When, for example, for a bird of 20 g, a mark is placed in the diagram at $r_s^*/r_i^* = 1.8$ it means: placed side by side in still air the model with $P = 20$ g, $r_s^*/r_i^* = 1.8$, and the real bird have the same heat loss H_{con}. These

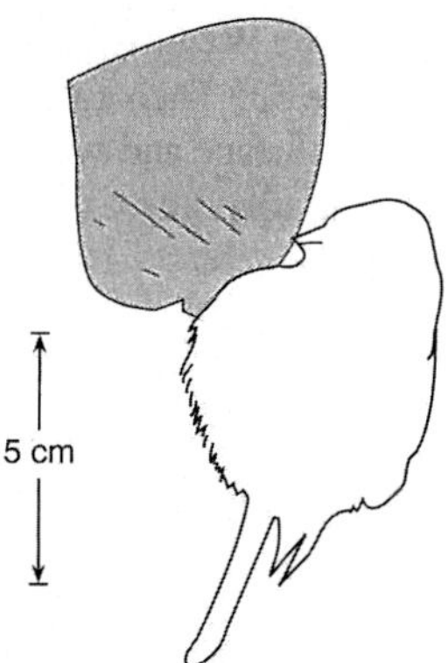

Fig. 58. Silhouette used to estimate the volume of a bird, in this case the blue tit (*Parus caeruleus*).

are winter-night values of northwestern European birds. As the diagram was meant to give an impression of the insulating properties of the plumage, the lower critical temperatures were recalculated and plotted as if the BMR was the standard BMR for a passerine bird and the difference between body temperature and critical temperature proportional to BMR. The actual BMR for the snowy owl (*Nyctea scandiaca*) for instance was about half the standard rate for *Passeriformes* which is very low, even for a non-passerine bird. The actual lower critical temperature is 4 °C. The diagram shows that the snowy owl is better insulated than the average bird of its size and in this case the best strategy would be to have a low BMR and to increase metabolism only as far as is needed, but we cannot say whether all owls have this wisdom.

For the non-passerine birds which have a lower BMR than the passerines, two plots were made in the diagram. The lower shows the insulating qualities of the plumage, the higher shows the actually observed lower critical temperature. For the passerines the difference between the standard BMR and actual BMR were small and only one value was plotted.

An attempt was also made to estimate the volume of the plumage of free-living birds. An estimate of the volume of the plumage was made by taking broadside photographs of free-living birds. A sphere of known size was photographed together with the bird or immediately afterwards at the same distance. (See Fig. 58.) From the silhouettes obtained in this way the volume of the bird could be calculated, assuming that rotation of the silhouette would give an acceptable approximation of the shape of the bird (Wallgren 1954). The volume of the 'naked' bird in cm^3 was taken to be equal to its weight in grams. Then the r^*_s and r^*_i values were calculated for the spherical model bird that had the same volume and weight as the real bird. The photographs used were of active birds during short moments of rest. Air temperatures were between –10 and –25 °C - far below the lower critical temperature, and feathers were fully fluffed so that the spherical approximation was not too far-fetched (cf. Fig. 58). At least for the smaller birds there is a good correspondence between the insulating properties estimated from the photographs and from measurements of BMR and critical temperature. From the diagram it becomes clear that smaller birds like tits (*Parus* spp.) which have LCT's near 20 °C must already spend at 0 °C twice the BMR to keep themselves warm. Even for larger birds the LCT is usually higher than 10 °C. It is clear that for smaller birds (and mammals) such shelter as snow can offer is of vital importance as well as such strategies as sleeping in natural or artificial nest holes and the huddling together of the short-toed treecreeper (*Certhia brachydactyla*) and the goldcrest (*Regulus regulus*).

With the larger mammals in the Arctic like the polar fox (*Alopex lagopus*), reindeer (*Rangifer tarandus*) and polar bear (*Thalassarctos maritimus*), the insulating properties of the fur are such that even at –40 °C they need not to raise their energy production above the BMR (Scholander et al. 1950).

By day a typical metabolic rate for a bird which interrupts its normal activities is 2 BMR. Even then at low air temperatures extra energy is needed purely as costs of thermoregulation. The great tit (*Parus major*), for instance, which at 20 °C loses heat equivalent to the BMR, loses 2 BMR at –1 °C, as the heat loss is proportional to $t_b - t_{air}$ (or $t_b - t_e$) (Fig. 58). With $t_{air} = t_e$ at –10°C in the shade, the metabolic rate is 2.43 BMR and a place in the sun may bring t_e to 0 °C and reduce the energy expenditure with 18%. The energy effect of sunbathing can be estimated from the following numerical data on R_{neta} and t_e: taken as an average over an approximately spherical shape R_{neta} can reach a value of 300 W/m^2 in summer for a black animal and even more on a bright winter day. The operative temperature is often 15 -20 °C and occasionally up to 30 °C above the ambient temperature. The extreme values are reached late in winter in sheltered sunny places where much reflected radiation is received from a steep snow-clad slope.

One item in the energy balance remained undiscussed. In order to provide the necessary energy a minimum metabolic intensity is required. Therefore an amount of O_2 has to be taken up via the respiration. Per gram oxygen used 15 kJ heat is produced on average quite independently of whether carbohydrate, protein or fat is used. One liter of air contains 0.29 g oxygen. The respiration air has to be warmed up to body temperature and saturated with water vapour. In this way we can estimate the amount of latent heat lost which is connected to the respiration intensity needed for the energy production required. In this case the heat loss is ca. 5% of the energy produced. (This can be checked with help of the data from Tables 3 and 11.) Our estimation agrees well with actual measured values (Shilov 1965; Evans & Moen 1975). This notion of evaporation as a minor item on the energy balance only holds when conservation of energy is the main concern.

In warm weather the bird or mammal faces the problem of getting rid of the metabolic energy it produces, the more so as even with a low activity level the metabolic rate easily rises to two to three times the BMR. In view of what was said on R_{neta} and t_e it can be understood that even with modest air temperatures, t_e can come so close to t_b in the sun that even with a thin plumage or pelage (a high α_{con}) the metabolic energy cannot possibly be dissipated by conduction alone.

Both the jackdaw (*Corvus monedula*) and the hooded crow (*Corvus corone cornix*) show signs of heat stress (open beak, panting) at t_e values of 33 - 36 °C (air temperature ca. 20 °C) and they seem to avoid being active for long periods in these conditions. Though no heat stress was observed, even the yellowhammer and the red-backed shrike, birds which characteristically perch on a branch in the sun, seem to avoid long periods above t_e = 30 °C and 33 °C respectively. Starlings reduce the length of their foraging bouts when it exceeds 31.5 °C (Clark 1987). For chicks of the Adélie penguins (*Pygoscelis adeliae*) in the Antarctic a t_e of 25 °C constitutes severe heat stress. The highest excess of t_e over t_{air} was 26 °C (Chappell, Morgan & Bucher 1990).

Alpine marmots (*Marmota marmota*) (Türk & Arnold 1988) diminish their activity above-ground with operative temperatures (t_e) over 25 °C which are often reached with air temperatures of 10 °C. The same is true for the North-American yellow-bellied marmot (Webb 1980; Melcher, Armitage & Porter 1990). For both species there is often no thermal equilibrium with the environment during above-ground activity. On the contrary, much energy is stored which means a rising body temperature. For the yellow-bellied marmot (Melcher et al. l.c.), which has a weight of 2 kg, a temperature rise of 0.075 °C /min means that energy is stored at a rate of 1.4 × the BMR. During bouts of activity the temperature rise can be 2 °C.

The examples given refer to birds and non-sweating mammals. Even then the energy release by evaporation can be quite important when the respiration rate is increased (panting). Evaporation rates for the great tit (*Parus major*) can be as high as 1g/h which is equivalent to 1.9 × its basal metabolic rate (J. Tinbergen pers. comm.).

The antelope ground squirrel (*Ammospermophilus leucurus*), a desert dweller can be active at operative temperatures of 60 °C for about 9 minutes. The heat gain from the environment, mainly by radiation, is then higher than the metabolic heat production and the body temperature increases with up to 5 °C. In this dry environment evaporation dissipates only 20% of the heat taken up, the stored energy being given off on retreat to a burrow. As Chappell & Bartholomew (1981) remark, the term endotherm seems inappropriate here.

So, we have reached a realistic estimation of the energy balance of a homoiothermic animal and of the influence of environmental factors, on the basis of a simple model.

Porter & Gates (1969) and others (see also Gates 1980) calculated for different kinds of homoiothermic animals under which combinations of temperature, wind and radiation the body temperature can be kept constant, on the basis of minimal and maximal values of the metabolic rate respectively. They arrived at diagrams showing in which part of the three-dimensional climate space (air temperature, wind, radiation) an animal can survive. Trites (1990) presented such climate space diagrams for two seal species on the basis of a more elaborate thermal budget model and found that pups of the northern fur seal (*Callorhinus ursinus*) on the Pribilof Islands, Alaska, can tolerate any normal combination of air temperature, wind speed and humidity, but that periods of very cold, wet and windy weather could be critical for the smaller pups. On the other hand, avoidance of heat stress is of importance for pups of the Galapagos fur seal (*Arctocephalus galapagoensis*) on the Galapagos Islands, and small body size is more adaptive in relation to the ability to seek shade and periodically to wet the fur.

4.2 Resistance and preference

We will now discuss the question of how plants and animals utilize the possibilities offered by the microclimate and how they are adapted to the limitations set by the microclimate. The most obvious case is the significance of extreme and possibly lethal temperatures, both maximum and minimum.

The highest temperatures to which mosses and lichens are subjected will lie close to the highest temperatures measured at the soil surface: we are dealing with dried-up plants differing very little from dry humus in a physical sense. Lange (1953, 1954, 1955) determined the temperatures tolerated by various mosses and lichens in a dry condition. Species of hot dry habitats such as *Cladonia pyxidata* and *Polytrichum piliferum*, could tolerate temperatures as high as 90 -100 °C. These species are thus well adapted to the extremes set by the environment.

This is even more so for seeds, where the lethal temperatures lie well above 100 °C, probably as high as 190 °C (Levitt 1972). We found that seeds of two annual plant species of dry coastal dunes in the Netherlands, *Erodium cicutarium* ssp. *dunense* and *Phleum arenaria*, survived a period of 2 h at 110 °C without any damage.

Most animals do not have such a dehydrated rest phase. But for those animals which have, resistance can be extreme as well. For instance the larvae of the African mosquito *Polypedilum vanderplanki* tolerate extremes of – 27 and + 100 °C; cysts of *Tylenchus*

(*Nematoda*) tolerate even 150 °C. Usually extreme desiccation cannot be tolerated that well, and lethal temperatures are not so extreme either. For many insects (imagines) the highest temperatures tolerated are between 40 and 50 °C (Cloudsley-Thompson & Chadwick 1964).

The eggs and larvae of the bark beetle *Ips typographicus* died at temperatures between 42 and 52 °C, which occurred in the cambium of pine logs (Schwerdtfeger 1963). For beetles the lethal temperature is between 47 and 52 °C (Thiele 1977), for the spiders *Pirata piraticus*, *Lycosa pullata* and *Pardosa chelata* (*Lycosidae*) at 35, 43 and 48 °C respectively; the eggs of *Theridion saxatile* die at 42 °C. (Nørgaard 1951, 1956; Hallander 1970). The Mediterranean cricket, *Gryllus bimaculatus* (*Orthoptera*) died after 3.5 h at 43 °C (Behrens et al. 1983). For the desert lizards *Agama savignii* and *Stellio stellio* the lethal temperatures are 47 °C and 45 °C respectively (Hertz & Nevo 1981).

Much is known about the temperature resistance of vascular plants. Lange (1959, 1961; Lange & Lange 1963) found that the leaves of many plants became damaged when subjected to 47 - 52 °C. For some desert plants the cooling through transpiration kept the leaf temperature below the lethal temperature (Lange 1959). More unexpected was the result of Gauslaa (1984) that in S Norway transpiration was essential for 14 of 69 species studied, to keep the leaf temperature below the lethal limit. Typical examples are *Petasites frigidus*, *Aconitum septentrionale* and *Rubus chamaemorus*. The leaves of *R. chamaemorus* probably came close to the lethal temperature (48.7 °C) under natural conditions. Heat damage was observed on the leaves of *Silene acaulis* after a day when the surface temperatures of this cushion plant reached 45 °C, 24 °C above the air temperature (Gauslaa l.c.). Kainmüller (1975) found that lethal temperatures for plants from the high mountains in summer were 50 - 60 °C. In winter heat resistance could even increase somewhat. This confirms the relation between cold and heat resistance (Levitt 1972). With succulent plants heat resistance can reach even higher values. So could leaves of *Sempervivum arachnoideum* tolerate temperatures up to 64 °C (Larcher 1980). In view of the radiation in the high mountains and the compact growth forms and succulent leaf forms of plants there, high leaf temperatures may be expected. Indeed, Larcher (1977) found temperatures as high as 55 °C.

In nature clear cases of high temperature damage are seldom found, although the differences between the temperatures actually found and the lethal ones are quite small. A classical example is the occurrence of foot ring disease, i.e. high temperature damage of pine seedlings (*Pinus sylvestris*) (Geiger 1961). We have never observed this ourselves. It is known from the forestry literature that the cambium of solitary beech trees with their thin bark can easily be heat-damaged at the south side of the tree, especially at the foot of the trunk and on trees along asphalt roads, which reflect much light and emit much heat radiation (Bernatzky 1978). At an air temperature of 32 °C the cambium temperature on the south side was 53 °C. Bernatzky also compared this with tree trunks painted white where the cambium has only 42 °C. In winter, temperature differences of 30 °C can appear between cambium at the south side and cambium at the north side (which was near the air temperature).

There is also much known about resistance to extremely low temperatures. Lichens are amongst the most resistant plants; extreme resistance is also shown by the shoots of higher plants from snow-free sites in the Alps, in winter *Loiseleuria procumbens* survived –70 °C and *Silene acaulis* even –180 °C (Larcher 1980). Antarctic lichens can tolerate –196 °C (Lange & Kappen 1972) where the absolute minimum temperature recorded there is – 88 °C, and the Mediterranean lichen *Cladonia convoluta* tolerates – 50 °C, which may be still further away, relatively, from its normal temperature range with – 10 °C as the absolute minimum and frost as a very temporary event. Some arctic and antarctic lichens can photosynthesise actively at –24 °C (Lange 1963). Still more extreme is the cold resistance of cysts of *Tylenchus* and *Polypedilum* (both mentioned above) –272 °C and –270 °C!

Woudenberg (1969) summarized the literature for fruit trees. For the usual species frost damage occurs at:

– 4 °C	bud phase
– 2 - – 3 °C	flower
– 1 - –2 °C	young fruit

For wild herbs and grasses minimum temperatures tolerated are lower. According to Kainmüller (1975) leaves of high mountain plants are not damaged until subject to –6 to –8 °C.

Probably the resistance level for lowland plants is similar. Reichelt (1954), in an Austrian grassland after a week in May with heavy night frost (with minima at 10 cm down to –9.5 °C), observed frost damage with the following grasses in order of decreasing damage:

Bromus erectus (thermophilous, i.e. favoured by warm conditions, avoids cold, occurring in southern and western Europe)
Alopecurus pratensis
Anthoxanthum odoratum
Dactylis glomerata
Bromus mollis (thermophilous)
Arrhenatherum elatior (thermophilous)
Trisetum flavescens (montane - subalpine)
Holcus lanatus
Poa pratensis (wide distribution range, up to N Siberia)
Festuca rubra (very wide distribution range).

Some of our own measurements follow on plants in coastal dunes. After a night (26 April 1959) with a minimum temperature of –8.8 °C at 10 cm, frost damage was recorded at the following species (in order of decreasing damage; species indicated with s are thermophilous with their main distribution in S and SE Europe; species indicated by w are woody): *Bryonia dioica*^{s}, *Rosa spinosissima*^{s,w}, *Rumex acetosa, Sambucus nigra*^{w}, *Ligustrum vulgare*^{s,w}, *Eupatorium cannabinum, Valeriana dioica, Hippophae rhamnoides*^{w}, *Crataegus monogyna*^{w}.

In all these cases we are dealing with species which show a capacity to recover, so that over a longer period of time the damage is not essential. Again, the difference between the minimum temperatures experienced and the lethal minimum temperatures must be small for several species. There are also species occurring at the northern limit of their distribution area where frost may be involved in keeping the northern limit where it is, or temporally moving it southward. An example is the southern-atlantic heathland bush *Ulex europaeus*, often killed by frost in the northeastern Netherlands, but not suffering from frost in the dunes of the southwestern Netherlands. The related shrub *Sarothamnus scoparius*, though occurring up to southern Scandinavia may also be severely frost-damaged in winter. A curious example is the cosmopolitan fern bracken (*Pteridium aquilinum*), occurring everywhere in boreal Scandinavia, yet still known as night frost-sensitive in the Netherlands and Britain, but here the development starts earlier. As far as the fern occurs in open vegetation it is restricted to microclimates with little night frost, such as east, south and west slopes, under the protection of trees and near to the sea and not in frost pockets.

The above-mentioned species are typical of heathland, a largely European-atlantic community type (e.g. Gimingham 1975). There are also some typical heathland species without any sign of frost sensitivity: *Calluna vulgaris, Deschampsia flexuosa* and the (boreal) *Empetrum nigrum*.

Table 34. Preferential temperatures (°C) of some poikilothermic animals (grasshoppers, unless stated otherwise).

Chorthippus albomarginatus	37.7[1]
Omocestus haemorrhoidalis	38.9[1]
Glyptobothrus brunneus	35.9[1]
Oedipoda coerulescens	44.7[2]
Two species of damp grassland	41.7 - 42.0[2]
Eleven species of moderately warm situations	42.6 - 43.8[2]
Five species of dry grassland	43.5 - 45.8[2]
Myrmeleon formicarius, larva (*Neuroptera*)	45.0[2]
Cicindela hybrida (*Coleoptera*)	44.0[2]
Phylan gibbus (*Col.*)	38.0[3]
Onymacris plana (Col.), Namibian Desert	40.0[4]
Lacerta vivipara (lizard)	37.5[2]
*L. agilis (*lizard*)*	38.5[2]
Agama savignii (lizard), Negev	40.4[5]
Stellio stellio (lizard), Negev	38.0[5]

[1]Gärdefors (1966); [2]Herter (1962); [3]van Heerdt, Isings & Nijenhuis (1956); [4]Henwood (1975); [5]Hertz & Nevo 1981.

The literature on cold and heat resistance is extensive (e.g. Levitt 1972; Precht et al. 1973; Larcher 1980). Here, we have only presented some examples showing the relation between the resistance level and extremes which may occur in the natural habitat in as far as the microclimate is involved; for many more examples see Gauslaa (1984). The only addition we may make here is that it has been shown that extreme temperatures, both low and high ones have a limiting effect on the photosynthesis of plants for many days after the extreme situation has occurred (Larcher 1980). This is damage in a sense, though not a visible one.

The non-active stages of insects and other arthropods are often resistant to very low temperatures, although not as low as those known for plants. Generally, the differences between extremes which occur and extremes which are lethal, are not large. For these stages recovery is not possible, even from partial damage. For instance, the Mediterranean cricket dies after 40 min at –5 °C (Behrens et al. 1983). Luff (1966) investigated the cold resistance of three beetle species, both in the field and in the laboratory. Under natural conditions in England a 100% mortality occurred at –8.5 to –10.5 °C with *Dromius melanocephalus* and *D. linearis*, and between 42 and 65% with *Stenus clavicorius*. At –5 °C mortality for all three species was ca. 20 %. A clear protection was provided by tussocks of *Dactylis glomerata*, in which the temperature did not sink below –4 °C. Only with *D. melanocephalus* some mortality did occur at –2 °C to –4 °C. Frost resistance in the laboratory was some degrees lower. (These were experiments in winter.) In April lethal temperatures in the laboratory were –5.8 °C for *S. clavicorius* and –3.2 °C with *D. linearis*. Clearly, low minimum temperatures in the field and the possibility of avoiding them, must be very important for such *Carabidae*. See also Bossenbroek et al (1977a,b).

As regards the preference for or a favourable effect of high temperatures, as can be found only in certain microclimatic situations, we may mention some interesting examples. First an example of effects on seeds. It has been known for a long time that a short exposure of seeds to high temperatures (oasting) promotes the germination of certain seeds. For instance, Grevillius (see Stoutjesdijk 1959) described this for *Calluna vulgaris* at 70 °C. Capon & Asdall (1967) found that several desert plants were promoted in their germination by temperatures of 50 °C.

Many microfungi in the soil develop best at temperatures of 40 to 50 °C. Such temperatures may very well occur regularly in the top soil in combination with sufficient moisture, as is found especially in organic material. Apinis (1965) found these fungi (in all stages) especially in hollow stems of dead grasses. It is also important to know that these fungi may grow in very dry conditions (a suction force of 200 - 300 bar, Lynch & Poole 1979).

Many poikilothermic animals have preferential temperatures of 30 - 40 °C (Table 34) which lie well above the actual air temperatures. Here we find the best examples of organisms which are absolutely dependent on warm microclimates. By their mobility they can move to the most favourable places. The basking of insects and reptiles drew the early attention of naturalists. Heimans & Thijsse, in 1904 in the Netherlands, reported on sunning flies (*Pollenia*), and adders (*Vipera berus*) in March; their English contemporary Richard Jefferies, 1889, wrote about 'a drone-fly on a sunny wall, January 20th' (see Barkman & Stoutjesdijk 1987 for references).

In some cases the function of basking on thermally favourable spots has been better investigated. Male adders search for warm spots as early as February, in order to reach a body surface temperature of almost exactly 34 °C, provided the combination of air temperature, radiation and wind make this possible. They find a position for the body as close as possible to, and perpendicular to the direction of the sunbeams. In this way their body temperature is often as much as 25 °C above the ambient temperature and sometimes even higher than the surface temperature in the immediate surroundings (Stoutjesdijk 1977b). Saint Girons (1975) made adders swallow mice with a miniature sender and found that the preferred internal body temperature was 31.3 - 32.5 °C and the maximum internal temperature tolerated 33.2 - 33.9 °C. These figures agree with the temperatures measured (with a radiation thermometer) on the outside of the adder.

Basking is of fundamental importance for spermatogenesis in the adder (Viitanen 1967) and this has also been shown for other reptiles (Dawson 1975). Later in the season the females also start basking, which is important for the development of the eggs of these ovoviviparous animals.

Clearly, a number of factors operate together here: the favourable microclimate with shelter for wind, a warm substrate (e.g. pine needles), and the ability of the animals to find the best spots. Later in spring, the adders choose less warm spots, for example partly shadowed ones, where again their preferred temperature of 34 °C may be realized.

For lizards the same situation occurs. *Lacerta agilis*, for instance, was found during sunny, windy weather in February, air temperature 8 °C, at about the only sheltered warm spot in the surroundings where it could develop a surface temperature of 29 °C.

The macrodistribution of such animals often indicates the importance of basking. As an example, *Lacerta agilis* does not occur north of the Mersey in England (near Liverpool), but in the much sunnier eastern boreal areas of Fennoscandia this species occurs up to 60° N (Jackson 1978; Berglind 1988). Similar relations have been found for butterflies (Higgins & Riley 1970).

The high temperatures reached on the south sides of an ant hill were mentioned already (see Ch. 2.7.2). From the middle of February onwards one may observe ants (*Formica rufa* and *F. polyctena*) concentrating on the south side of the mound. They are searching for the most favourable places, for example holes made by woodpeckers. Here shelter for wind and the optimal exposition are found. Because the ants form a dense layer their temperature can be measured with a radiation thermometer. They reach temperatures between 28 and 32 °C, which is 15 - 20 °C above the air temperature. With higher temperatures the ants become so active that they disperse over a wider area. Higher surface temperatures than 42 °C are avoided. During their stay in the sun all kinds of physiological processes start (Kneitz 1970), which can also be shown histologically.

Ants can also act as carriers of heat (Zahn 1958). The heat collected during basking is released again deeper in the nest. In view of the numbers of ants and their temperature a substantial contribution to the heat transport into the nest seems possible.

Caterpillars of the butterfly *Euphydryas aurinia* hibernate and become active in early spring. They search for food in dense vegetation and bask on warm spots such as dead leaves and grass. They can reach temperatures up to 30 °C above the air temperature. The conversion of food is optimal with 35 °C (Porter 1982).

The hairy caterpillars of the moth *Macrothylacia rubi* can be found basking in November, on sunny, sheltered spots in heathlands. They need high temperatures for pupation. The pupae can move up and down an oblong web in order to reach the most favourable position towards the sun. Spiders (*Lycosidae*) are active in February on favourable spots such as the south side of dry grass tussocks. Their egg sacs are also put in the sun. D. C. Brandt observed how *Pardosa lugubris* moved continuously so as to obtain a temperature of 28 - 30 °C for the egg sacs. The speckled wood butterfly (*Pararge aegeria*) basks to reach a temperature of 33 °C, which is essential for mate-locating flights. At low air temperatures flights are short and basking is done on a substrate which is much warmer than the air. Basking by females is probably important for egg production and oviposition (Shreeve 1984).

Solbreck (1976) described how the bug *Lygaeus equestris* is dependent on a rather high starting temperature for its migration flights. This is reached by basking on a suitable surface such as a thatched roof. At an air temperature of ca. 10 °C body temperatures of 40 °C were reached and the body temperature at which the flight started was above 30 °C. Short periods of sunshine during which a high level of activity is possible, are decisive here. The same holds true for prey catching by *Lycosidae*, insects and reptiles (Dreisig 1980; Avery, Bedford & Newcombe 1982; Kaufman & Bennett 1989). Clearly, short periods of sunshine can be used most effectively on a substrate which warms up quickly (see Ch. 2.7.2).

In many cases the possibility of searching for the most favourable situation is linked to a spatial differentiation of the microclimate on a smaller scale, where the animal can take in the most favourable position. Lensink (1963) studied the behaviour and microdistribution of three grasshopper species in the dry dune grasslands of Voorne. The vegetation consisted of a small-scale mosaic of seven types which differed clearly in structure and microclimate. In spring and early summer the animals migrate from the warm, open, sunny types to the denser, cooler and moister types. Later, the females return to the first-mentioned types in order to lay their eggs. During cool summers the animals stay longer in the open sunny spots. The highest densities were found where spots with a favourable microclimate for egg laying and hatching of larvae are found close to spots with the most favourable microclimate for larvae and adults.

Adult heather beetles (*Lochmaea suturalis*) start flying massively in spring, but only if the sun shines and there is no wind, and the average day temperature is above 16 °C. As soon as the sun is covered by clouds or the wind starts blowing they stop their flight (van Schaick Zillesen & Brunsting 1983). To make use of favourable conditions is also a matter of avoiding unfavourable periods by bringing the life functions to a very low level.This could be compared with what lichens and mosses do during dry periods.

As to the preference for certain microclimates, the situation for plants is much more complicated than for animals, because the below-ground and the above-ground parts live in completely different environments. Because of their immobility plants cannot make use selectively of certain favourable situations. This is why a certain stage of development, notably the germination, may be decisive for the rest of the plant's life, including the habitat of the adult plant.

Temperature fluctuations, with their variation in amplitude and period, are essential properties of the climate in general and of the microclimate in particular. The significance of daily temperature fluctuations for the germination of seeds has been demonstrated by various authors. For many legumes a treatment with alternating temperatures, with a high maximum (60 °C) makes the seed coat permeable, so that germination can take place (Williams & Elliott 1960; Quinlivan 1966; Baskin & Baskin 1974). Seeds of *Lycopus europaeus* do not germinate at a constant temperature but germinated very well with a strong alternation of day and night temperatures, especially when the day temperature is over 25 °C (Thompson, Grime & Mason 1977). This is in agreement with the temperatures measured at the surface of wet mud, the usual germination site of this European plant species. The germination percentage of *Holcus lanatus* increases strongly in the dark with the amplitude of temperature fluctuations (Thompson, Grime & Mason 1977; Grime 1979). In a dense sward this species germinates poorly; but in open spots abundantly. The temperature fluctuations near the ground are much smaller in dense grassland as compared to open grassland.

Hardly anything is known about the significance of short-term fluctuations, especially in the lowest air layers and at the surface, over a period of say 30 min. Such fluctuations occur with shifting clouds and are strongest where soil and organisms respond very fast and strongly to direct solar radiation, i.e. when their specific heat by volume and their albedo are small. With the Mediterranean cricket *Gryllus bimaculatus* population growth, as determined by the speed of development and the number of eggs laid, was strongest with a temperature change every second hour. However, a comparison was made only with changes over a longer period (with the same average) and with constant temperatures (Behrens et al. 1983). When the larvae of this cricket could voluntarily bask under an artificial sun shining for 12 h a day, their time of development was shortened enormously. During basking, about 10% of the total development time, the animals had their preferred temperature of 34 °C. Their development rate was 2 - 3 times as high as could be expected from the day-degree viewpoint (Remmert 1985).

Little is known about the significance of the rapid changes in temperature brought about by thermal convection and measured on a time scale of seconds. Medical physiology has provided indications that rapid temperature fluctuations are perceived (Wachter 1976). Temperature fluctuations of ca. 10 °C/s can be measured near to the ground above a warm surface. Stamens of plants and other fine organs such as antennae of insects are doubtless able to follow such rapid changes. Short peaks of a super-optimal temperature with a duration of 30 - 60 seconds can damage *Colias* butterflies; they cause the animals to live at an average temperature which is below the optimal (Kingsolver & Watt 1983). Kappen & Zeidler (1977) provide evidence for the significance of short temperature shocks. A treatment of 40 °C during 15 s was lethal for leaves of *Populus deltoides*, while a longer stay at higher temperatures did not damage them.

It has long been known of certain grasshoppers that e.g. 39% more of the eggs laid hatch under fluctuating temperatures as occurring in nature, compared with a constant temperature provided in an incubator, even if this were the optimum temperature. For other animals the effect of fluctuating temperatures can be zero or even negative (Schwerdtfeger 1963). Eggs of the salamander *Amblyostoma punctatum* develop equally rapid at alternating temperatures of 5 °C and 21 °C (6-h period) or a constant temperature of 13 °C, but much

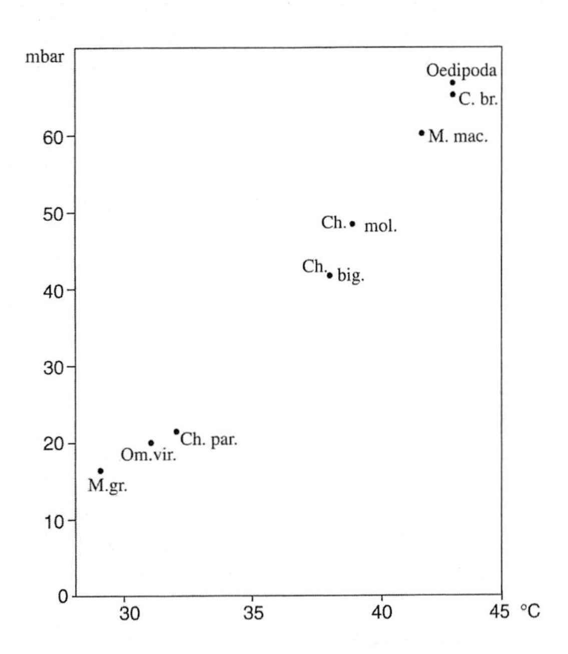

Fig. 59. Temperature and humidity conditions for grasshoppers in a transect between tall grassland on humid soil and bare, dry sand. From left to right: *Mecostethus grossus, Omocestus viridulus, Chorthippus parallelus, C. biguttulus, C. mollis, Mecostethus maculatus, Chorthippus brunneus* and *Oedipoda coerulescens.*

faster if the fluctuation period is 12 h or 24 h instead of 6 h (Buchanan 1940).

Preference for high air humidity is very common. Some examples may illustrate this. The sow-bug *Porcellio scaber* is in equilibrium with an *RH* of 95% (den Boer 1961). At a higher humidity the animal takes up water. On clear, still nights it searches for higher levels, for instance higher up a tree stem, where the air is not entirely saturated with water vapour (see Ch. 2.9.2). As a result it looses an eventual surplus of water taken up at too high air humidity levels. Also races of the mosquito *Anopheles maculipennis* show subtle responses to variation in *RH*. Within the range 95 - 100% *RH* three races showed different preferences (Hundertmark 1939). Such differences indeed occur in nature, for instance during a still night moving from inside a forest to outside.

Of course, temperature and humidity preferences have to be considered together in order to obtain a biologically meaningful picture. Thiele (1977) has investigated such combinations for a large number of carabid beetles. Four combinations are recognized:

1. Cool - moist. Species adapted to this combination occur in forests, particularly montane beech forest. An example is *Molops piceus*: preferential temperature 16.6 - 17.8 °C, 74 - 79 % *RH*.
2. Cool - dry. Only one species preferred this combination, the forest species *Agonum assimile*: 7.8 - 12.5 °C, 45 -54% *RH*. The preference for low humidity may be unexpected; indeed it is subordinate to the temperature preference of this beetle.
3. Warm - moist. This group, including various species of the genus *Bembidion*, prefers sunny open moist riverbanks. These beetles are like a miniature *Cicindela*; they run on long legs but also fly often. Such a preference can be understood easily in view of what was said earlier about the temperature of a wet surface in the sun (Δt = 10 °C).

4. Warm - dry. Here we find many grassland species. As an extreme case a preference temperature of 40 - 50 °C is mentioned for *Callister lunatis*. It should be added that there were also many species without a clear preference for temperature and relative humidity.

The temperature and humidity preferences of grasshoppers were studied in the field by Jakovlev (1959). The diagram shown in Fig. 59 is based on his data. The temperatures (measured with mercury thermometers) are from unshaded, freely-exposed sites; they probably closely approach those of animals in their natural situation. On the mbar scale the difference $e_b - e_{air}$ is given, where e_b is the maximum vapour pressure at body temperature.

4.3 Phenological aspects

"If the oak is out before the ash Then you'll only get a splash
But if the ash beats the oak Then you can expect a soak"

The importance of temperature differences in the microclimate for the development of plants and animals is illustrated by various phenological aspects, i.e. aspects related to seasonal development. In Ch. 2.12, some examples have already been given. Some more will follow here. On the whole, our knowledge of this subject is still fragmentary. Generally, differences in microclimate will increase differentiation in time of the phenological phases.

Flowers of the hawthorn (*Crataegus monogyna*) on branches near the ground open much earlier than flowers on branches higher up. On the other hand, sprouting of branches of oak bushes (*Quercus robur*) near the ground proceeds slower than that of branches on the top of the bushes, and also more slowly on heathland, where the nights are relatively cold (Chs. 2.9, 3.5), than on inland dunes, which remain relatively warm at night (this may be a difference of two weeks!). Oaks on heathland as well as on inland dunes are always later than oaks in woodlands and parks and there may also be genetical differences involved, or/ and differences in soil fertility.

Table 35. Threshold temperature values below which four species of *Idaea* do not feed, moult or pupate. (After Ryrholm 1989.) ave = *I. aversata* L.; bis = *I. biselata* Hufn.; dil = *I. dilutaria* Hb.; hum = *I. humiliata* Hufn.

Temp. °C	Feeding last-instar larvae	Feeding half-grown larvae	Moulting	Pupation
13				dil
12				hum
11				bis
10			dil	ave
9			hum	
8				
7			bis, ave	
6				
5				
4				
3		bis		
2		dil, hum		
1	dil, hum			
0				
–1				
–2	ave, bis			

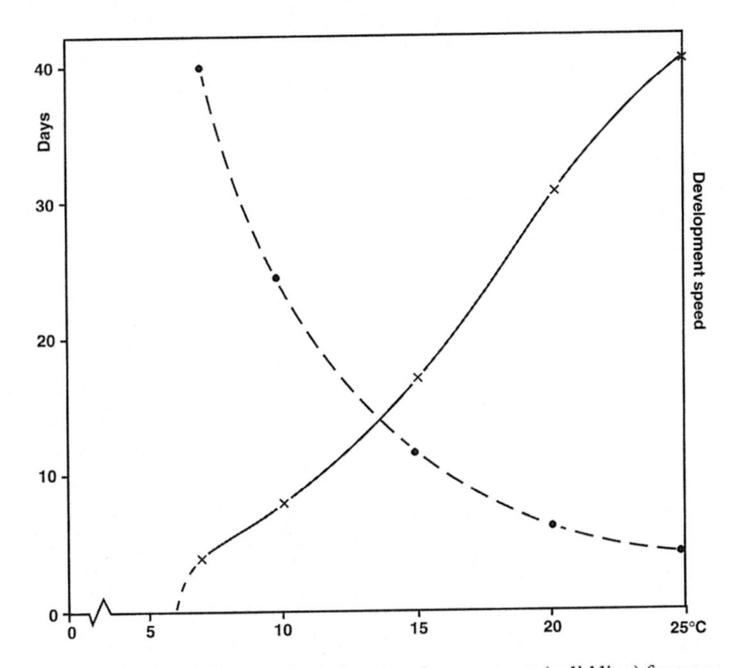

Fig. 60. Development time in days (interrupted line) and relative development rate (solid line) for eggs of the fruit moth *Orthosia* as a function of the temperature. After de Fluiter, van de Pol & Woudenberg (1963).

These two examples show that both day temperatures and night temperatures may be important in triggering development. Other examples here are the tomato (*Solanum lycopersicum*), the growth of which is mainly determined by night temperatures (Went 1944), and the spruce (*Picea abies*) which grows during the night, but is dependent on the six warmest hours during the preceding day (Dahl & Mork 1959).

We also know of cases where temperature has hardly any influence on the speed of development. For example, the growth of peas (*Pisum sativum*) appeared to be independent of temperature within the range 4 to 20 °C (Went 1957). Judged from its behaviour in the field, the common weed *Stellaria media* is also rather temperature-independent.

Still, important phases in growth and development usually start at temperatures which are species-specific, and are also determined by the development phase (threshold temperature). Table 35 from Ryrholm (1989) shows threshold temperature values for several developmental stages in four moth species of the genus *Idaea*.

From the starting point the speed of the growth or development process may increase progressively with temperature, but in other cases the relation is linear (Fig. 60). At the optimum temperature the rate of the process is maximum and it will usually decrease again with further increase in the temperature. In some cases the optimum range is very narrow; in other cases there may be only small changes over a broad temperature range.

We are dealing with complicated processes here, determined by a number of elementary processes, each with its own temperature dependence (Went 1944). In those cases where the rate of a process increases progressively with temperature, one may expect a stronger effect from short periods with an increased temperature, as compared to a smaller continuous temperature increase with the same average. This was the case in the experiments on spruce (*Picea abies*) of Dahl & Mork (1959).

When the temperature fluctuates within an interval where the relation between growth

Table 36. Threshold temperature (Thr) and temperature for development at 90% of the maximum rate (90) for some butterflies in orchards (de Fluiter, van de Pol & Woudenberg 1963).

Species	Thr	90
Adoxophyes reticulata, egg	10.5	21
A. reticulata, larva	15	25
Orthosia sp., egg	6	22
Ibid., larva	9	21
Enarmonia pomonella, egg	10	21
Paniurna argyrana, egg	10	19

and temperature is linear, a constant temperature has the same effect as changing temperatures with the same mean. This was shown by Arnold & Monteith (1974) in experiments on the growth of the grass *Festuca ovina*. Here the relation between growth and temperature was linear from 5 to 25 °C. With four other grasses, Peacock (1976) found a linear relation for the intervals 0 - 20 °C.

The relation between temperature and the development rate of insects has been investigated extensively. Within a certain temperature interval, this relation appears to be linear or sigmoid, but not very deviant from a linear relation (Thiele 1977; Joosse, Brugman & Veld 1973; de Fluiter, van de Pol & Woudenberg 1963).

The range between the threshold value and the optimum value of the temperature is often very short, for insects not more than 10 °C (Table 36). For plants, the range is usually much larger, as shown by the data on grasses (Arnold & Monteith 1974). This means that small differences in temperature may have a big effect, especially when the threshold is exceeded in one case and not in the other. On the other hand the optimum may be exceeded easily when the temperature is greatly increased, in cases where we are dealing with static or scarcely mobile development stages.

We also know of more complicated examples. For instance, Richards (1957) showed that eggs of the bug *Oncopeltus fasciatus* developed well at 17 °C; the larvae only developed into imagos, if during the development of the eggs the temperature was above 20 °C for some hours every day. (The average temperature could then be under 17 °C.)

We do not know much about the importance of temperatures below the threshold. In Richards' experiments a period of several days at temperatures below the threshold prevented normal development later on at higher temperatures. However, more generally the effect of temperatures under the threshold cannot be shown and development may occur at alternating temperatures, even if the average is below the threshold (Behrens et al. 1983).

In phenology the concept of heat sum is often used. This implies the summing up of average 24-h temperatures (macroclimate) over a period starting with the first day a development threshold is passed. A similar temperature sum is known as Cumulative thermal time (Cannell & Smith 1983); in the formula suggested by Cannell & Smith the threshold temperature (often 5 °C) is subtracted from the 24-h temperature before summing up. Finally one speaks of degree-days. Such temperature sums often give good results, although it is hard to understand how the air temperature at 2m, as the only meteorological parameter can give useful data, where the microclimatic conditions for the living organism deviate so strongly from the macroclimate. Maybe this is due to the fact that many phenological observations have been made on trees and shrubs where the differences in temperature between the perceptive organs and the air are rather small. Incidentally, other climatic factors may disturb this correlation. Thus Mork (1941, see Dahl & Mork 1959) and Caprio (1974) found that the day-degree approach could not be used when areas with different amounts of sunshine were compared. Therefore, Caprio used a so-called heliothermic index, taking into account the number of sunshine hours. This is related to the species-energy and the climatic energy concept (Wright 1983; Ryrholm 1989).

Lauscher (1953) mentions a case related to the previous one. The nun moth *Ocneria monacha* was very abundant when the number of sunshine hours in the critical period in spring was extremely high. The eggs, laid on the trunks of pines (*Pinus sylvestris*), developed unusually rapidly, especially at forest edges, compared with its parasite which develops in the soil.

Lauscher, Lauscher & Printz (1955) point out that in Norway the flowering periods of trees were advanced over the years 1928 -1952 as compared with the period 1897 - 1926, but retarded as far as herbs were concerned. They suggest that this could be due to a difference in microclimate. One could think of milder and more cloudy weather in the second period leading to lower temperatures near the ground affecting the herbs negatively.

Finally, though the rhyme cited above -there is also a German version - may not have great predictive value, it shows how two trees may react differently to the same weather conditions.

4.4 The influence of microclimate on texture and structure of vegetation

4.4.1 Introduction

In the preceding sections we have discussed the influence of microclimate on organisms, with an emphasis on their physiology, behaviour and phenology. Morphological adaptations such as size, form, thickness and special features of the surface such as hairs, feathers etc. were mentioned more occasionally.

We will now discuss in somewhat more detail how microclimate influences biotic communities as a whole. Two major aspects can be distinguished, the floristic-faunistic composition of the community, and its structure. It is well-known that the microclimate has a great influence on the species composition of biocoenoses. It is not possible to go into any detail here. It may be sufficient to say that each species of plant, animal and micro-organism can occur within a certain range of the environmental factors making up an organism's environment (ecological amplitude) and that characteristics of the microclimate play their part in determining whether the organism is within the ecological amplitude. So, microclimate influences species composition. But, species composition influences microclimate as well: the individuals of each species (genets or ramets, see Begon, Harper & Townsend 1986) have a certain architecture, which to a large extent determine the structure of the entire community and thus the microclimate (see Ch. 3.3). The restriction 'to a large extent' means that individuals (especially ramets) may vary in size and life form, depending on the community they form part of. Further, their number and arrangement, both horizontal and vertical, may vary. Instead of looking at all individual organisms, we may study the biotic community as a whole. On this level of abstraction it will be even more difficult to judge the extent to which the community and the microclimate are primary or secondary.

The structure of a biotic community is usually determined by its vegetation component (exceptions are, for instance, coral reefs, oyster beds, termite nests). Plants form over 90% of the above-ground terrestrial biomass and, because of their immobility, a more-or-less fixed matrix within which the mobile animals have to move. In the following we will therefore restrict ourselves to the vegetation component and follow the division by Barkman (1979) between structure in a strict sense and texture.

Texture is the sum of all morphological characters of the vegetation, including their frequency and share in the biomass, apart from the species composition and disregarding their arrangement in space and time. Structure is the arrangement of textural elements in space (vertical stratification, horizontal patterns) and time (diurnal and seasonal periodicity).

Of the textural characters, leaf size, leaf consistency and leaf inclination will be considered. Such parameters are usually divided into classes and the percentage occurrence of the classes of a parameter form a so-called spectrum.

Regarding the study of structure on the level of the plant community very little is published and here we can only give some examples. This section is not a systematic treatment of the topic, but rather a stimulus to further investigations!

4.4.2 Leaf size

To summarize some basic facts discussed earlier: leaf size is a measure for warming up by the sun; leaf size is, via the size of sun - and shade- spots, related to the maximum possible LAI, amount of leaf area per m^2 of vegetation; leaf size determines the structure of litter, among other things the amount of suffocation of soil cryptogams and the creation of moist air-cavities in the litter. Conversely, leaf size is determined in part by the (micro)climate, but the average leaf size of a certain vegetation type is also positively correlated with fertility and moisture status of the soil. The largest leaves are found in warm-moist (equatorial) areas (for references see below). In the microclimate temperature and humidity have opposite effects on evaporation. In a cool, dry forest the leaves of the herbs are on average smaller than the leaves of the trees; in a cool moist forest, on the other hand, they are bigger. In a dry, warm grassland the leaves are smaller than in a neighbouring dry cooler forest.

A detailed analysis of leaf size in relation to altitude was made by Werger (not publ.). He found that in the Alps the average leaf size of *Homogyne alpina* in the same habitat decreases from 4.9 to 3.6 cm^2 at between 2200 and 2600 m a.s.l.; for *Arnica montana* leaf size decreased from 30.1 to 12.6 cm^2 at between 1900 and 2450 m. Here we have to realize that temperature decreases with altitude, but nevertheless the potential transpiration in-

Table 37. Contrast between open *Sedum* vegetation and tall *Juniperus* scrub on the alvar of Öland, regarding microclimate and leaf characteristics.

		Open *Sedum* vegetation	*Juniperus* scrub
Max. temp. at 0 cm (°C)		44.5	21.2
Min. temp. at 0 cm (°C)		2.0	6.4
Temperature amplitude (°C)		42.5	14.8
Max. saturation deficit at 0 cm (mbar)		73.2	3.6
Leaf type distribution (%)	Size		
Bryophyllous	< 4 mm^2	50	16
Leptophyllous	4 - 20 mm^2	16	14
Nanophyllous	0.2 - 2 cm^2	24	26
Microphyllous	2 - 20 cm^2	11	35
Mesophyllous	20 - 100 cm^2	0	9

Table 38. Comparison of roadside vegetation as to the relation between sclerophylly index (ratio between dry and fresh biomass), leaf size - both measured in June 1980 - and duration of potential sunshine in h/d over the months June and July. Drenthe, the Netherlands.

No. of measurements	2	4	5	2	4	3
Potential sunshine (h/d)	0.5	2 - 3.5	4 - 5	6.5 - 9	11.5-12.5	14
% sclerophyllous leaves	0	0	0.8	0.2	4.6	25.1
Sclerophylly index	0.15	0.19	0.20	0.22	0.23	0.10
% leaves < 2 cm^2	9.0	2.7	3.7	1.5	20.0	47.0

creases, certainly above 2000 m, because of stronger solar radiation, lower air pressure, lower vapour pressure and more wind.

Leaf size is related to the danger of overheating. Large leaves must transpire strongly to avoid lethal temperatures (Ch. 4.1.3, Gauslaa 1984). There is also a relation with water use efficiency in comparison with the CO_2 fixed by photosynthesis (Stoutjesdijk 1970a; Parkhurst & Loucks 1972; Givnish & Vermeij 1976; Zangerl 1978; Gauslaa 1984). An analysis of the relation between leaf size and the moisture status of the soil for *Saussurea alpina* along a wet-dry gradient on a south slope in southern Norway, showed that leaf size increased from 1.5 × 3 cm to 3 × 10 cm between dry and wet. In the wet habitat the leaf size was such that only strong transpiration could prevent the leaves from reaching a lethal temperature (Gauslaa 1984).

On the sunny and windy island of Öland (SE Sweden) we find a sequence of plant communities from low, open vegetation with *Sedum* species on the flat limestone via closed grassland to tall juniper scrub (see Sjögren 1988 for a recent account). The difference between the open *Sedum* vegetation and the *Juniperus* scrub is considerable, as follows from measurements carried out in June 1978 (Table 37). During the measurements the weather was cool and windy and the contrast could have been greater still. In this sequence there is a clear trend towards larger leaves and a milder microclimate. The larger leaves are especially related to a lower saturation deficit.

Table 38 shows the relation between sunshine and leaf characteristics. The measurements concern roadside vegetation differing in the number of hours of potential sunshine per day during June and July. Up to 10 h/d there is little correlation with leaf size; for higher values the percentage of small leaves increases suddenly.

In this analysis 20 sites were involved, situated in pairs or trios, each along particular roads. Four groups were distinguished: sunny sites (su); country roads with rows of trees

Table 39. Leaf size spectrum in relation to microclimate in roadside vegetation. su = sunny sites; ds = dry shade; sh = dense shade; os = open shade.

Leaf size class	Leaf size	su	ds	sh	os
Bryophyllous	< 4 mm^2	0.2	-	0.1	-
Leptophyllous	4 - 20 mm^2	0.5	-	-	-
Nanophyllous	20 - 60 mm^2	5	-	0.1	-
Subnanophyllous	60 - 200 mm^2	22	0.1	10	3
Microphyllous s.s.	2 - 6 cm^2	39	54	46	21
Submicrophyllous	6 - 20 cm^2	23	42	31	53
Mesophyllous	20 - 100 cm^2	11	4	13	22
Macrophyllous	100 - 500 cm^2	0.1	-	0.6	-

across open fields, thus dry and wind-exposed: dry shade (ds); roads through woodland: dense shade (sh); roadsides at the northern edge of woodland: open shade (os). The distribution of leaf sizes over the four types is shown in Table 39. In the order su-ds-sh-os there is a shift towards larger leaves. It will be clear that os is at the end of the series: the microclimate is more humid (despite more diffuse light and more wind as compared with the woodland), there is more rain, more dew and, because it is cooler, less evapotranspiration.

4.4.3 Leaf consistency

Several types of leaves can be distinguished in addition to orthophyllous (normal) leaves: malacophyllous (very thin) leaves; sclerophyllous (hard, leathery) leaves, and succulent (thick, juicy) leaves. The differences between these leaf types can be quantified with the help of the following parameters: leaf size, leaf thickness, fresh weight and dry weight. Sclerophylly is taken here as the ratio dry leaf weight/leaf area.

Sclerophylly occurs especially in cold and dry climates and with a shortage of nitrogen and phosphorous. In the alpine zone sclerophylly increases with altitude (colder and drier), in *Homogyne* (see Ch. 4.4.2) from 0.84 to 1.12 and in *Arnica* from 0.45 to 0.64 (Werger unpubl.). If we compare vegetation with different duration of snow cover but at the same altitude, we find that sclerophylly increases with decreasing duration, meaning increasing effects of drought and frost, from 0.20 in the *Salicetum herbaceae* (9 -10 months of snow cover: snow bed vegetation) to 0.40 in the *Caricetum firmae* (5 months of snow cover) (Pümpel 1977). In woodlands the sclerophylly of the herb layer decreases from winter to summer.

In the analysis of roadside vegetation discussed above, the sclerophylly index (here defined as the ratio dry/fresh biomass) increased from 0.15 to 0.37 with increasing duration of potential sunshine. Also, the percentage of sclerophyllous leaves increased strongly. Both parameters increased particularly in the range 10 - 15 h/d. In the Öland example the percentage of sclerophyllous leaves decreased from 26 to 14 % from the open herb to the closed shrub vegetation.

Succulent vegetation is found particularly in dry and hot climates and also in saline environments. Succulence increases with drought, unless night frost occurs often (temperatures of < –4 °C at 10 cm). Succulence can be expressed as the ratio volume/surface.

In *Homogyne* (see above) succulence increases from 2200 to 2600 m from 2.8 to 3.7 (Werger unpubl.). On the other hand, Pümpel (1977) failed to find a relation with snow cover duration. On a limestone hill, Werger (unpubl.) found the highest degree of succulence (0.90 - 0.97) on the driest soils against only 0.54 on moist colluvial soil down the hill. On the other hand, in the roadside vegetation discussed above, degree of succulence was positively correlated with soil moisture and there was no correlation with the dryness of the microclimate (sunshine hours per day).

4.4.4 Leaf inclination

Leaf inclination, the angle of the leaf with the horizontal plane, has a strong influence on the vertical partitioning of the incoming sunlight, as we have seen in the above sections. Conversely, the inclination can be considered as an adaptation to the microclimate. Through changes in the position of the leaves a plant can influence the amount of radiation received, the temperature and the transpiration. Horizontal leaves catch a great deal of light. They

Table 40. Leaf inclination types according to Barkman (1979).

	Name	Leaf characteristics	Inclination
e	erect	almost vertical	60 - 90°
ep	erectopatent		30 - 70°
pa	patent		10 - 40°
h	horizontal		– 20 - 20°
d	decumbent	pending	– 50 - 0°
pe	pendent	pending abrupt	– 90 - – 40°
Special cases:			
sf	spherical	all inclinations	– 90 - +90°
hs	hemispherical	all inclinations between	0 - + 90°
sp	spreading	lower half ep, upper half h	
r	recurvate	lower half h, upper half pe	
ar	arcuate	lower half ep, upper half d - pe	

predominate in places with little light, where there is much shade or a short growing season. In the high mountains the growing season is short indeed. The *Salicetum herbaceae* is free of snow for only 2 - 3.5 months and this vegetation has the lowest inclination, 24.3 ° (Pümpel 1977). The *Caricetum curvulae*, 3.5 - 5 months snow-free has an average inclination of 45.2°. Furthermore, the inclination increases with height above the ground, which is directly parallel with the amount of light coming in.

In Dutch grasslands the tall grasses usually have leaves with a high inclination, while species with more horizontal leaves, especially rosette plants, such as *Bellis perennis*, *Hieracium*, *Hypochaeris*, *Plantago* all occur in the lower layers. This also holds for woodlands. In an alder carr, *Alnetum*, in the northeastern Netherlands the average inclination in the canopy was 23°, in the herb layer 5°. Even on the same plant, leaf inclination increases higher up the stem.

According to Barkman (1979) we distinguish six standard leaf inclination types as well as some special cases (Table 40). Table 41 gives the percentage distribution of these leaf inclination types over the four types of roadside vegetation. On the sunny sites most of the leaves are erect or erectopatent, in the open shade a large proportion are horizontal, while decumbent leaves are mainly found here. Arcuate leaves occur mainly in the dense shade. This type is common in forests, as may be expected, with both forbs such as *Digitalis*

Table 41. Leaf inclination spectra in relation to microclimate in four types of roadside vegetation. For the site types see Table 39.

Leaf inclination type	su	ds	sh	os
Erect	38	8	2	2
Erectopatent	30	51	25	14
Hemispherical	16	1	-	-
Patent	6	1	13	23
Spreading	4	3	6	1
Arcuate	3	7	30	2
Horizontal	3	33	20	40
Recurvate	-	1	-	-
Decumbent	1	6	4	18
Pendent	-	6	-	-

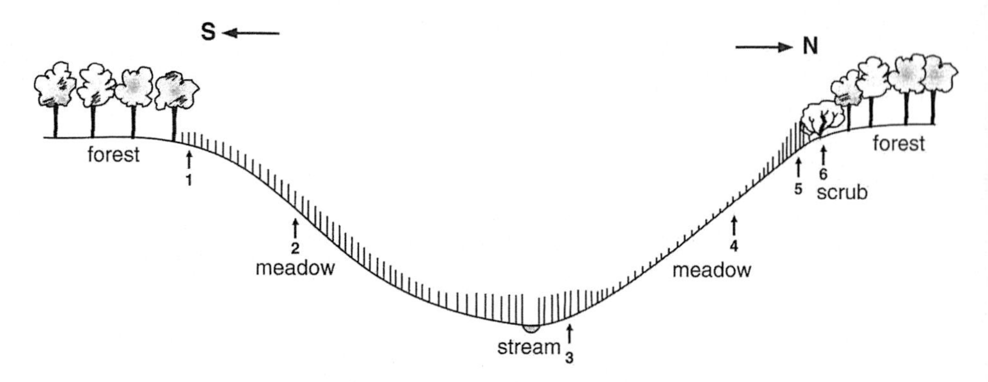

Fig. 61. Schematic north-south transect across an east-west running rivulet valley in the Spessart (S Germany). Numbers 1 to 6 refer to relevé numbers in Table 42.

purpurea, and *Orchis purpurea*, and tall graminoids such as *Milium effusum*, *Melica nutans* and *Carex sylvatica*. However, the arcuate type is also characteristic for sunny, but wet places with reeds, including *Phragmites australis* itself and the tall grass *Calamagrostis canescens*, further *Carex* species such as *pseudocyperus*, *pendula* and *paniculata* (the phytosociological alliance *Magnocaricion* in Europe). It is not yet clear what the function of this leaf type is.

Fig. 61 and Table 42 show how the leaf inclination type spectrum can also be used to characterize the relation between leaf type and position of vegetation types in the landscape. Through a rivulet valley running east-west a north-south transect from one bank across the valley to the other bank was analyzed with six phytosociological relevés, representing the following vegetation types (numbers refer to numbers in Fig. 61).

1. Hay meadow in the shade of a woodland to the south: *Arrhenatheretum* with *Anemone nemorosa*, *Cardamine pratensis* and *Vicia cracca*;
2. Sunny hay meadow: *Arrhenatheretum typicum*;
3. Wet forb vegetation: *Junco-Menthetum longifoliae*, with dominance of *Mentha longifolia*;
4. Dry hay meadow on south slope: *Arrhenatheretum brometosum erecti*, with *Bromus erectus* and other indicators of dry grassland;
5. Forb fringe between hay meadow and scrub: *Trifolion medii*, with *Agrimonia eupatoria*, *Origanum vulgare* and *Trifolium medium*;
6. Scrub bordering woodland belonging to the *Prunetalia spinosae* with dominant *Prunus spinosa*.

The distribution of leaf inclination types resembles that of Table 40. The high leaf inclinations (e, ep) are most common on the grassland sites, but there is a clear difference within the grasslands between sunny sites (2 and 4) and open shade (1) sites, with erect leaves dominating in the sunny and erectopatent leaves with a lower inclination dominating in the open shade sites. In the wet vegetation arcuate leaves dominate strongly. In the fringe community hemispherical leaves make up more than half of the leaves, while in the scrub decumbent leaves together with horizontal and pendent ones dominate.

Table 42. Leaf inclination spectrum for six sites along the transect of Fig. 61 based on phytosociological relevés. Percentages are quantitative proportions of leaves from a sample taken in each relevé.

Analysis	1	2	3	4	5	6
Aspect	N	N	-	S	S	S
Inclination °	7	15	0	15	5	5
Leaf type						
Erect	12	45	-	42	15	-
Erectopatent	44	19	5	22	19	4
Hemispherical	-	5	-	27	52	-
Patent	1	4	10	1	1	3
Arcuate	14	1	66	-	5	1
Horizontal	6	15	17	8	8	24
Decumbent	12	9	2	-	1	40
Pendent	-	-	-	-	-	29
Spherical	10	2	-	-	-	-

5. The analysis of the microclimate

"It is just as difficult to do everything wrong as it is to do everything right"
W. Schmidbauer (1985)

5.1 Methods of investigation

5.1.1 Objects of investigation

The microclimate should be investigated particularly when we assume that it will deviate strongly from the macroclimate, and moreover when it seems to affect the local distribution of plants and animals and their communities, and the local survival of populations. It is a matter of experience to judge the field situation on these aspects. Deviant microclimates develop especially when the weather is clear and dry (much radiation in the day time, strongly negative net radiation at night) and little wind. On the coast and mountain tops (strong wind!) and in outspoken oceanic climates (clouds!)the microclimate will not be much differentiated. Convex landscape forms have less strongly-developed microclimates because they are more windy and there is a run-off of cold air by night. Thus, microclimatic extremes will be well developed in concave landscape forms such as basins and depressions, open places in woods, especially when sheltered and situated in continental climates.

Regarding the vegetation component, mosses and lichens living on the ground are subjected to more extreme climatic conditions than herbs and taller plants. For shrubs and trees there is the additional difference that they usually root much deeper. On the other hand, all higher reaching life-forms are subjected to near-the-ground conditions as juvenile plants.

The microclimate is less important as a primary factor if other environmental factors have an extreme impact. An example are salt marshes, where factors such as salinity, redox potential and inundation frequency are overriding in determining the local distribution of organisms. Here it is less meaningful to investigate the role of microclimate for vegetation. Of course, the converse question is much more interesting here too: how does the vegetation developing under these extreme conditions create and influence the microclimate, and consequently the fauna.

If we wish to understand vegetation gradients observed in the field, microclimatic research may be rewarding, especially if we assume variation in microtopography and physical soil properties such as texture, humus content, lime content and phreatic level. Microclimatic parameters such as albedo, specific heat and thermal conductivity are directly dependent on such soil characteristics.

When we are dealing with differences in the understorey of stratified vegetation we should concentrate on structural variation in the higher vegetation layers, especially in spatial gradients or mosaics of different structural types, for instance a zonation from grassland to woodland or a mosaic of heathland, grassland, moss vegetation and bare spots. Again, we are confronted here with the priority question: did the variation in vegetation come first or the variation in microclimate? And, in the case of the latter sequence, the next question may be: has the vegetation which developed under the influence of a certain microclimate changed the original microclimate, and has variation originally present been increased or decreased? In such a situation it may be useful to remove the vegetation in part and compare the microclimate on the bare spot with that of the surrounding untouched vegetation. It is always useful to consider those influences which are independent of the local vegetation, e.g. inclination and exposition, wind shelter, possible sunshine.

In the following sections our starting point will be the search for different microclimates in different habitats. Of course with another starting point one would develop other methods in order to answer other questions (see also Barkman 1977).

5.1.2 Choice of measurement points, location and number

First of all there should be a relation between the grain of the vegetation pattern and the density of measurement points. For each element of a spatial complex (zonation or mosaic) a number of points must be selected (stratified sampling). The sampling density within each element is dependent on the apparent or presumed internal variation and also on the desired statistical accuracy. It is difficult to give fixed rules here; one should rather establish sampling density empirically from case to case. Still, some general guidelines should be considered:

1. A minimum of three measurement points per element is necessary in order to obtain at least an impression of the variation.
2. In the case where one wishes to characterize the microclimate of a vegetation type, one should select large stands assigned to that type and locate measurement points in the centre of each, to avoid any edge effects. Because wind, with different humidity or temperature (obtained in neighbouring vegetation), may disturb the situation, and also because there is a strong gradient in wind velocity near the ground, measurements in windy weather may give quite deviant results. It is therefore best to measure when there is no wind at all. As a rule of thumb one would measure up to a height above the ground (or if there is a closed vegetation cover, above this cover) which is 5-10% of the diameter of the homogeneous stand one wishes to typify. So, with a stand diameter of 10 m in 30 cm tall grassland, a total measurement height of 100 cm, i.e. 70 cm above the dense vegetation, would be realistic, or with stand diameter of 20 m in an open shrubland 1.50 m tall, a total height of 2 m. In situations where there is wind one should reduce this height to 0.5 - 1 %, or reduce it by a factor of 10.
3. If, in apparently homogeneous vegetation, microtopographical variation is observed one should stratify the sample according to this variation.
4. If these three guidelines are satisfied measurement points should be selected at random.

If the measurement points are located, they should be drawn on a (sketch) map of the area under investigation and a sequence of measuring points established (and drawn as well). It is wise to keep to this sequence in order to minimalize disturbance of the vegetation, and also to obtain constant time intervals between measurements. The shorter the walking route the better, but one should approach the measurement points from a direction opposite to the wind and the sun. In this way one can prevent disturbance of parts of the vegetation which are included in what is considered typical for the microclimatical conditions. Because one should also measure at night, the markers along the walking route should be indicated with luminous paint! As a general basis for the measurements one should make notes on the vegetation and the soil (preferably a complete stand description) and in general make an analysis with a microclimatic eye (see Ch. 5.2.8).

5.1.3 Time of measurement

In order to obtain a comprehensive picture of the microclimate one should measure over the entire growing period, meaning all year round for evergreen vegetation, which includes various types of temperate vegetation such as peat bog, heathland, grassland, juniper scrub, and coniferous forest. For moss- and lichen-dominated vegetation one should concentrate the measurements in the winter season. For winter annuals (i.e. annual species germinating in early autumn and flowering in the early spring) the measurements can be restricted to the winter half of the year, for macrofungi summer and autumn, etc.

Measurements should cover an entire day at least once per season, but even better is once per month. To increase objectivity one could do such a monthly measurement systematically, on a fixed day. Random choice of days is rather unrealistic because in that case one cannot cope with unexpected bad weather. Bad includes here strongly varying cloud cover during the day! As said before, the best results are obtained on clear, still days. In order to obtain data on special aspects it may be necessary to take extra measurements on days with a special type of weather. For instance, in order to know the maximum daily evaporation per site, one should measure on windy, warm days. In order to know the influence of the wind, one should measure on windy days varying in sunshine and temperature.

The selection of days and their number can be based partly on general patterns in the macroweather; one could restrict measurements to one day of each of the major weather types within each season (Wilmers 1968). Again: the more one knows about the situation the fewer measurements are necessary.

The above remarks concern incidental measurements in which the observer is directly involved. Nowadays automatic measurements can also be made, so that more continuous measurements are possible. However, the amount of data collected will soon become unmanageable, so that a restriction from the beginning is useful. Also, various measurements are integrated, for instance rain meters, Piche-evaporimeters, anemometers, temperature sum devices and electric light sum meters, or for minimum and maximum measurements (temperature). Such measurements can be quite useful for comparing sites, e.g. light sums or minimum temperatures inside and outside a woodland. They increase in importance when one has a good insight into the situation: 'what can be expected', and can link the results to the general weather situation: 'this extreme minimum could only occur on this particular night'.

A further point of interest is the time of day of the measurement. With constant bright weather the maxima of temperature and saturation deficit are usually reached around midday. Minima of temperature and saturation deficit are usually reached shortly before sunrise. In order to obtain a first idea about the daily course of weather factors, one should take a first measurement before sun-rise and then at 8.00, 13.00 and 18.00 h (solar time). It would be much better, though, to measure each hour during a 24 h-period. Together with measurements of continuously-operating instruments such hourly measurements give an accurate picture of the microclimate of that particular day.

According to our experience one can manage measurements of all the important microclimate factors at 16 - 20 points (within a radius of 100 m) with six observers. In case one wishes to obtain a 24-h picture three such teams, working 8 h each, would be needed.

As an addition to relatively intensive measurements at a relatively small number of points one can do rapid, more extensive measurements at a large number of points simultaneously. In that case the best time is 13.00 solar time, when the spatial differences are maximum and the change in time minimal. Nowadays we have very fast working instruments available for the measurement of surface temperature (radiation thermometer), light intensity (photocells), wind velocity (hot-wire anemometer), air humidity and soil moisture (see Ch. 5.2.3). In this way one person can do 100 - 200 observations per hour. Since the microweather varies greatly both in time and in space it is recommended that one combines time-intensive with space-intensive methods.

With automatic, self-registering instruments one has the advantage of being able to measure very frequently, for instance every 5 min, and, moreover, of being sure that the microclimate is not repeatedly disturbed, which inevitably happens when one measures manually and visually. Each observer produces heat and water vapour, forms a wind-break, casts shadow and, this being the main problem, tramples vegetation. On the other hand the automatic devices have serious disadvantages as well. There is the danger of theft and damage, and the risk of undiscovered technical defects causing the loss of hours, sometimes days of expensive observations. Moreover one is limited to a relatively small area and usually also to a limited number of measurement points. The rapid measurement of hundreds of points cannot (yet) be automatized, certainly not when the measurement programme is to be continued the day after in a different type of vegetation. Finally, we repeat that an understanding of what really happens can never be obtained without direct personal contact with the local situation in the field.

As to the duration of measurements, the monitoring of some phenomena does not demand measurements to be taken all day long. For instance, light measurements on mosses and lichens are no longer necessary as soon as they are dry (see Ch. 3.8.3).

5.1.4 Measurement height

Each vegetation type has a characteristic microclimate profile, both above-ground and below-ground, which changes with the hour of the day and type of weather. If we are interested in the vegetation as a whole, we have to study the entire microclimate profile, including the top of the vegetation canopy, or at least determine the type of profile and the place and value of the cardinal points: maxima and minima. If we wish to compare vegetation types of different heights it is of little use to measure on any standard height.

5.1.5 Choice of parameters and locations

Often quite trivial errors are made regarding the sort of parameter measured and/or the location of measurement points. One does not need to measure light in winter if one is interested in the relation between light and photosynthesis of deciduous trees; it is of no use to measure soil moisture at 50 cm depth if the vegetation we are interested in roots in the upper 20 cm.

If one investigates the microclimate as an ecologist, one has to look at things from the 'viewpoint' one is interested in. Plants should be looked at from the 'plant's eye', and the same holds true for animals and plant communities. One has to understand how and why the microclimate of a plant, animal or community is as it is and functions as it does, and to be able to predict it on the basis of relief, soil type, soil moisture and vegetation structure.

5.1.6 How to measure

Each measurement influences the microclimate. The very installation of instruments creates disturbance. At each new visit there is a new if only temporary disturbance. It is therefore useful to keep the following in mind.

1. Always approach the measurement points from the same side, preferably from the northeast (see above).
2. Approach open vegetation along forest edges from the forest and avoid casting shadows and impeding the influence of the wind.
3. Always measure against sun and wind.
4. With electrical instruments, use a relatively long wire between sensor and galvanometer, but not so long that resistance or induction currents will create disturbances, or that wires get broken.
5. If there is a regular inspection around fixed measurement points, first measure the parameters which do not imply disturbance, for instance, first net radiation and surface temperature with a radiation thermometer (no disturbance at all), the air humidity (Vaisala) or psychrometer and wind velocity (hot-wire anemometer), and finally light intensity (see Ch. 5.2). The most damaging disturbance is usually the placement of the instruments in the vegetation.

A special problem can be caused by the inertia of some instruments. A mercury thermometer, for instance, has an adjustment time of several minutes. With such an instrument one cannot do many measurements in a short time. On the other hand, some instruments are too fast. Very fine thermocouples linked to sensitive volt meters respond so quickly that no constant temperature can be recorded over a realistically long time. Fluctuations per second are often not interesting because soil, plants and animals also have a certain response time. In practice, one often has to compromise: a slower instrument often has a 'radiation error': a large sensor warms up to a level above the air temperature, because the heat transfer coefficient is smaller than with a fine sensor.

One would of course wish to adapt radiation error, size and response time of the measurements to the special features of the organism(s) one is interested in. As to the temperature, this can be realized simply by putting needle-like thermocouples directly in the organism. It is also possible to construct sensors which physically resemble the object to be measured (see Ch. 5.2.7). In vegetation with many species such adaptations are impossible, however. Here we have to find an intermediate approach between, on the one hand measurement of fundamental physical parameters such as radiation balance, water balance, wind, air temperature and humidity, and on the other hand specifically biological values of microclimatological parameters.

To understand the relationship of the organism with its environment we need measurements which characterize this environment, but the parameters used can be of a composite nature i.e. evaporation, operative temperature, black bulb temperature (Ch. 5.2.7).

5.2 Instruments and investigation techniques

5.2.1 General remarks

First some general remarks before we start discussing the various instruments. It is good to know how the measuring instrument works and what errors it may have. It is also important to understand the nature of the phenomenon one wishes to measure, especially complicated matters such as long-wave radiation.

The instruments should be calibrated regularly. This can be done very simply, for instance a mercury thermometer can be compared with a calibrated thermometer. If this is not possible and one has several instruments of the same type, one should calibrate them with each other.

For almost all parameters it is possible to use instruments which transform the parameter into an electric signal. This has the advantage that one can record at some distance from the place of measurement and that only one recorder is needed for different parameters. Digital meters are to be preferred to analogue ones (having a pointer). If the sensitivity of an analogue instrument is high, the interval to be measured often falls beyond the range indicated on the meter. A digital volt meter, for instance, can indicate voltages between – 10 mV and +10 mV with an sensitivity of 2 μV. An analogue instrument with the same sensitivity would have a range of, at most, 0.5 mV. The sensors must be relatively robust when used for continuous measurements, which is often hard to combine with other requirements.

It would be beyond the scope of this book to present detailed instructions for all measurements. Generally speaking, the possibilities for effective measurements in the field are being improved all the time. The old-fashioned analogue recorder, representing the results of several sensors graphically as a point cloud, was difficult in practical usage, and moreover not very suitable for field work. Nowadays we have the datalogger, storing data on magnetic tape or diskette and printing data on paper. This instrument offers more possibilities but it is usually not made for field work. Recently, smaller and relatively cheap instruments of this type have become available which are especially designed for fieldwork and offer possibilities for the recording and processing of various signals.

5.2.2 Temperature measurement

Liquid thermometers

For many temperature measurements the traditional mercury thermometer is still very useful. Due to its simplicity very little can go wrong, and it records its own temperature very accurately. In a slightly different form it can also be used as a maximum thermometer. Because of its fragility it cannot be used in hard soil. Moreover it has a long response time and large radiation error as a result of the large mercury reservoir, and the temperature one wishes to know is often not the one really indicated. One cannot measure the temperature at, say, 2 mm above the ground. One cannot take any measurements at a distance and one cannot make it self-registering either. All these disadvantages are shared by the alcohol thermometer which in addition is less accurate (deviation through air bubbles). However, this type can be used as a minimum thermometer if placed in a horizontal position.

Thermo-elements

Measuring with thermo-elements, also called thermocouples, is a very flexible method as to the size and shape of the sensor. Where two metals touch each other a potential difference arises, the size of which is dependent on the temperature. With the metals usually used in electric connections such as copper, silver, gold or brass, the potential difference is small, but still large enough to give disturbances with very accurate measurements. For temperature measurements special alloys are used with a strong thermo-electric effect.

Fig. 62A illustrates the principle of a measurement with a thermocouple using the metals copper (Cu) and constantane (Const). If A and B have the same temperature, the potential difference between Cu and Const at A is equal to that between Const and Cu at B, but of a contrasting sign, and the instrument indicates a net voltage of zero. If A and B differ in temperature the net potential difference is not zero any longer but increases with increasing

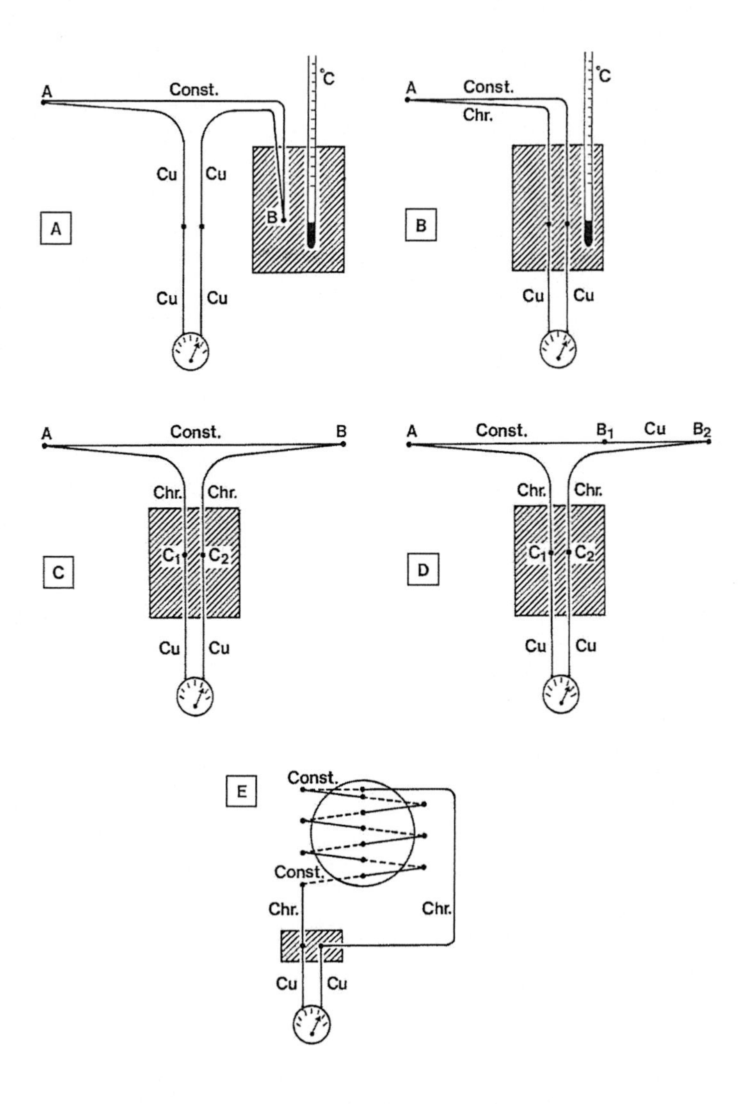

Fig. 62. Thermo-electric connections. A. Combination of copper, Cu, and constantane, Const. The active junction A measures the temperature relative to the reference junction B with known temperature. Hatched: metal block or thermos flask. B. The same for constantane and chromel, Chr. The transitions to copper both have the reference temperature. C. Connection to measure temperature differences. D. If B_1 and B_2 have the same temperature the nature of the interconnected material is not important. E. Connection for a thermopile to measure radiation. Only the junctions within the circle receive radiation to be measured. The two transitions Chr - Cu are kept at the same temperature (hatched).

temperature difference. For the combination Cu-Const the voltage created is ca. 40 μV/°C.

The thermo-element is by nature very suitable for the measurement of temperature differences. In order to measure absolute temperatures, one of the junctions has to be given a known temperature. This is called the cold junction or reference junction: the other junction is called the active junction. One approach is to bring the reference junction in a thermos flask filled with water the temperature of which is measured with a mercury thermometer. If the water is about air temperature it will change little. Another possibility is to put the reference junction in melting ice. This is a constant temperature of 0 °C, but usually far from the temperature one wishes to measure and it is advantageous to keep the difference between reference and active junction as small as possible. If there is a one-percent error in the calibration factor, its effect is of course smaller with small temperature differences. If the use of a thermos flask is inconvenient, one can also put the reference junction and the thermometer in an isolated block of metal, for instance a copper cylinder of 5 cm diameter, and read the temperature there with every measurement.

Fig. 62 shows other possibilities as well. With chromel and constantane, which both differ thermo-electrically from copper, a connection can be set up according to Fig. 62B. Another possibility is shown in Fig. 62C, provided the junctions C_1 and C_2 are kept at the same temperature, which need not to be known. This construction is particularly useful if temperature differences are concerned. If the respective thermocouple wires are not connected directly, but via a metal wire, then the result is the same as if the connection were direct, provided B_1 and B_2 have the same temperature (Fig. 62D). The material with which the two wires are soldered is not important either. Thermocouples can also be welded, a refined technique is described by Zanstra (1976).

Thermocouple wire can be obtained in gauges ranging from some mm in diameter to very thin, 0.02 mm. Heavier wire (diameter > 0.1 mm) with suitable insulation is usually sold as pairs: copper - constantane (40 μV/°C), chromel - constantane (60 μV/°C) and chromel - alumel (40 μV/°C). With a digital Volt meter with a sensitivity of 1 μV we can easily reach an accuracy of 0.1 °C, without problems from spurious thermo-forces.

If we start from a reference junction of 0 °C the thermo-force generated is not entirely linearly proportional to the temperature difference. This can be adjusted by calibration or by using correction tables provided by the manufacturer. Small differences may arise between material from different suppliers. Here, calibration is recommended, especially when the temperature difference between reference and active junction is large.

There are instruments which compensate for the temperature of the reference junction electronically and show the temperature of the active junction directly. For recorders and dataloggers this is a very good system, but where this implies long, expensive thermocouple wires, a separate reference junction, for instance one deep in the soil, is still to be recommended.

For mobile measurements it is often easier to use only one indicating instrument to which sensors for all kinds of parameters can be connected and to accept the separate thermometer for the reference junction. Nowadays electronic cold junction compensators for different types of thermocouples are available in the form of a small IC (chip, integrated circuit) that can be included in the circuit. In that case one can read the temperature of the active junction directly.

Thermocouples can easily be connected in series to build a thermopile (Fig. 62E), which is suitable for measurement of small temperature differences and for radiation measurements. Such a circuit is also suitable for measuring average temperatures over a larger space.

Thermocouples have the advantage of being cheap and also easy to construct; they can measure at a distance and be connected to registration apparatuses. Moreover, if the junctions are small the radiation error is very small. On the other hand, this gives a very short response time, as a result of which it may be difficult to take a reading of rapidly changing temperatures, for instance near a strongly sun-heated soil surface. However, there are digital measuring instruments recording maximally once per second, or slower.

Resistance thermometers and semiconductors

The resistance of pure metals increases with temperature, for platinum by ca. 0.4 %/°C. This is the principle of the resistance thermometer. Usually, this instrument takes the form of a little cylinder some mm in diameter around which platina wire is wound with a resistance of 100 Ω at 0 °C (Pt 100). The resistance is usually measured by means of a bridge circuit. Without individual calibration it is suitable for measurements with an accuracy of 0.1 °C. When the wire is wound on to a cylinder of 5 mm diameter the radiation error can be 6 °C. A thin wire (30 μ) stretched in a zigzag over a square frame, making as little contact as possible with the frame, has a radiation error of maximally 0.3 °C.

Advantages over thermocouples are measurements via a plane surface (which is more representative) and smaller fluctuations which make it easier to take readings. A disadvantage of the resistance thermometers is that they are more fragile and cannot be used on a very uneven surface, if we want to measure near that surface. In dense grassland it cannot be used either.

Resistance thermometers with a negative temperature coefficient, NTC resistances or thermistors, are made of sintered metal oxides. The resistance decreases by about 4% for each degree of temperature increase. They are made in the form of small flakes, bars or spheres. Their minimum size is ca. 1 mm. If specially prepared, identical thermistors can be obtained which give an accuracy of 0.2 °C without calibration.

Finally temperature can be measured with semiconductor diodes. If a current passes through a diode the voltage over the diode is linearly proportional to the temperature. The output is more than 1mV per °C.

Thermistors and diodes have a larger radiation error than platina wire sensors. All resistance and semiconductor thermometers do not need a reference temperature; the 'output', the voltage per °C, is higher than with thermo-couples and increases in the sequence resistance thermometer, thermistor, diode. They require a less sensitive and cheaper indicating instrument, which may have advantages with registration and special applications such as the driving of a summarizing counter. On the other hand they need an accurate voltage source. And, as to their manufacturing and size, they are less easy to adapt to divergent situations. Semiconductor thermometers, which allow maximum-minimum readings as well, are sold cheaply as complete units with a liquid crystal display. Usually they are quite accurate and have the advantage that readings can be taken at a few m distance.

Air temperature

For the measurement of the air temperature mercury thermometers are used in classical meteorology. When placed in the sun the temperature indicated may be more than 5 °C above the real temperature. On the other hand, at night the temperature indicated may be too low.

For measurements at some height above the surface of the earth, this 'radiation error' can be avoided by placing the thermometer and other instruments in a white-painted and well-ventilated instrument shelter. The standard arrangement here is a height of 1.50 m above short-cut grass.

In a smaller, different form, instrument shelters are used in official meteorology, viz. for measurement of the minimum temperature at 10 cm. During the day, they are only suited to measurements at higher levels, for instance in a forest (Kramer, Post & Woudenberg 1954). For the measurement of minimum temperatures a freely exposed thermometer may be preferred, because in this way the temperature closely approaches that of the vegetation at night. Use of shelters is only advised when one wishes to compare results with data from the official meteorological services though, for instance in England, a freely exposed thermometer is used (grass frost thermometer).

Another method is that of the ventilated thermometer, where air is sucked along the thermometer, while this is protected from radiation influences by a double nickle-plated cylinder. This principle is applied in the well-known Assmann psychrometer with which air humidity is measured at the same time (Ch. 5.2.3). As such, this is an accurate instrument. A big disadvantage, however, is its long response time: it is some minutes before one can take a new measurement. With measurements of vegetation there is a risk of disturbance of the local microclimate because of the ventilation and a still higher risk of disturbance to the reading. Also, air measured near the ground comes from a higher level than that at which the inlet opening is located. The closer to the ground the less mobile the air (Nyberg 1938).

Micrometeorological demands often make it necessary to measure with a small sensor, placed so as to be as free as possible, where the measurements can be made at some distance. The radiation error of a sensor is smaller, the smaller its dimensions are, because the heat transfer coefficient is higher. With a thermo-element of wire with a diameter of 0.06 mm, the radiation error was lower than 0.5 °C (Stoutjesdijk 1961; Fritschen & Gay 1979), with a diameter of 0.03 mm less than 0.3 °C.

Even with somewhat larger sensors the radiation error can usually be kept within reasonable limits by using a small aluminium screen intercepting direct sunshine. Such a small screen must be constantly adjusted to the position of the sun. With a permanent arrangement a larger screen would be necessary, but that implies a disturbance of the microclimate e.g. with measurements near a surface in the sun.

Surface temperature

In view of the steep gradients both in the lower air layers and in the top layers of the soil, measurements with a mercury thermometer can only give an approximate, usually too low, value of the temperature of a surface directly and strongly warmed up by the sun. The best approximation is obtained when the bulb is just under the surface. A thermocouple of thin wire is better, but deviations are still possible, especially when the material used has a low heat conductivity. Vaartaja (1949) found that the error can be kept small by using thermocouple wires as thin as possible and leading them some cm along the surface.

Leaf temperatures can be measured with a fine thermo-element, preferably in the form of a very thin (0.02 mm) band of flattened wire. Reliable measurements are possible, but great care is required to ensure a good contact with the leaf (Pieters 1972). On leaves with hairs or an irregular surface good measurements are almost impossible.

The best method is measurement with a radiation thermometer. The principle can be understood from Fig. 63. At the bottom of a cylindrical hole in a copper block a thermopile

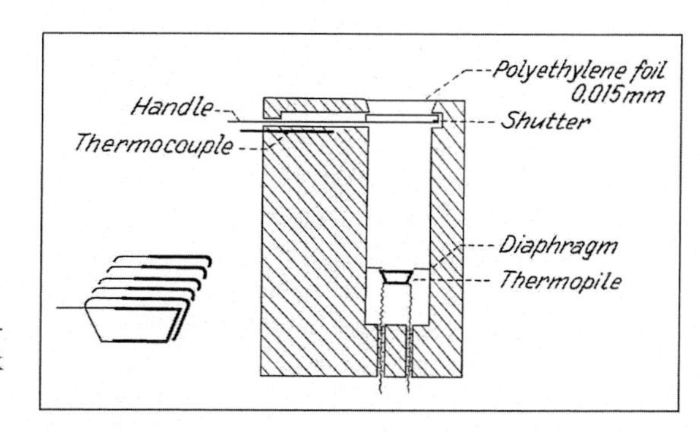

Fig. 63. Construction of a simple radiation thermometer. After Stoutjesdijk (1966). Left: detail of thermopile.

is placed as a radiation detector. At its upper end the hole can be closed with a shutter (temperature T_2). If the shutter is open and the instrument directed towards a black surface with temperature T_1 the thermopile receives long-wave radiation from this surface: $f_1.\sigma.T_2^4$, where f_1 is a proportionality factor which is dependent on the opening angle of the instrument. If J_1 is the output in mV from the thermopile with the shutter open and J_2 the output with the shutter closed, we have:

$$J_1 - J_2 = f_1 \cdot f_2 \ (\sigma T_1^4 - \sigma T_2^4) \quad (5.1)$$

where f_2 is another proportionality factor, which is dependent on the sensitivity of the instrument. The shutter has the temperature of the copper block, which is measured with a thermocouple or a small mercury thermometer. In this way T_1 can be calculated. A complication arises when the object not only emits long-wave radiation but also reflects solar radiation. A glass shutter transmits solar radiation and ideally, the sensor should receive the same amount of solar radiation with the shutter open or closed. Another solution is to paint the sensor white. In that case the solar radiation is not taken up, but the long-wave radiation is. Because neither of the methods is perfect, the best solution is to use them in combination (Stoutjesdijk 1966, 1974). The copper block with the thermopile is isolated from the surroundings completely, and a thin polyethylene window (0.02 mm) just below the shutter prevents disturbance by air movement without intercepting long-wave or solar radiation.

One measurement thus consists of a reading with the shutter open, one with the shutter closed, and a reading of the thermometer or the output thermocouple.

Commercial radiation thermometers function according to the same principle, only the shutter opens and closes at a high frequency as a result of which an alternating current signal arises. They usually have a germanium lens concentrating the long-wave radiation, and filters intercepting the visible light and restricting the transmitted long-wave radiation to a certain wave-length interval, e.g. 8-12 mm. They allow a direct reading of the temperature and they work very fast, also because the adjustment time is very short. The newest types are supplied with a sight (finder lens) and are able to measure the temperature of an object only 2 mm across.

As with previously discussed instruments, working with the radiation thermometer is not entirely without problems. In the air between the object and the instrument emission and absorption of long-wave radiation takes place. With a measurement at a distance of 1 m, the error can be up to 1 °C if the air temperature differs 10 °C from the surface temperature. This concerns the first type of radiation thermometer. When the sensitivity is restricted to the wave-length interval 8 - 12 µm, where the atmosphere is nearly entirely transparent, this problem does not arise, but such an instrument is, of course, not suited to measurements of the long-wave radiation of the atmosphere which is mainly in wave-lengths < 8 and > 12 µm.

Another problem occurs when the surface to be measured is not 'black'. If the emission factor ε is smaller than 1, a reflection of the long-wave radiation from the surroundings occurs, which is complementary to the emission. If the temperature of the surroundings differs Δt from the surface temperature, the error is $(1 - \varepsilon) \cdot \Delta t$, as a first approximation. Of course, by the temperature of the surroundings is meant the average radiation temperature of the surroundings as far as it can be 'seen' by the surface (cf. Ch. 2.3.3).

A large error may occur in the following case. Quartz has a low emission and thus a strong reflection in the interval 8 - 12 µm. This is the wave-length interval where a clear sky hardly produces any long-wave radiation. If one measures with an instrument which is sensitive in this very interval, as is usual with commercial instruments, then a considerable error will arise (Büttner & Kern 1965). The cold sky dome is, as it were, reflected in the surface. For an instrument sensitive over the entire spectrum the error is much smaller (Stoutjesdijk 1974c).

5.2.3 Measurement of air humidity

Many methods exist here, see Sonntag (1966-1968); Fritschen & Gay (1979) and Schurer (1981). A basic method is the determination of the dew point, but this is usually too complicated for our purpose.

The psychrometer is both accurate and generally applicable. It is based on the difference in temperature between a dry thermometer indicating the air temperature, and the same thermometer surrounded by a wet wrapping. The latter cools off through evaporation. In the ideal case when no heat is supplied by radiation or conduction an equilibrium situation arises where the energy, removed as latent heat from evaporation, is compensated for by the sensible heat taken up from the air.

Because in both cases the transport takes place through the same boundary layer, albeit in opposite directions, we can write (see Ch. 2.6):

$$\frac{\alpha}{\gamma}(e_w - e_{air}) = \alpha(t_{air} - t_w) \qquad (5.2)$$

Here, e_w is the maximum vapour pressure at the temperature of the wet surface, e_{air} the vapour pressure of the air, t_{air} and t_w the air temperature and the wet bulb temperature respectively. The symbols α and γ are the heat transfer coefficient and the psychrometer constant.

From Eq. 5.2 follows:

$$e_{air} = e_w - \gamma\,(t_{air} - t_w) \qquad (5.3)$$

It follows that the indication of an ideal psychrometer is independent of α and thus of wind velocity. Evaporation increases with wind velocity, but the amount of heat taken from the air increases as well and proportionally, so that the equilibrium temperature attained is the same.

This ideal situation does not happen often in practice. The wet thermometer is always colder than the objects in its surroundings as a result of which it receives more long-wave radiation than it emits. There is also heat conduction through the thermometer tube to the colder bulb. The influences of radiation and conduction decrease as α increases (Ch. 2.6). With the Assmann-psychrometer a ventilation is applied with a wind speed of some m/s and both mercury reservoirs are protected from solar radiation by double nickel-plated tubes.

A larger value of α may be obtained by diminishing the size of the wet thermometer as much as possible. Nowadays electrically driven ventilated psychrometers with small electrical thermometers exist, which are considerably faster. With the Assmann-psychrometer the adjustment time is ca. 5 min, while in this new instrument only ca. 20 s. Moreover, in this way the risk of sucking air with a different temperature and humidity from elsewhere is diminished. Also, the dry and wet bulb temperature, vapour pressure, relative humidity, and dew point can be displayed digitally,without calculations.

Still faster are psychrometers with very fine wet and dry thermocouples. Thin cotton

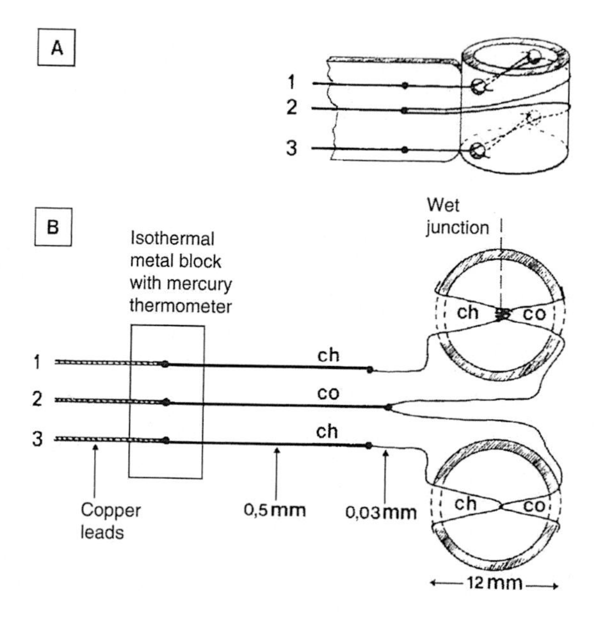

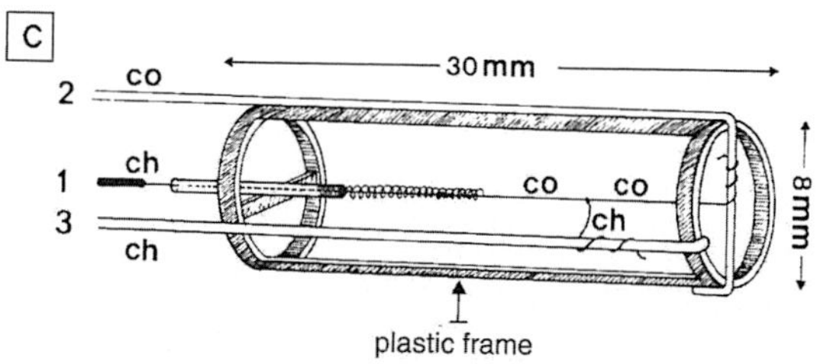

Fig. 64. Psychrometers with thermocouples. Sketch of the construction (A) and the electrical connection (B) and connection scheme of a miniature psychrometer. At (C) the wires at either side of the wet junction are wrapped with thin cotton thread taking up water from a capillary. After Stoutjesdijk (1980).

thread is wound round the wet thermocouple. If the thermocouple wire is thin enough and the wrapped part long enough, the error through conduction becomes small. Due to the small size α is so large that no artificial ventilation is needed. In Fig. 64 two variants of a thermo-element psychrometer are sketched. Type c, but with wires of 0.06 mm diameter and twice the size shown in the figure, functioned almost like an ideal psychrometer (Stoutjesdijk 1961). A very small air movement was sufficient to maximize the difference between dry and wet junction temperature. Even in completely stagnant air the difference was near the maximum. For such a psychrometer, with a little air movement (as is usually present in vegetation) we can, for calculating the vapour pressure from the dry and wet junction temperatures, use a value of γ which applies to a laminar boundary layer, ca. 0.57

Table 43. Difference between dry and wet bulb temperature for different types of psychrometer. Air temperature 22 °C. The letters A and C refer to Fig. 64.

Type of psychrometer		Temperature difference (°C)	
		With ventilation	Without ventilation
Assmann-psychrometer		11.4	
Thermocouple psychrometer A		10.8	10.8
Thermocouple psychrometer C	small type	11.6	11.3
	large type	12.0	11.8

mbar/°C (van der Held 1937; Sonntag 1968). Unwin (1980) constructed a small thermocouple psychrometer with very weak forced ventilation.

For the Assmann-psychrometer the value of γ generally used is 0.66 mbar/°C. The values of γ mentioned are valid for an air pressure of 1000 mbar; the value of γ is proportional to air pressure. The use of a higher value of γ with the Assmann-psychrometer has to do with the higher wet bulb temperature, not because the latter would function differently, but because conduction and radiation still play a part in making the temperature of the wet bulb higher than when a thermocouple is used. Apparently, heat conduction through the wires already had some effect on the small thermocouple psychrometer.

Fig. 64 shows a construction where both the dry and wet junction are placed in a perspex cylinder 12 mm in length and of 10 mm internal diameter. Wires of chromel and constantane are bent into a v-form and hooked in each other. With the soldering of the wet junction only the contact point may be covered with tin. The wrapping is a knot in a thin cotton thread which can keep a drop of water of 1.5 mm. The circuit schemes for both types of psychrometer are presented in Fig. 64B. In the circuit 1-3 the temperature difference between dry and wet bulb is measured, both chromel - copper transitions are placed in the same metal block where the constantane - chromel reference junction and a thermometer are also placed. The temperature difference with the reference junction is measured in the circuit 2 -3. This ‘drop psychrometer’ gives slightly lower values for the psychrometric difference than the others, as might be expected, but it is almost entirely insensitive to air movements (Table 43).

The drop psychrometer is very suitable for measurements in dense vegetation and has been proved to be resistant to mechanical damage in the field. This type of psychrometer requires much care during the construction. The other types, where the wires at both sides are wrapped with cotton thread, are much easier to construct and can be reproduced accurately. The cotton thread should be as thin as possible and carefully scoured with a wetting agent. With measurements around freezing point one should check whether the water on the wet bulb is frozen or not. Ice has a lower vapour pressure than water of the same temperature and water can be supercooled several degrees below 0 °C. If ice has been formed, the vapour pressure over ice should be used in the formula (Table 3).

For moistening a psychrometer one should use distilled water. This is important, especially with continuous measurements, otherwise the water round the sensor changes into a salt solution, which has a lower vapour pressure than water. With continuous measurements it is difficult to construct a well-functioning water supply and to screen the instrument from the sun without hindering air movement. Stigter & Welgraven (1976) have described a larger model thermo-element psychrometer suitable for continuous measurements.

The di-electrical sensor (called Vaisala sensor after the manufacturer) is based on a hygroscopic effect. A small plate of a special plastic (4 mm × 8 mm) absorbs moisture from the air dependent on the relative humidity. Thin layers of gold which have been vacuum-deposited at either side of the plate make a condensor of the plate with a capacity which is proportional to its moisture content. The instrument is accurate and rapid, but has its disadvantages: it collects dirt easily and is difficult to clean. Further it should be calibrated regularly, especially after contact with liquid water. This type of instrument measures primarily the relative humidity, based on the assumption that the plate has the temperature of the air. If the temperature is higher, for example through radiation, the relative humidity measured is too low. The latter complication can be made use of by electrically heating the plate under a very high relative humidity and then checking how much lower the measured relative humidity is. All in all, the Vaisala sensor is a reasonable instrument giving good results, not in need of ventilation or water supply, but on the other hand it does not tolerate wetting. As with other instruments, it gives the best results if the user knows the functioning of the instrument and the situations for which it is best suited.

The classical hair hygrometer is still used, usually in the thermohygrograph. This instrument can be placed in the instrument shelter at 1.50 m and used to relate the measurements to the general weather data. It is also suitable in the forest, but not for measurements on the micro-scale. The hair hygrometer has to be calibrated regularly.

A real micro-method has been devised by Corbet et al. (1979), for, among other things, the measurement of the air humidity in flowers. A salt solution takes up water from the atmosphere or releases water, until the concentration is in equilibrium with the relative humidity of the air. A small drop of potassium citrate is placed in an eye of copper wire and the concentration is determined refractometrically.

Finally, some remarks are made on the measurement of moisture in the soil. The classical method of weighing and drying of a fixed volume of soil, taken with a steel cylinder, is all right but time consuming and tending to cause disturbance. A new device has been developed by the Technical and Physical Service at the Agricultural University at Wageningen, the Netherlands, which measures the di-electrical constant of the soil, which is largely determined by the moisture content. There is also a version for measuring the moisture at the soil surface. An older instrument, giving a rough indication (on a scale 0 - 100) of the soil moisture in the top layers, is commercially available under the name Aucon.

5.2.4 Measurement of radiation

Three methods are known: the non-selective method where radiation is transformed into heat, the photo-electrical method transforming radiation directly into electricity, and the photochemical method, where some chemical reaction is caused by the light.

If we are dealing with the heat and water budget we are looking for a method for measuring energy fluxes in energy units independent of the wave-length. Then a thermal method is obvious, and when long-wave radiation is concerned, it is really the only method. The principle of this method is based on a temperature difference and can best be explained by starting with the measurement of the net radiation (Fig. 65A).

We start from a horizontally placed black plate. If the upper surface receives more (or less) radiation (R_1) than the underside (R_2) a temperature difference arises. This difference is proportional to $R_1 - R_2$. This system functions well in still air, but wind makes the differences smaller and leads to instability in the recording. In order to diminish the effect of wind, a dome of polyethylene film is used nowadays, which is kept in shape by means of over-pressure. (Fig. 65B). For permanent use, when disturbance could arise because dew may be precipitated on the dome, a variant is available where dry air is blown over the dome. In practice, the 'plane plate' may consist of a small plate of plastic, e.g. perspex, some mm thick, and the difference in temperature between upper and under side is measured by a thermopile, i.e. a number of thermocouples connected in series; the thermopile is made of galvanically copper- or silverplating constantane wire (Stoutjesdijk 1966); both sides of the plastic plate are then covered with thin copper foil which stabilizes the temperature.

A simple form of this radiation meter with plane polyethylene film is shown in Fig. 64C. Here radiation from a low angle is screened off. Especially at night this construction gives useful values for sample measurements. For $r / h = 4$ (Fig. 65C) the error is less than 5 %. On the basis of the principle in Fig. 65C different variants are possible. First, a one-sided measurement of incoming radiation is possible by replacing the lower polyethylene cover

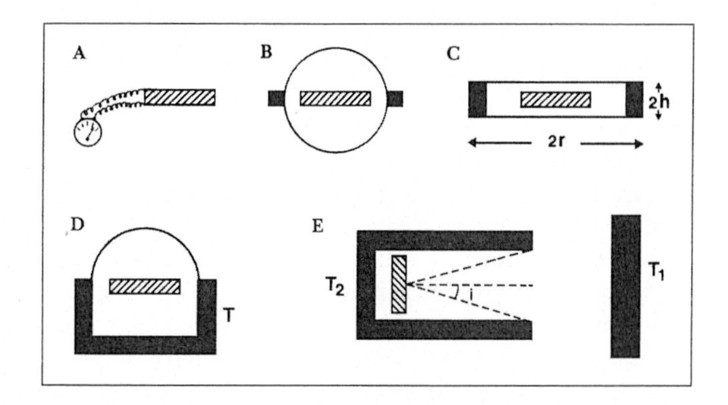

Fig. 65. Radiation meters. A. Free placed meter for net radiation. The temperature difference between upper and underside of the plate is measured with a number of thermocouples connected in series. B. As A, but sheltered from the wind by domes of polyethylene. C. As B, plane model. D. One-sided model for solar radiation + long-wave radiation (polyethylene dome), or for solar radiation only (glass dome). E. Arrangement for calibration of radiation meters. Black: aluminium.

(Fig. 65B) with a black metal cover, of which the temperature is measured (Fig. 65D). The radiation received can be calculated according to:

$$\sigma T_1^4 - R = f \cdot J \tag{5.4}$$

Here, T is the temperature of the metal cover, R the radiation intensity to be measured, f a calibration factor, and J the 'output' of the sensor.

This construction can also be used for the measurement of solar radiation by replacing the polyethylene hemisphere by one of glass, transmitting only short-wave radiation. Usually Moll-Gorczinsky's solarimeter is used for this purpose, which is based on the principle that a thermojunction placed free in the air attains a higher temperature under radiation than the same junction when (thermally well-)connected with the metal case.

Another method makes use of the fact that a white surface is heated much less by solar radiation than a black one, while both have the same absorption of long-wave radiation. With Dirmhirn's (1958) black-white-pyranometer a thermopile is formed from a number of black and white sectors. Szeicz (1965) constructed a tube-shaped instrument according to this principle for taking measurements in vegetation.

A fundamental calibration of radiation meters is obtained by placing the element in a situation such as that sketched in Fig. 65E. If the black metal plate has a temperature T_1 and the cylindric metal tube a temperature T_2, and the meter has an opening angle i the element receives a net radiation:

$$(\sigma T_1^4 - \sigma T_2^4) \sin^2 i \tag{5.5}$$

For the calibration of radiation balance meters a calibrated solarimeter is often used. Both instruments are placed in the same position towards the sun and direct solar radiation is intercepted with a small screen. Both instruments receive equally less radiation, and for the solarimeter this amount is known. For the measurement of short-wave radiation causing specific biological or photochemical effects the following approaches are possible. For solar radiation, a separation can be made between ultra-violet and photosynthetically-active radiation, $\lambda < 700$ nm (PAR) on the one hand, and radiation with wave-lengths > 700 nm, on the other hand, by placing a radiation meter side by side with the usual solarimeter of which the outer cover is replaced by filter glass which only transmits radiation > 700 nm.

Within a certain wave-length interval the photo-electrical method offers great advantages. The sensitivity is high and there is no disturbance from the effects of temperature. On the other hand, the sensitivity is heavily dependent on wave-length. The much used silicium photo-element (usually called a photocell) is sensitive to radiation between 400 and 1100 nm, but has its maximum sensitivity at 750 nm. With suitable filters certain wave-length intervals can be isolated, for instance red and far red. Also, meters can be constructed with a quite uniform sensitivity to photosynthetically-active radiation (van der Hage 1984). By using filters, an instrument can be made the indication of which runs exactly parallel to that of a solarimeter. Of course, this is due to the circumstance that the energy in the > 1100 nm wave-lengths is small and roughly proportional to that in the short wave-lengths, because the Si-cell does not perceive radiation > 1100 nm (Kerr & Thurtell 1967). Monochromators can also be built with interference filters and photocells, with which narrow wave-length bands can be selected. An oblong filter is used in this kind of instrument; the transmitted radiation is dependent on the position of the filter (Sauberer & Dirmhirn 1958; Stoutjesdijk 1972a).

The smallest silicium cells have a diameter of about 4 mm, enabling measurements to be made in dense vegetation. They have to be used with a diffusely dispersing filter of milk glass or plastic in order to make the sensitivity independent of the angle of radiation. Such 'point measurements' are not very representative. A better average may be obtained by a long 'light bar', being a number of narrow sensors connected in series, parts of which can be screened off. If the situation allows, for instance in a forest, we can also obtain average values over a larger surface by directing a sensor with a limited opening angle towards a matte white surface. The measurements in woodland (3.4.3, Fig. 38) were made in this way using the above-mentioned interference monochromator. For comparative measurements small deviations from the 'ideally white' surface are not important. Incidentally, types of white paint are available reflecting no less than 88% of all visible wave-lengths.

Photocells produce a current which, within certain limits, is independent of the external resistance. This resistance should not be too large (ca. 100 Ω). When a modern high-ohm measuring instrument is used, a resistance must be connected parallel with the photocell.

Measurements with photocells are often performed comparatively and expressed in relative units. Calibration can be done in W/m^2 or in light quanta. This requires special arrangements and expertise and is always based on non-selective methods.

Earlier measurements were often done with selenium cells, adapted to the sensitivity of the eye, which has a maximum of 550 nm and decreases strongly at either side. The unit used here is the lux (lx). Transformation to energy units is only possible for radiation of a certain spectral composition. For sunlight the approximate relation is $1\ kW/m^2 \approx 107.5$ klx.

Photochemical light measurement is seldom done. The advantage of this method is that integrated measurements can be done at many places. However, the sensitivity is predominantly in the ultra-violet and measurement results must be interpreted with care (Lagrew & Baskin 1975).

5.2.5 Measurement of heat flow

Heat flow measurement is applied when bodies take up or release heat, like the soil, for instance. The principle can be understood best in comparison with an electrical circuit. If one wishes to measure the current, one can include a small resistance with known ohm value in the circuit and measure the voltage over this resistance (the total resistance in the circuit has hardly been changed).

A heat flow is measured in roughly the same way. If, for instance, a thin plate with known heat conducting capacity is put onto the soil surface, the heat flow can be calculated when the temperature difference between upper and underside (the analogue of the voltage difference in the resistance) is known. In order to keep the heat resistance small, the plate must be quite thin and of well-conducting material. Then the temperature difference is small. It can best be measured with a number of thermo-elements connected in series. By wrapping a small perspex plate of some mm thickness with thin constantane wire which is half coated with copper (see Fig. 63 and Stoutjesdijk 1966) such heat-flow plates can be easily constructed. For calibration, which is necessary, one can use commercial, calibrated plates.

If there is no need for a high sensitivity, heat-flow plates can easily be made of so-called print plate, which is covered with a thin layer of copper on both sides. One can then connect the copper surfaces with a thin constantane wire (Stoutjesdijk 1978).

This method can be used for many different measurements. One example is the measurement of the heat flow through the walls of a nesting box with a breeding great tit (*Parus major*) (Mertens 1980).

5.2.6 Measurement of wind

Wind velocity is traditionally measured with rotation anemometers consisting of a vertical axis with three horizontal bearers on which aluminium domes are mounted. In the usual type the axis drives a counter so that the total distance per unit of time over which the domes move can be calculated. For measuring an average wind speed this type of instrument is very suitable. However, in its original version it had a rather high starting speed, at least 0.5 m/s. Now, there is a greatly improved version, without a mechanical counter, but which uses an intermittant light beam and pingpong balls in stead of aluminium domes. Here the starting velocity is not more than ca. 0.1 m/s (Bradley 1969).

By nature of their construction these instruments measure only the horizontal component of wind movements. Because the anemometers are rather large they can hardly be used in vegetation - also of course because wind speed is often under or around the starting velocity of the instrument. With very variable wind direction and speed the data are often difficult to interpret. In such cases a non-mechanical instrument will give better information, especially on wind speed fluctuations.

For this situation one often uses the hot-wire anemometer, which essentially measures the heat transfer coefficient of an electrically heated platinum wire. The minimum wind speed to be indicated is some cm/s; with higher wind speed, from 5 m/s onwards, the accuracy decreases rapidly. For the study of air currents in boundary layers a subtle and vulnerable version of this anemometer with extremely thin, freely exposed wire is used. Another, more robust version with a short wire protected by a frame is suitable for measurements in vegetation. There are also hot-bulb anemometers using a small bulb instead of a wire which are very suitable for fieldwork.

Some of these instruments take into account the temperature of the air. In other instruments the temperature of the hot wire is so high, up to 300 °C, that the temperature of the air can be disregarded. The sensor may also take the form of a bar, but this makes the meter sensitive to wind direction and this can be a disadvantage for use in vegetation with unpredictable air currents.

5.2.7 The use of models

If one is interested in the significance of environmental factors for plants or animals, one first has to know how these factors effect the organisms. Two approaches are possible. The first is to measure directly the temperature, transpiration, etc. of the organism and then relate the results to the actual value of the environmental factors. The second is to establish the relations between the occurrence of organisms and the environmental factors, and then deduce the characteristics of the organisms from the general measurements.

In many cases with an autecological problem, the biological object can be replaced by a model responding in a similar way to the decisive environmental factors.

Bakken & Gates (1975) made casts of lizards, the temperature of which was measured in the situation of the living organism. The model lizard was painted so as to approach the reflection of a living lizard. When transpiration is small, as is the case in many reptiles and insects, the model attains the same temperature as the real animal. Brandt (1974) studied a lycosid spider sunning its cocoon and placed a similar cocoon with built-in thermocouple at its side (see Ch. 4.2). Especially for insects, the use of intact dead specimens as sensors for temperature measurements is promising.

In medical physiology the katathermometer and frigorimeter are used. The first type is a thermometer with a large reservoir (ca. 10 cm^3) which is warmed up to 37 °C. Then the time needed for the reservoir to cool off to 36 °C is measured. This instrument integrates, as it were, the effects of radiation, wind and air temperature as a human being experiences them in the open air. Still better for this purpose is the frigorimeter, a sphere wrapped in a dry or moist cover, kept at a temperature of 34 °C. Then the energy added by an electric heater is measured (Fanger 1972). Metal casts of the bodies of birds and mammals can be covered by the original plumage or pelt. When the cast is provided with a thermostat and a heating wire, a realistic model of the energy exchange with the environment is obtained (Heller 1972; Byman, Hay & Bakken 1988). An unheated metal model can also give valuable information on the operative temperature (see Ch. 4.1.5).

The classical Piche-evaporimeter is suitable for the simulation of the evaporation of a wet soil, plants, or poikilothermic animals with a moist skin (for instance slugs and amphibians). This instrument consists of a disc of wet green blotting paper fed with water from a glass tube with a graduated scale so that the evaporation can be measured (de Vries & Venema 1954; Roth 1961).

The evaporation of a wet surface is determined by the net absorbed radiation, the wind, and the saturation deficit of the air (see Ch. 2.6). A leaf reacts to these factors as well, but the radiation has relatively more impact here than wind and humidity. While the evaporation of a wet surface always increases with increasing wind velocity, the evaporation of a weakly-transpiring leaf will decrease, on the other hand. A good correlation is to be expected between Piche-values and the evaporation of wet moss and of wet bare soil. As soon as the top soil layer has dried up, this correlation does not hold any longer. Thus the Piche-evaporimeter indicates the potential evaporation and, provided it is used with insight in the actual situation, it can give valuable indications, especially when a summation value is concerned. Walter (1960) presented many examples of the use of the evaporimeter.

Disadvantages of the Piche-evaporimeter are that it is easily damaged by frost, and that in sunshine after rain the air in the reservoir expands and water is pushed out. In dry periods the instruments are a source of water for wasps.

Generally, the use of models cannot replace the measurements of simple situations with uniform physical sensors measuring only one factor at a time. However, models can form a valuable additional approach, particularly because they can easily be used over longer periods of time and in many places simultaneously.

5.2.8 Tools for site descriptions

As has become clear from this book, much about the microclimate can be derived from the visual observation of landscape and vegetation structure. With the description of sites from a microclimatological viewpoint some, often quite, simple tools can be used.

The orientation of slopes, forest edges etc. can be determined with a compass. An inclination meter is often part of the more sophisticated compasses but it can easily be improved.

It is important to describe which part of the sky dome or the sun ecliptic is screened off by obstacles rising above the horizon. The screening determines when the direct solar radiation is intercepted and it also has an influence on the reflected short-wave radiation and the daily long-wave radiation received from the surroundings.

A simple tool to be used here is the horizontoscope (Geiger 1961). This is a segment of a perspex sphere in which one observes a mirror image of the surroundings. If a circular piece of transparent paper is placed below the perspex and looked at from above one can estimate which part of the sky dome is screened by trees, hills and other objects. The daily track of the sun has been printed on this paper for the different seasons. It is easy to draw the contours of all obstacles on a duplicate transparency. The sketched contours indicate the projection of the screened-off and the free part of the sky dome respectively. In order to obtain a picture of the radiation climate over the entire year it is necessary to differentiate for parts of the dome which are only screened in certain periods of the year, notably deciduous shrubs and trees which screen differently in summer and winter.

The globoscope has a convex mirror which is photographed from above. The photograph can be interpreted with the help of transparencies for the same purpose as the globoscope (Lee 1978).

Finally, fish-eye photographs with a suitable overlay can be used; in fact the globoscope produces fish-eye pictures too (Anderson 1964; Watson & Johnson 1988). Fish eye photographs are especially valuable in complicated situations.

Barkman (1988) developed a method for the quantitative description of vegetation structure, which we will summarize here. Horizontal and vertical denseness of (low) vegetation is estimated by means of a simple apparatus consisting of two aluminium laths, each 1m long, with a horizontal aluminium plate at 15 cm from the pointed lower end, which is driven into the soil up to the basal plate. The laths are perforated at regular intervals. They are placed at 1m distance in the sample plot. Cylindrical pointed aluminium bars (9.5 mm diameter) are thrust horizontally through the holes from lath to lath. Then vegetation is trimmed with a pair of scissors in a vertical plane along the bars. The next step is to place the laths 10 cm backwards and put the bars in place after having glued orange and blue plastic ribbons across them at 10 cm distance. At each height one can estimate both the vertical covering of the bar by vegetation, as seen from above, and the horizontal covering as seen from the side through 10 cm of vegetation. In this way x values for each height v'_x for the vertical covering and values h_x for horizontal covering are obtained. Then the bar is replaced at the highest position between the laths, and two auxiliary laths are placed 10 cm behind the

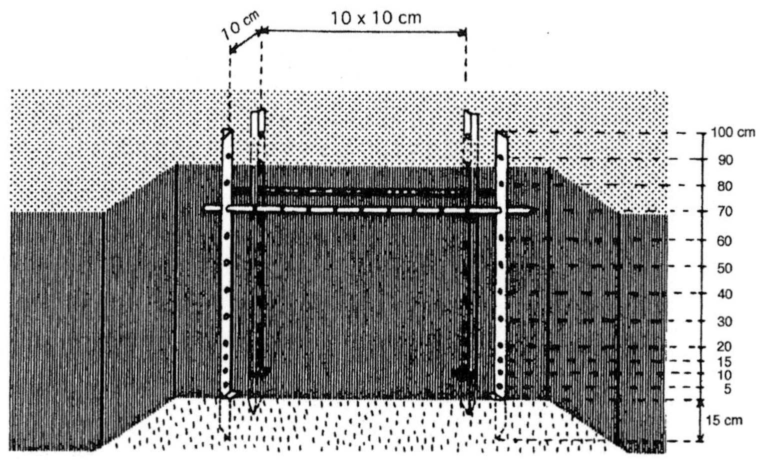

Fig. 66. Apparatus for estimating horizontal and vertical denseness of short vegetation. After Barkman (1988).

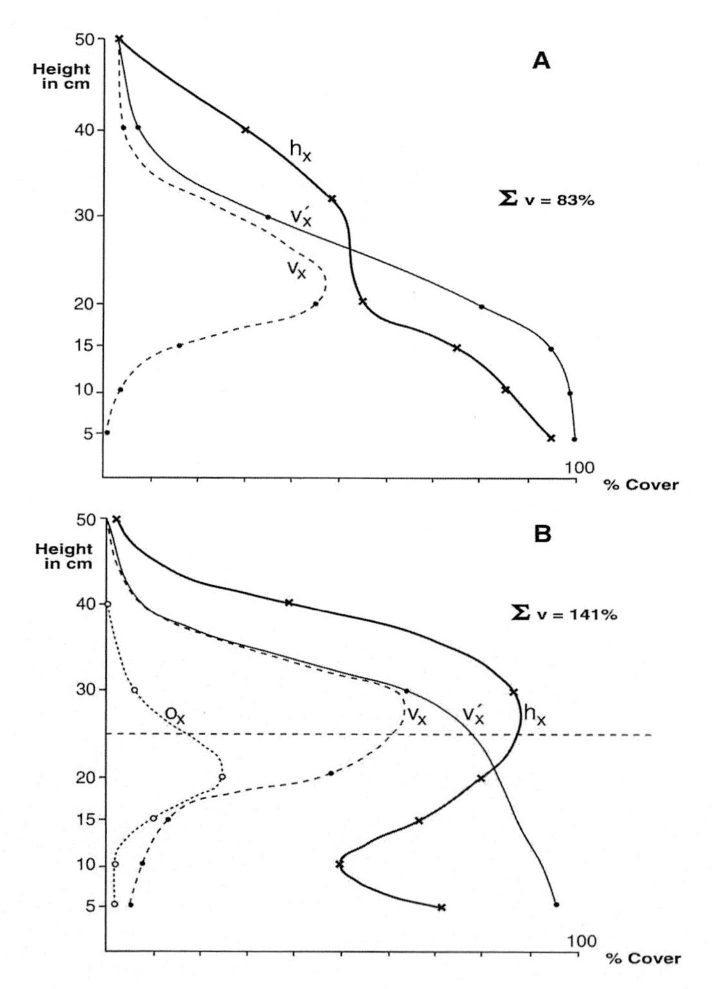

Fig. 67. Vegetation-structural characters of two different types of short vegetation. A. Hayfield at a sunny site (*Lolio-Cynosuretum*), Drenthe, the Netherlands, 9 June 1979; B. Dry *Calluna* heath, Drenthe, the Netherlands, 30 August 1979.

main laths and connected by a bar at the same height. All parts of plants between the two bars and in front of the first bar, exceeding this level, are cut off. (They can be collected for a stratified biomass determination.) Then the two bars are lowered 10 cm in position and the new vertical covering v_x of the first bar is estimated. Fig. 66 shows the apparatus in outline and Fig. 67 illustrates what kind of denseness profiles may be obtained. In grassland at a sunny site (Fig. 67a) with mainly erect and erectopatent leaves, the three curves show a regular course with high densities below 20 cm, with h_x always greater than v'_x. In a dry heathland (Fig. 67b) with many horizontal leaves, v'_x exceeds h_x below 20 cm and v_x shows a clear maximum between 20 and 30 cm.

Evidently, such detailed structural measurements and estimations can be combined with microclimatological measurements, and of course also with observation of the distribution and patterns of movement of animals. In this way we have demonstrated once more how vegetation, fauna and microclimate can, and should be investigated in an integrated approach in order to understand the biological significance of microclimate and the interactions between microclimate, vegetation and fauna.

6. Birds in the sun

6.1. Energetics and energy flows of birds

6.1.1. Introduction

You will agree with me that the sun is an essential source of energy for the animal world. I think your own experience can be a good starting point for a more quantitative approach. It is easy to do a simple experiment, on a sunny day with a north-eastern wind and an air temperature of 10 °C. Seat yourself on a sheltered spot with your back towards the sun. You'll feel very comfortable, after 10 minutes almost too warm in your solid outdoor clothing, probably light cotton undershirt, a warm shirt, a thick sweater of fleece and a jacket lined with some kind of "tex" that is ventilating but impermeable to the wind, (or so they say). Under these conditions I have measured the temperature in the sun on the surface of my jacket with a fine thermocouple (0.05 mm) that was as it were part of the surface. The surface temperature was 67 °C, that is 57 °C in excess of the air temperature and 30 °C warmer than the skin. In the shadow, sheltered from the wind, there is little left of that comfortable feeling, not really cold, but certainly not warm. On the surface of the jacket I measure 42 °C, that is 32 °C above ambient air temperature, but not more than 5 °C above the body temperature. The third possibility: in the shadow and in the wind. I feel how heat of my skin is sucked in spite of the wind-tight membrane in my jacket. The surface temperature is now 22 °C, that is 12 °C above the air temperature, i.e. there must be a considerable heat flow leaving your body.

In summary: the sun is a source of energy which, when absorbed under natural conditions it can cause high temperatures, i.e. be of great importance to the living organisms. The great importance of the wind respectively shelter in this aspect is stressed. In certain situations solar radiations and wind are decisive and the air temperature is of secondary importance.

6.1.2. Energetics of the bird

Birds and other hot-blooded animals always have a basal metabolic rate (BMR), that keeps the vital processes going (Gavrilov, 2013). The ambient temperature at which the BMR is just sufficient to keep the body temperature at about 40 °C is called the lower critical temperature (LCT) (Fig.68).

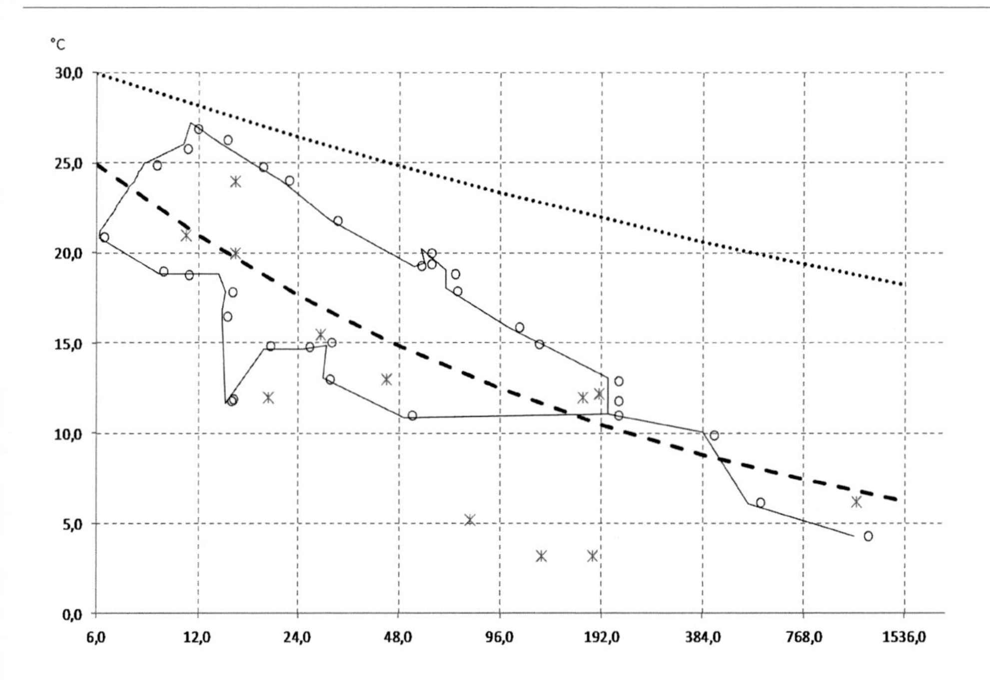

Fig. 68. Data on the lower critical temperature of passerines in relation to weight. o: Collected by Kendeigh from various authors, obtained by metabolic measurements. Some values have been calculated from relation between BMR and C after fig. 69. x: Various data calculated from the plumage conductances derived from the photographically obtained volume of free-living birds, Stoutjesdijk c.f. in: Stoutjesdijk & Barkman 1992. The lower dotted line was calculated by Kendeigh c.s. as the best fit for these data. The upper dotted line was obtained in the same way for non-passerines. The virtual values not shown, compare fig. 70.

The simple model of a core of, say, 40 °C and an insulating mantle is a good approximation, not only a theoretical one. You can simply construct one and study the thermal behaviour, either from a theoretical viewpoint or experimentally (Stoutjesdijk, 2009). In many cases, a thermal model of a real bird can be made by means of a bulb of thin metal either covered with an insulating layer, or simply bare. A blackened beer-can may often be suitable to estimate operative temperature of a jackdaw. A body temperature of 40 °C means that heat is given off to the air and that this heat loss (H) is proportional to the difference body temperature-air temperature (as a first approximation). Where sun and wind play a role we can for not too complicated cases calculate an operative temperature which takes the place of the air temperature and often can be estimated with an acceptable approximation, either theoretical or by simple measurements (remember the beer can and the jackdaw).

It is more complicated when solar radiation penetrates the plumage (Stoutjesdijk 2002, 2003, Wiersma & Piersma, 1994). In fig. 69 the simple black bulb model is compared with various models that are partially penetrable for solar radiation. Note that the fluffy models may reach Δt values of about double those of a black bulb of the same size. Although the results obtained are completely counter-intuitive they can be readily understood from the viewpoint that solar radiation penetrates into the "plumage" and is gradually absorbed. See also under "Operative temperatures in winter".

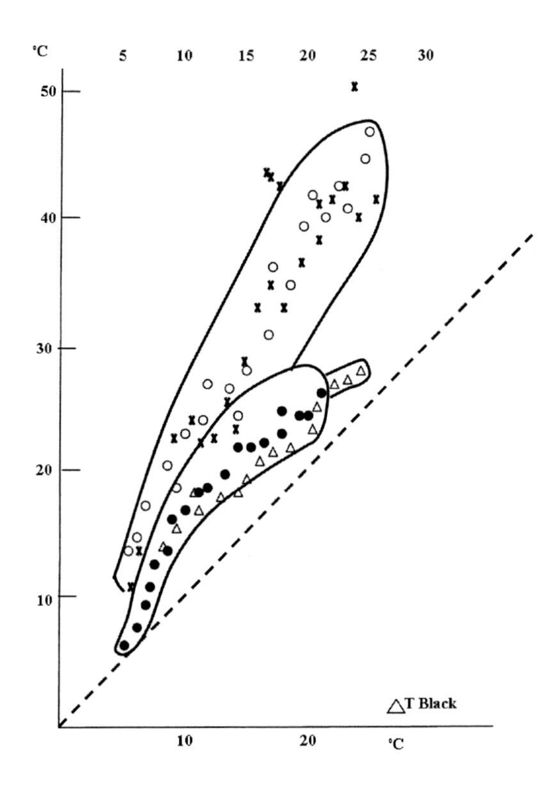

Fig 69. Difference core temperature – air temperature (ΔT) for models with a coat of, respectively, grey wool fibres (x), white artificial fibres (o), swanpelt (•) and down (Δ) in relation to ΔT of a black bulb. The symbols give average values for each 1 °C interval of black bulb ΔT. The enveloping curves contain all the data for the models with grey and white fibres (except a few extreme values) respectively, those with swan pelts. The results for the down feathers are only shown as averages, they lie mostly within the range of the swan pelts (from Stoutjesdijk 2002.)

In this case it is obvious to use a pelt drawn over the carcass and to observe the thermal behaviour of this imitation bird in the sun. Somewhat more refined you can include a thermostat and measure the dissipation of the energy. This may of course deliver useful data even when the sun does not shine. This method is in the first place useful when the same species is considered throughout. Here I keep things as simple as possible to obtain an overview of the thermal aspects of birdlife based on observations, simple measurements and calculations. I hope to build a simple model without discrepancies between the various approaches to the thermal aspects of birdlife. Where necessary, I have more detailed estimates of the thermal properties of the plumage.

On the other hand, I calculated the conductance from photographs of birds in the winter looking like feather spheres, or at least egg-shaped bodies (Photo 1, *Parus montanus*). By comparing with objects of known dimensions the volume can be calculated. We assume furthermore that the volume in cm^3 of the living core is equal to the mass of the bird in grams (fig. 70).The measurements were made in direct sunlight, usually it is assumed here that the radiation is evenly distributed over the sphere.

Photo 1. Willow tit (*Parus montanus*) in the cold (photo P. Stoutjesdijk) and its appearance in summer on the right (photo 123RF, John J. Henderson).

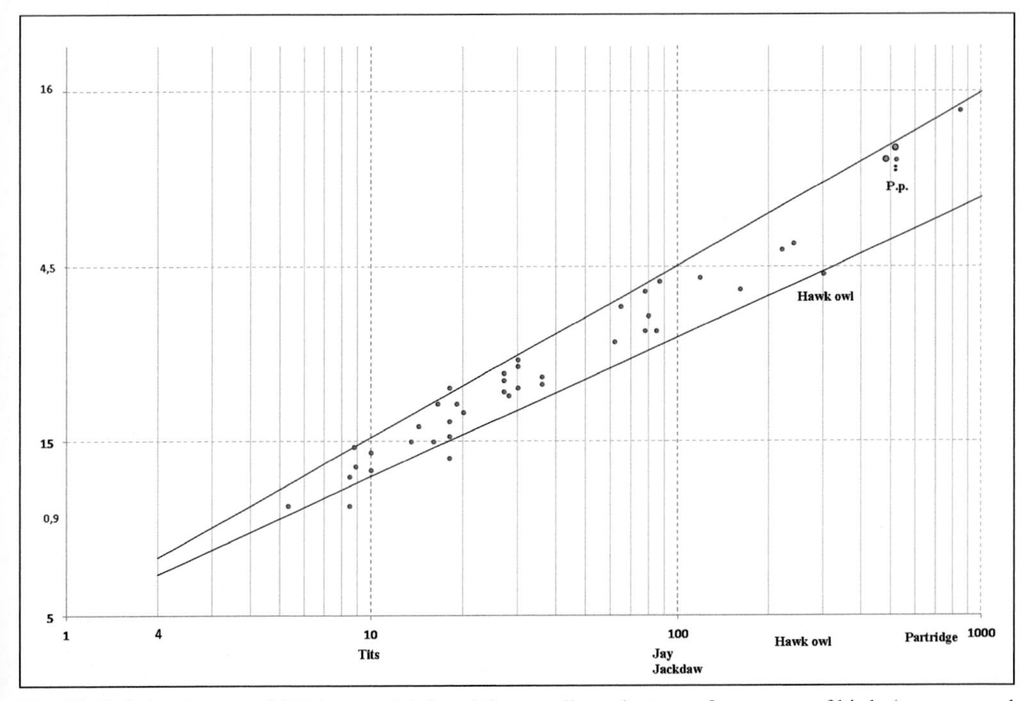

Fig. 70. Relation between M (Mass=weight) and C, overall conductance for a range of birds (some named along M axis). Data from autumn and winter, taken from same observations as in fig.57, p148). Conductance written as a function of M (Mass in gram) on double logarithmic scale gives a line with a slope of less than 0,50. P.p. = partridge.

My aim is here to arrive at a temperature in which the various thermal aspects of the environment viz. air temperature, radiation and wind are combined into a so-called operative temperature (Top) (cf. Stoutjesdijk 2003). Note that a black bulb is sensitive to the above mentioned factors in a similar but not exactly the same way as a bird and that the difference of a fluffy from a closed plumage may be enormous. The operative temperature is thus the product of the environment as well as of the properties of the bird.

6.1.3. Production and flow of energy

The uptake or delivery of energy is dependent upon the thickness of the plumage and the difference between body temperature and operative temperature. It is clear therefore that with a certain operative temperature there is a heat loss (H) that is exactly compensated by the basic metabolic rate (BMR): i.e. the limit below which the metabolic rate does not sink with a healthy bird (Fig. 70). This temperature is called the lower critical temperature (LCT) (see p.143).

Put precisely: H=C(Tb-Tc), H = C (Tb-LCT), where H is the heat loss in Watt and C the thermal conductivity in Watt per degree centigrade, both taken for the whole bird. For the body temperature (Tb) usually 40 °C is taken.

When BMR is approximately proportional with $M^{0.67}$ and C is proportional to $M^{0.50}$, (where M is Mass) then we have:

LCT = 40 - BMR/C is proportional with $M^{0.17}$. For the same bird this means that the LCT is very sensitive to small errors in the power ($^{0.50}$) of M which is used for the conductance. E.g.: when the conductance is not $M^{0.50}$ but $M^{0.60}$, BMR/C is proportional, not to $M^{0.17}$ but to $M^{0.07}$. After Kendeigh c.s., C can only be exactly proportional to $M^{0.5}$. To me, this is a bit far-fetched, but I can understand that $M^{0.5}$ is a rather sharp upper limit, though it should not be written as $M^{0.5000}$ as Kendeigh c.s. do!

Above LCT often a new trajectory is defined. I add here the definitions of Kendeigh c.s. of the "zone of thermoneutrality" where body temperature is regulated primarily by changing the effectiveness of the body insulation and where the BMR is maintained. The term "upper critical temperature" is used at the point where heat stress starts (Beintema c.s., 1995). The changes in heat exchange are effectuated by small changes in the structure of the plumage, an increased respiration or a somewhat changed posture. A minor air movement gives an increased heat transfer, each of these not clearly visible. One must realise here that the bird usually takes up energy on the sun side, while there is a heat-loss on the shadow side. There is a subtle integration of incoming and outgoing energy fluxes. On the sun-side the operative temperature can, with air temperatures below 0°, easily rise over 40 °C, above LCT anyway. Note that the fluffing of the feathers on the sun-side can reduce the total heat load.

After Marchand (1991) the transition from physical to metabolic temperature regulation is much sharper for mammals than for birds, as the mammals are not restricted by plumage which for birds in the first place is made for flying.

When the heat loss is equal to BMR, then BMR/C = Tb-LCT, i.e. the difference between the body temperature and the lower critical temperature. Said otherwise the operative temperature (top) is then equal to LCT. In fact we have gradually reduced the bird to a sphere covered with a homogeneous layer of insulation, in homogeneous surroundings, what concerns radiation and air temperature. The temperature inside the sphere, say 40 °C, is maintained by metabolic processes. The energy produced is higher than BMR when the operative temperature is below the LCT. The temperature range immediately above the LCT is often called the neutral range, probably because symptoms of heat stress cannot be observed. Still, there must be an increased heat loss achieved by small changes in the structure of the plumage and increased respiration, or a somewhat changed posture. A minor change of air movement may facilitate heat transfer, which indeed usually remains unobserved.

In this connection it must be realised that a bird in the sun always absorbs energy on the sun side and dissipates heat on the shadow side. In other words, there is a subtle integration of incoming and outgoing energy fluxes (Fig. 71). Note that the BMR $\approx$ to $M^{0.66}$, while C $\approx$ to $M^{0.45}$. This means that LCT sinks with increasing M, as confirmed by direct measurements.

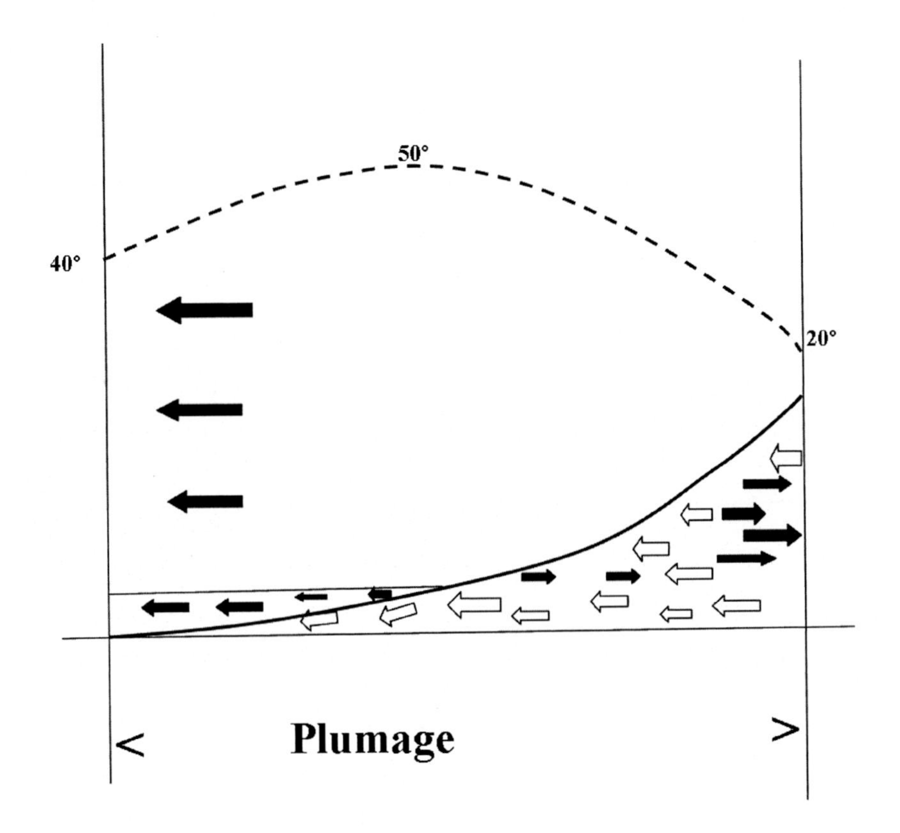

Fig. 71. Schematic section of a bird's plumage in the sun. Heat flow shown by black arrows, solar radiation by open arrows

I must note that passerines have a higher BMR, thus a lower LCT than the non passerines and higher than for all other homeotherms (Fig. 62): 65% after Schmidt Nielson (1997). After the recent literature the difference is 30-50%. Gavrilov (2013) and McNab (2009) give an extensive treatment on the relationship with taxonomical and ecological position. The Passerines become thus less isolated, although a significant difference with non-passerines persists. For clarity I must stress that the differences between the groups depend upon the numerical factors. The proportionality with $M^{0.67}$ is not changed, that is, the BMR of passerines first written as 0,054 x $M^{0.50}$ should become approximately 0,040 x $M^{0.50}$.

Summarising the above we get: 40-BMR/C = LCT.

After Kendeigh c.s. (1977) LCT = 0,3981 $M^{-0.24978}$

Here 0,3981 is a numerical factor, it has nothing to do with the body temperature.

Sixty five years ago our physics professor said: "don't write too many decimals, it makes me think you have not yet fully understood the matter". When we take this advice to heart, we arrive at two different expressions for LCT.

1) (after Kendeigh) : LCT = $40/M^{0.25}$, here 40 is unrelated to body temperature.

2) I propose: LCT = 40° - BMR/C, where 40 stands for the body temperature.

Both expressions look fairly similar, but the first says: LCT = 40/ divided by something. Note that here LCT cannot be below 0 °C. The second expression says: LCT is forty degrees °C minus something.

The first expression (Kendeigh) is based on measurements of metabolic processes by various physiologists. It is thus the most direct one, and not derived from physical principles, but rather correlative i.e. which curve fits best through the collection of data.

The second expression is obtained by taking together the thermal properties of the plumage and a standard expression for the BMR. Here the LCT can be below 0 °C as is the case in real life. The second expression says that the LCT is "40 °C/minus something".

After Kendeigh c.s.1977, the conductance in Watt/C/bird must be exactly proportional with $M^{0.5000}$ for a standard bird. This is equivalent to: $M^{0.1666}$ x $M^{0.667}$ / $M^{0.33}$.

The reasoning of Kendeigh goes as follows (somewhat simplified): the overall conductance C $\approx$ the surface: $M^{0.66}$ and inversely proportional to the thickness of the plumage.
Divided by the specific conductance of the plumage: $1/M^{0.33}$ that would give: C $\approx$ t $M^{0.33}$ LCT = 40 - $\approx$ $M^{0.67}$ / $M^{0.33}$ the thickness of the plumage must vary, as $W^{0.33}$ because the weight of the plumage is proportional to ($\approx$) the weight of the whole bird. Said otherwise: the overall conductance C $\approx$ M 0.3333. It is now assumed the relative thickness of the plumage does not change with the mass of the bird, but that the plumage becomes courser and of a higher specific conductivity for a heavier bird in such a way that C $\approx$ $M^{0.50}$. For the somewhat larger birds the differences in relative thickness of the plumage are in fact rather small, but for the smallest tits etc. the relative thickness of the plumage may reach a value of 1.8 – 2.0 (Fig. 57) The same applies to the haw.

In Fig. 70 the calculated values of C in Joule per degree C per bird are plotted on double logarithmic paper. The average C value is here proportional with $M^{0.44}$, the upper boundary with $M^{0.49}$, which means that doubling the weight (mass), the conductance increases with a factor $2^{0.5}$, the basal metabolic rate with the factor $2^{0.65}$ and the LCT sinks. With weight doubled, C would also increase with a factor two, the LCT would become higher, theoretically impossible and at variance with the experimental evidence. All this together does not help us to explain the expression LCT = $40/M^{0.25}$ which probably is a correlative relation obtained from metabolic measurements provided by various authors (see Fig. 68). LCT is clearly defined here, unfortunately the term cannot be found in the index of the most recent handbooks, at most accidently in the text (Carey, 1996, Whittow, 2000).

In a first approximation the BMR is proportional to $M^{0.66}$ and the conductance of the plumage proportional to $M^{0.50}$, that means the larger the bird, the lower the LCT, as is in accordance with the experimental evidence. For details see BOX.

6.1.4 Operative temperatures in winter

In the following, we leave the "Spielerei" with home-made or theoretical birds alone for a while. By a concrete example from southern Sweden (Fig. 72) (Stoutjesdijk 2003) I first show what is possible as concerns the operative temperature under winter conditions. This example also applies to a "real" winter in the Netherlands. The measurements are made on a sheltered patch covered with snow. On the branches of the surrounding wood every now and then a light air movement can be seen. On the spot where the measurements were taken, the air is stock-still. The sky is cloudless. Scattered groups of juniper bushes with a closed structure are found. The figure gives an idea which (operative) temperatures can possibly be observed by means of measurements on dummies.

With an air temperature of 0.5 °C, Fig. 72 also gives an idea which energy fluxes are responsible for the temperature of the models allowing a realistic estimate of the operative temperatures of small passerines on a bright still winter day with snow. The solar elevation is around 20 °C. On a plane perpendicular to the direct solar radiation, more than 1000 Watt/m^2 solar energy is received, on a horizontal plane about half that amount.

The white surface of the snow reflects about 80% of solar radiation. The "sky", i.e. the clean and cloudless atmosphere has the effect of a cold dome (-32 °C) that emits much less long-wave, (thermal) radiation than the earth in opposite direction. When you add the incoming and outgoing radiation fluxes, long-wave and short-wave, you see that the earth surface receives on

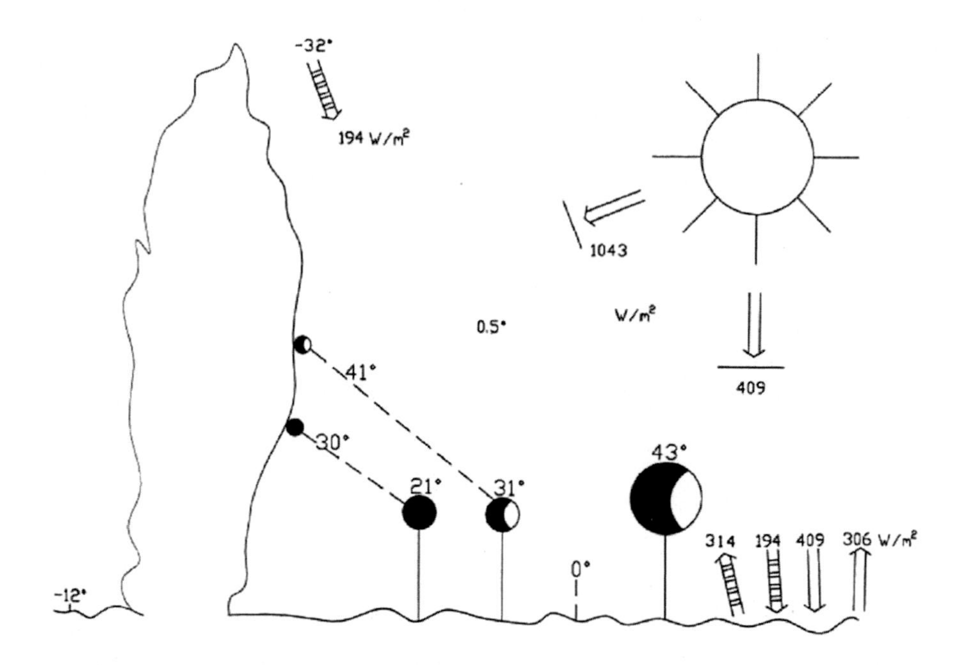

Fig. 72. (from: Stoutjesdijk 2003, Fig. 4.) Radiation conditions and model temperatures on a cloudless and still winter day (air temperature 0.5 ℃). Temperatures inside black bulbs (thin Al, 0.06 mm) and of double bulbs with insulation in between. The frontal part has been removed. The radiation perpendicular to the sun is shown as well as the global radiation on a horizontal plane: 1043W/m^2 resp. 409W/m^2. Also shown long wave (thermal) radiation from the clear sky: 314W/m^2, operative temperature -32 ℃, 306W/m^2 and in the open shadow -12℃, 263W/m^2. The open arrows refer to shortwave radiation (sunlight) and the hatched arrows to long wave (heat radiation).

Note that the temperature excess of the large bulb was 12° higher than that of the small one. The temperature of the small bulbs was 10° higher when close to the Juniper, though not actually touching it, than when freely exposed. This is in the first place due to the fact that the cold sky is replaced by the dense needlework of the Juniper which is about 40° warmer. Furthermore, the needlework reduces transfer by moving air.

the balance almost nothing, but a little bird above the snow surface receives solar energy from all sides.

In the "open shade" the surface temperature is -12 °C (Stoutjesdijk 1974). The overall balance of the surface is negative.

With an air temperature of 0.5 °C, Fig. 72 also shows the temperatures of freely exposed models in this environment under a clear sky. A black single-walled bulb with a diameter of 5.5 cm has a temperature of 20° C. A similar model of the same size, double-walled, is covered with an insulating layer of white artificial fibres. Of the outer surface a segment of 4.4 cm diameter is removed. Even when the fibres and the bulb are seen to be "white", the overall reflectivity of the frontal part is estimated to be about 50% of the received radiation and the temperature of the core is still 30° C, and 10° C above that of the single-walled bulb.

A bulb of double size (13 cm) reaches a temperature of 43 °C, which is not exceptionally high. For a larger object the heat transfer coefficient in W/cm^2 °C becomes smaller. A smooth black bulb of the same size under these conditions would have reached 30°C. The difference between a fluffy and a smooth black surface is the radiation which is absorbed on the black surface and immediately returned as heat to the air. The radiation energy that penetrates into the fibre material is transformed into sensible heat and then "slowly", via conduction transferred to the outer surface (Fig. 69, Fig.70, Stoutjesdijk, 2002).

On the sun-side of the juniper (Fig.72) the "cold dome" of sky (-32 °C) is replaced by a wall of dense needles with a surface temperature of 8 °C that also reflects much more radiation than the weak diffuse skylight that is intercepted. The two small bulbs situated close to the juniper (without touching the branches) here reach temperatures 10° C higher than the exposed bulbs , viz. 30 °C respectively 41 °C, due to the better radiation conditions and inhibited air movement by the presence of the shrub (Stoutjesdijk, 2003). Most "real" birds would not support these temperatures for more than a few minutes.

6.2. Observations on behaviour

6.2.1 Field observations on sunning birds

Fig. 73 is based on my field observations on sunning birds in Sweden and The Netherlands. This diagram shows incidental observations ranged along the vertical scale (temperature). It includes casual field notes, mostly of several species of titmice. The birds were not only sunning but foraging as well. The sun can save metabolic energy but not really replace it. When you observe over a certain stretch of time, you often see a sunning bird alighting towards another sunning place. Usually the sunning is not for longer than 5 minutes, rarely 15 minutes or more. This kind of sunning is in the first place exhibited by tits that try to enhance the effect by seeking the vicinity of dense branches of juniper or spruce. Even with virtually still weather, this behaviour may increase the effect of the sun by up to 10 °C. The vertical scale gives a first impression of the effect of sunning over the trajectory of air temperatures between -20 ° and +30 °C. The strong decrease of sunning behaviour of the titmice above 0 °C is striking. I can imagine that the weather above freezing point is often cloudy or overcast. I have seen gold-crests on the south-side of a pine or spruce which were clearly foraging only, not sunning. Haftorn, (1986) writes that in fact gold-crests, apart from foraging in the sun, never show any real sunning behaviour and he cites other authors who have observed the same. Dots (Fig. 73) along the vertical axis represent the air temperature at which a sunning flock was observed. As all these flocks were observed under the same conditions it seems acceptable for all observations to add 20 °C above the air temperature.

These measurements were made with a black bulb of 5 cm diameter. On the typical sunning spots the temperature of this bulb was always between 18 ° and 24 °C above ambient air temperature.

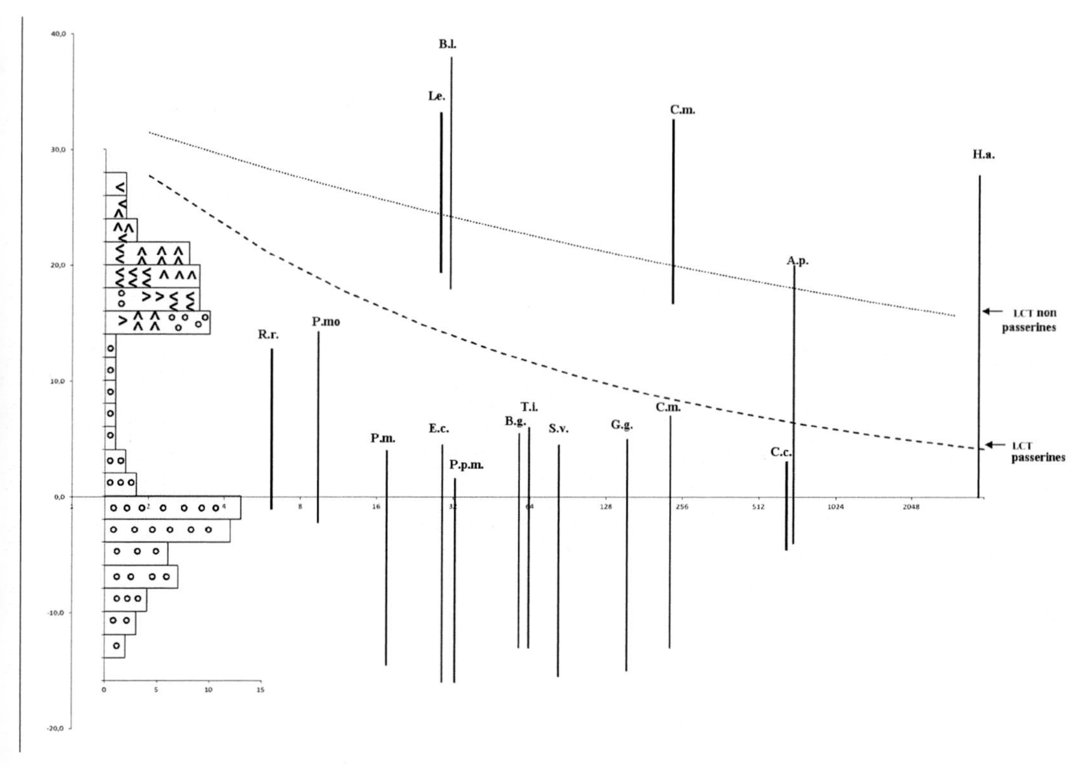

Fig. 73. Various aspects of birds in the sun with air temperature (base of vertical lines), operative temperatures (top of vertical lines) in relation to the Mass in grams of the bird (horizontal axis). Incidental observation of sunning behaviour along the temperature axis (vertical axis). from left to right: Goldcrest (R.r.); 2 Willow tit (P.mo.); Great tit (P.m.); Yellowhammer (E.c. 2x); Lanius cubitor (L.c.); Bullfinch (P.p.m.); Bohemian Waxwing (B.g).; Redwing (T.i.); Great spotted woodpecker (D.m.); Jay (G.g.); Jackdaw (C.m)., 2 x); Starling (S.v.);Black crow (C.c.); Wigeon (A.p.); White-tailed-eagle (H.a).
o: sunning of small birds
x: LCT Passerines
●: LCT non-passerine
In the transverse bars along the vertical axis are shown the various types of heat stress:
<: open beak (gaping) as sign of heat stress
Δ: extreme sunning against parasites (see text). horizotal bar: nr. of observations, see bottom on left side

In the field "temperature-mass" (Fig. 73) I have compiled the data of a number of separate observations as a function of both the air temperature and the mass of the bird. Every observation is represented by a vertical line of which the base indicates the air temperature and the top the operative temperature.

A sunning tit with an air temperature of 0 °C has an estimated operative temperature of around 20 °C, that is an LCT where its metabolism is equal to the BMR at 0 °C, the metabolic rate is at 2 BMR. A jackdaw, roughly ten times the weight of a great tit , can already at 10 °C maintain its body temperature with an energy use of roughly 1 BMR. That means that at -20 °C the metabolic rate is 2 BMR. What matters is the distance in degree C from the body temperature: 40 °C. For the great tit the energy use at – 20 °C is equal to 3 BMR.

A yellow-hammer sunning at -16 °C almost during a full hour reaches an operative temperature which can be estimated to be about 10 °C below LCT. About the same applies to a Swedish bullfinch, (which is much heavier than a Dutch one). In the same category are waxwing, blackbird, redwing, fieldfare, jay and jackdaw.

We are here concerned only with observations longer than 30 minutes. Where I did not observe either the beginning or the end of the sunning, or where low temperatures prevented the operative temperatures to reach the LCT, in principle sunning could last endless. Yellow-hammers occasionally bask under a cloudy sky.

The larger birds are less sensitive to wind (see page 30), and for a black crow (C.c.) the operative temperature can be 8 °C above ambient, with a well visible breeze in the treetop.

Due to the lower BMR for the non-passerines this group feels heat stress only at higher temperatures. The spotted woodpecker is with light frost at a sheltered spot still clearly below its LCT. A wigeon sunning at – 4 °C on a south slope can do that without getting too hot, with an operative temperature of 15 °C and a LCT of about 20 °C.

De Vries & van Eerden (1995) demonstrated that ducks lose much more heat when swimming than on dry ground because the plumage is compressed (that is thinner and thus with an increase conductance). I want to add here that the heat-loss from the feet is greatly reduced thanks to the counter current mechanism: this should not be underestimated. Through measurements on copper "feet" I conclude that the heat transfer in water must be estimated to be roughly 50 times as much as in air with the same temperature difference. This means that a temperature difference of one degree centigrade between the webbed feet and the water would be equivalent to en extra heat loss of roughly one BMR.

What to say of two white-tailed eagles sunning for 45 minutes in an enormous willow tree. There was a light wind with an air temperature of 20 °C. I estimated by means of a dummy the operative temperature to be 30 °C and the LCT 17 °C with a weight of 4 kilogram. There must have been an incoming heat-flow of an estimated conductivity of 30-17 x C Watt/ bird and a temperature rise of about 0.4 °C per hour.

The Spanish author Carrascal c.s. (2001) noted that with air temperatures over 9 °C, tree creepers in the Spanish mountain-woods avoid the sun-side of the pines and retreat more and more into the forest, where they are not easily visible for the sparrow-hawk. The authors speak of a "trade-off" between radiation of the sun and the higher risk of predation by the sparrow-hawk there. I thought more in terms of "too warm": a model of a tree-creeper easily reaches 25 °C above air temperature on the thick bark of an old pine. The surface of the bark is often more than 30 °C above air temperature, e.g. 44 °C with an air temperature of 8 °C! (Stoutjesdijk , 1977). That was the state of affairs when on a cloudless clear day in spring I discovered a tree-creeper in an open patch of old Scotch pine. The bird remained meticulously on the shadow side of the trunk. Reaching the sharp transition shadow-sun it retreated every time, as if an invisible electric wire was drawn there. When I left the wood a sparrow-hawk flashed through my field of view.

It is clear that in the sun and with a somewhat higher air temperature the operative temperature can rise so high that even with the metabolic rate at the lowest level (BMR) the body temperature starts to rise. "Stay out of the sun" the obvious advice would be, but apparently that is not always possible. The first visible signal of heat stress is the open bill (Photo 2). Sometimes it can be seen that the pouch is pumping vigorously (photo 3, grey heron). I once could observe this beautifully on a hazel hen that, accompanied by her chickens, walked into my Swedish garden to eat red currants. I mention in passing that a wet surface, such as the inside of the throat, can lose much more heat than a dry surface, about 6 times as much in a typical case where both surfaces are at 35 °C.

We have now arrived at the conclusion that sunning can be an unavoidable heat-load. You can see jackdaws forage with an ambient temperature of 20 °C in open grassland where the

Photo 2. Jackdaw with minor heat stress (photo P. Stoutjesdijk)
Photo 3. Grey heron (*Ardea cinerea*), pumping gular pouch (photo 123RF, Vladimir Mucibabic)

temperature is several degrees higher, and the radiant temperature still more. It is clear that the metabolic rate is here above BMR anyway and the operative temperature must be so high that the body temperature could exceed 40 °C and certainly does so with an active bird. For a jackdaw of 200 g I calculate, at an air temperature of 20 °C and an operative temperature of 50 °C, a temperature rise of 1 °C in about 30 minutes. When this is the highest acceptable, shadow would be the only solution. This is what we see with jackdaws. They are commuting often between sun and shadow, foraging in the sun for 15 minutes, followed by 15 minutes in the shadow. For a little bird of 12 g the critical sunning time would be about 6 minutes.

Barnacle geese (*Branta leucopsis*) with downy young were observed at an air temperature of 20 °C to switch between sunning and swimming and back about every 5 minutes. The question is here what initiated these switches: probably both the cold water and the sun. Anyway, models covered with a pelt of a young swanskin attain in the sun an excess temperature of 23 – 28 °C above air temperature and a gooslet will easily experience an operative temperature of over 40 °C when the air temperature is 20 °C.

On a warm day (25 °C) in the midday hours I observed that starlings avoided sheltered dry grassland completely, to come back later in the afternoon. Normally they had a strong preference for windy sites for foraging.

6.2.2. Sunning behaviour against parasites

There is a special way of sunning which my wife and I first observed when we with a pair of borrowed grandchildren walked in the extensive semi-natural city-park of Värnamo, Småland. We saw a jackdaw on a branch sunning in a posture that suggested extreme heat-stress, in a con-torted position, hanging wings and head askew; everything suggested that a complete succum-bing was nearby. My wife pointed out the dangers of excessive sunbathing to the children. The elder girl was so impressed that she immediately started to alarm the whole jackdaw community. I saw this type of sunning more often in the garden of the main "konditori" of Växjö, Sweden, where they sat with spread wings on the roof (Photo 4). Orthey sat with some difficulty on the iron supports of the lanterns on the terrace. I thought it could be explained as a risk connected with greed for stealing unguarded Wienerbröd or Princess Torte.

Photo 4. Jackdaw spread on roof (photo P. Stoutjesdijk)

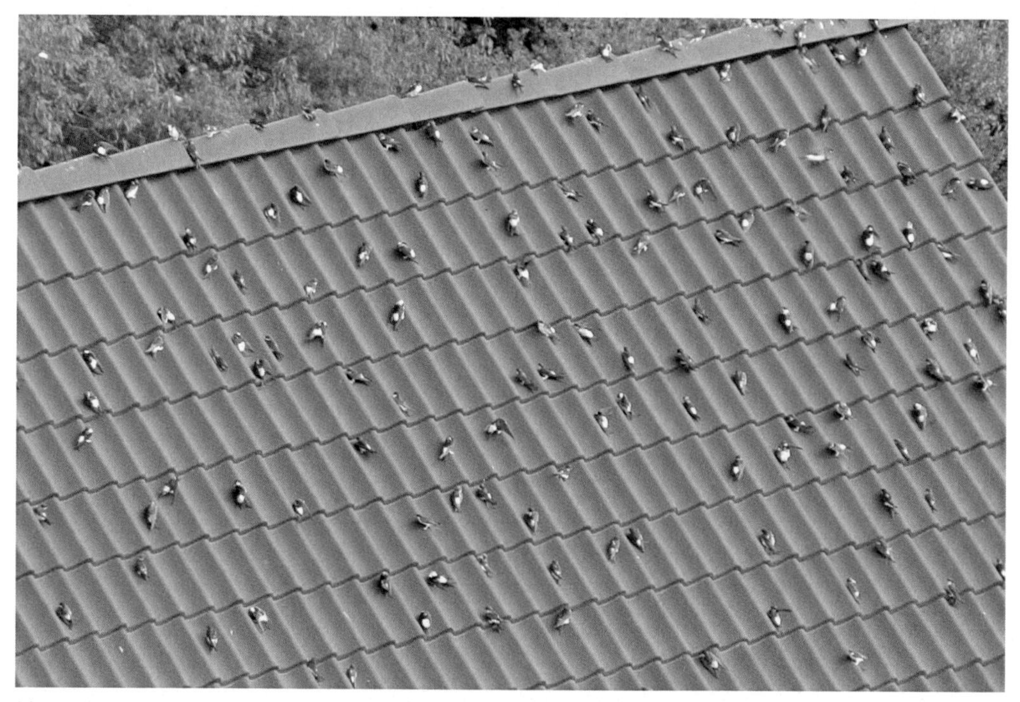

Photo 5. Martins (*Delichon urbica*), sunning on a roof (photo Yvonne Stoutjesdijk)

Photo 5a. Martins, sunning on a roof (photo Yvonne Stoutjesdijk)
Photo 6. Blackbird, male, sunning (photo P. Stoutjesdijk)

cess Torte. See also photo 5, sunning martins. At first I could not find anything like an explanation in the literature and I satisfied myself with the idea that human sunbathing also may run out of control. Finally I found in the dictionary of ornithology (Landsborough Thomson, 1965) a section under the head "feather maintenance" describing these forms of sunning (p.282). Summarising: "when performing the more extreme postures in the full heat of the sun, birds may gape and often appear to have collapsed in distress." The same author remarks elsewhere that the solar heat may well cause ectoparasites to become active in the plumage. Waldbauer (1998) writes that "the heating of the plumage makes lice and other ectoparasites uncomfortable, causing them to move and making them more accessible to the bill of the preening bird" (Photo 7).

Münch (2003) describes for crossbills "ein hochritualisierte Handlung" of which once started by one bird, more and more birds take part and that lasts up to 5 minutes. With the back to the sun, the head bent forward, the bill wide open, permitting the radiation to penetrate into the corners of the mouth and through the erected plumage to the skin. The bird remains 5 minutes motionless, the eye always open directed to the sun with only every now and then a movement of the eyelid. The excessive fainting scenes by jackdaws were performed in Sweden.

A similar behaviour was seen with blackbirds, in the years 2010 and 2011, more rarely in 2012 and 2013, from the kitchen window in Ossendrecht. It could be seen (Photo 6) how various essential parts were bathed in the sun and also that the treatment of the whole bird did not last more then 2-5 minutes. This probably was enough and would not be sustained longer. I got the impression that the same bird came back several times, sometimes I was sure because the bird had recognisable marks. A typical series would last about one hour in which the sunning lasted 2-7 minutes with interruptions of a few minutes. Usually the bill was opened after 2-3 minutes. On the same sunning place and in the same way, sunning behaviour was observed by hedge-sparrow, robin and occasionally great tit, always during short periods, indicating it to be an action in order to extricate parasites from their coat. (Note also the passage on crossbills, Münch p.175). The blackbirds were always sunning on the same spot, sheltered by the house and dense shrub. The soil was covered with a thick layer of dry leaves with temperatures (in the sun) of 20-25 °C above that of the air. For the underside of the bird this would mean an increase of the operative temperature of 10 °C as compared with the substrate at air temperature. A grey heron was also observed sunbathing in a similarly characteristic posture (photo 8).

About the temperature inside or on plumage I did not find reliable information in the literature. I made some measurements in the centimeters deep down on the skin of the rump of a cock (*Gallus gallus*) that was glued upon a thin walled vessel filled with water at 40 °C. With a fine thermocouple I measured a temperature of 65 – 69 °C at about 3 mm from the surface of the skin, with an air temperature of 20 °C. At the outer surface of the plumage I measured under the same conditions up to 60 °C. Here it must be realised that the body temperature, said otherwise the temperature of the "heat sink" under the insulating layer, is 40 °C.

A paper by Dubska c.s. (2011) that recently came to my attention fits well into the above. Common city-birds such as blackbirds, song thrushes and to a lesser degree robin and great tit are important hosts for the tick *Ixodes ricinus*, in its turn bearer of *Borrelia afzelii* transmitting Lyme disease.

The evident importance of sunning in this connection is not mentioned, neither is it referred to in publications by Waldbauer (1998) and Allander and Dufva (1997), which both discuss extensively the importance of common songbirds as hosts to *Ixodes*.

Photo 7. Blackbird removing parasites (photo P. Stoutjesdijk)

Photo 8. Grey heron, sunning (photo 123RF, Ronald IJdema)

Epilogue

"Nothing so difficult as the beginning in poetry except perhaps the end" Byron

Looking at this book on microclimate with as much of an open mind as we can, we are inclined to say it has become an attempt to understand the links between climate and life. (This includes man, although he was left almost entirely outside this book.) The first step was to understand how the 'minor weather' in the immediate surrounding of plants and animals is connected with large scale weather. Approaching the problem from the opposite direction means studying the relations of plants and animals with their immediate environment.

On the way from microweather observations to microclimate descriptions we have made only a few steps. General trends can be recognized e.g. in the situation with certain weather types and in extremes to be expected.

F. Wilmers has made a valuable attempt to give a climate description with the microclimate in mind. He characterized the climate of NW Germany by the frequency of the occurrence of several weather types, chosen especially because of their obvious relations with the situation in the microclimate. Simplified, the following weather types are distinguished.

Radiation type
Strong solar radiation, without extremely high intensities. Few clouds, weak winds. Dry air. Typical for high-pressure situations.

Squall type
Radiation intensity changeable with very high peaks through reflection by high Cumulus clouds. The high radiation intensities are usually combined with a wet surface. Wind rather strong and changeable.

Neutral type
Overcast with Stratus clouds. Low radiation intensity. Often very high air humidity.

Cyclonic type
Cloudy, low radiation intensity. Windy.

Dependent on the weather type, an enormously differentiated pattern of microclimates is formed near the soil. This is mainly through the effect of the vegetation but is also influenced by soil and landscape types. It hardly seems possible to characterize this complex in a simple way, which at the same time, affords an overview. Only rigorous simplification and schematization may perhaps disclose some general traits.

A first attempt might be to look at the climate-vegetation complex as a whole, i.e. follow the most characteristic traits of the microclimate for each biome from desert via steppe, mediterranean evergreen vegetation through deciduous forest, taiga to tundra. At the moment we cannot do more than illustrate this idea with a few examples. As a first simplification we start with radiation weather. We can say that in the arid zones microclimate is invariably dry and warm. 'Dry and warm' are taken in a relative sense i.e. drier and warmer than in the free air.

It seems to be a general rule that where there is a lack of water, even if this is of edaphical origin, this is aggravated in the microclimate.

In the deciduous forest zone the predominant microclimate seems to be humid and cool or with, at most, a small temperature excess, but the range is wide here and warm and dry microclimates are found.

Near the alpine tree limit the direct sunshine is great, open shade effects are strong, high surface temperatures are common because the park-like landscape gives rise to sunny wind-sheltered spots with much organic litter which warms up quickly and to high temperatures.

More details could be added here about the range of situations that are realized and about the frequency of different types of microclimate. Furthermore the pattern of differentiation may be coarse or fine-grained.

Of course, much natural diversity has been destroyed, also in this respect, and new situations have been created as well. Often, conditions are forced upon the landscape which are originally strange to the climatic zone in question or very rare under natural conditions. Quite common are the creation of steppe-like conditions in the temperate zone. On the other hand, we have irrigation under arid conditions. We might speak here of extrazonal microclimates.

As for the role of animals, we have long been inclined to draw a parallel with the statement of a well-known Dutch psychiatrist, who said "Ordinary people build castles in the air, but the fools live in them (and the psychiatrists collect the rent)". So it seems that the plants make the microclimates while animals, with a few exceptions like mound building, have only a passive role. The role of large and small herbivores in determining vegetation structure is however more and more recognized as an important one and so we may assume that in the natural landscape even the differentiation of the microclimate may owe much to the fauna (from cattle to caterpillars).

Considering microclimate from the view point of plants and animals it may be stated that some of them often live close to the limit of what they can endure. Microclimate extremes are thus a meaningful aspect of their living conditions. On the other hand, some plants and animals can only live in special microclimate conditions or they need them in decisive moments of their lives.

Finally the effect of the local variation of the microclimate is such that development rates are quite different and the phenological stages are reached with great differences in time from place to place with the result that the spatial diversity creates diversity in time as well.

References

* Textbooks with a general bearing on the topic of this book

Anderson, M. C. 1964. Light relations of terrestrial plant communities. *Biol. Rev.* 39: 425-486.
Anderson, M. C. 1966. Some problems of simple characterization of the light climate in plant communities. In: Bainbridge, R., Evans, G. C. & Rackham, O. (eds.) *Light as an ecological factor*, pp. 77-90. Blackwell, Oxford.
Apinis, A. E. 1965. Thermophile Microorganismen in einigen Dauergrünlandgesellschaften. In: Tuxen, R. (ed.) *Biosoziologie*, pp. 290-303. Junk, The Hague.
Arnold, S. M. & Monteith, J. L. 1974. Plant development and mean temperature in a Teesdale habitat. *J. Ecol.* 62: 711-720.
Aulitzky, H., Czell, A., Fromme, G., Neuwinger, I., Schiechtl, H. M. & Stern, R. 1961. Beschreibung des Gurglertales (hinterstes Ötztal in Nordtirol). *Mitt. Forstl. Bundesversuchsanst. Mariabrunn* 59: 33-52.
Avery, R. A., Bedford, J. D. & Newcombe, C. P. 1982. The Role of Thermoregulation in Lizard Biology: Predatory Efficiency in a temperate diurnal Basker. *Behav. Ecol. Sociobiol.* 11: 261-267.
Baker, R. R. 1978. *The evolutionary ecology of animal migration.* Hodder & Stoughton, London.
Bakken, G. S. 1980. The use of standard operative temperature in the study of the thermal energetics of birds. *Physiol. Zool.* 53: 108-119.
Bakken, G. S. 1990. Estimating the effect of wind on avian metabolic rate with standard operative temperature. *Auk* 107: 587-594.
Bakken, G. S. & Gates, D. M. 1975. Heat transfer analysis of animals. In: Gates, D. M. & Schmeil, R. B. *Perspectives of biophysical ecology*, pp. 255-290. Springer, Berlin.
Bakken, G. S., Vanderbilt, V. C., Buttemer, W. A. & Dawson, W. R. 1978. Avian eggs: thermoregulatory value of very high near-infrared reflectance. *Science* 200: 321-323.
Bakker, D. 1960. *Senecio congestus* (R. Br.) DC. in the Lake IJselpolders. *Acta Bot. Neerl.* 9: 235-259.
*Bannister, P. 1976. *Introduction to Physiological Plant Ecology.* Blackwell, Oxford.
Barkman, J. J. 1949. Notes sur quelques associations épiphytiques de la Petite Suisse Luxembourgeoise. *Arch. Inst. Gr.-Duc Lux. N.S.* 18: 79-94.
Barkman, J. J. 1951. Impressions of the North Swedish Forest Excursion. *Vegetatio* 3: 175-182.
Barkman, J. J. 1958. *Phytosociology and Ecology of Cryptogamic Epiphytes.* Van Gorcum, Assen.
Barkman, J. J. 1965a. Die Kryptogamenflora einiger Vegetationstypen in Drenthe und ihr Zusammenhang mit Boden und Mikroklima. In: Tüxen, R. (ed.) *Biosoziologie*, pp. 157-171. Junk, The Hague.
Barkman, J. J. 1977. Die Erforschung des Mikroklimas in der Vegetation. Theoretische und methodische Aspekte. In: Dierschke, H. (ed.) *Vegetation und Klima*, pp. 5-20. Cramer, Vaduz.
Barkman, J. J. 1979. The investigation of vegetation texture and structure. In: Werger, M. J. A. *The Study of Vegetation*, pp. 123-160. Junk, The Hague.
Barkman, J. J. 1985. Geographical variation in associations of juniper scrub in the central European plain. *Vegetatio* 59: 67-71.
Barkman, J. J. 1988. A new method to determine some characters of vegetation structure. *Vegetatio* 78: 81-90.
Barkman, J. J., Masselink, A. K. & de Vries, B. W. L. 1977. Uber das Mikroklima in Wacholderfluren. In: Dierschke, H. (ed.) *Vegetation und Klima*, pp. 35-81. Cramer, Vaduz.
Barkman, J. J. & Stoutjesdijk, Ph. 1987. *Microklimaat, Vegetatie en Fauna.* Pudoc, Wageningen.
Barkman, J. J. & Westhoff, V. 1969. Botanical evaluation of the Drenthian District. *Vegetatio* 19: 330-388.
Barton-Browne, L. B. 1964. Water regulation in insects. *Annu. Rev. Entomol.* 9: 63-82.
Baskin, J. M. & Baskin, C. C. 1974. Some ecophysiological aspects of seed dormancy in *Geranium carolinianum* L. from Central Tennessee. *Oecologia (Berl.)* 16: 209-219.
Begon, M., Harper, J. L. & Townsend, C. R. 1986. *Ecology - Individuals, Populations and Communities.* Blackwell, Oxford.
Behrens, W., Hoffmann, K. H., Kempa, S., Gässler, S. & Merkel-Wallner, G. 1983. Effects of diurnal thermoperiods and quickly oscillating temperatures on the development and reproduction of crickets, *Gryllus bimaculatus. Oecologia (Berl.)* 59: 279-287.
Bengtsson, K., Prentice, H. C., Rosén, E., Moberg, R. & Sjögren, E. 1988. The dry alvar grasslands of Öland: ecological amplitudes of plant species in relation to vegetation composition. *Acta Phytogeogr. Suec.* 76: 21-46.
Berdowski, J. J. M & Zeilinga, R. 1983. The effect of the heather beetle (*Lochmaea suturalis* Thomson) on *Calluna vulgaris* (L.) Hull as a cause of mosaic patterns in heathlands. *Acta Bot. Neerl.* 32: 250-251.
*Berenyi, D. 1967. *Mikroklimatologie.* Fischer, Stuttgart.
Berglind, S. Å. 1988. Sandödlan, *Lacerta agilis* L., på Brattforsheden in Värmland (with summary). *Fauna Flora* 83: 241-255.
Bernatzky, A. 1978. *Tree ecology and preservation.* Elsevier, Amsterdam.
Billings, W. D. & Morris, R. J. 1951. Reflection of visible and infrared radiation from leaves of different ecological groups. *Am. J. Bot.* 38: 327-331.

Bongers, F. & Popma, J. 1988. *Trees and gaps in a Mexican tropical rain forest.* Diss., Univ. of Utrecht, Utrecht.
Booth, T. H. 1990. Mapping regions climatically suitable for particular tree species at the global scale. *Forest Ecol. Manage.* 36: 47-60.
Bornkamm, R., Lee, J. A. & Seward, M. R. D. (eds.) 1982. *Urban Ecology.* Blackwell, Oxford.
Bossenbroek, Ph., Kessler, A., Liem, A. S. N. & Vlijm, L. 1977a. The significance of plant growth-forms as shelter for terrestrial animals. *J. Zool. Lond.* 182: 1-6.
Bossenbroek, Ph., Kessler, A., Liem, A. S. N. & Vlijm, L. 1977b. An experimental analysis of the significance of tuft structure as a shelter for the invertebrate fauna, with respect to wind velocity and temperature. *J. Zool. Lond.* 182: 7-16.
Boyko, H. 1962. Old and new principles of phytobiological climate classification. In: Tromp, S. W. (ed.) *Biometeorology*, pp. 113-127. Pergamon Press, Oxford.
Bradley, E. F. 1969. A small sensitive anemometer system for agricultural meteorology. *Agric. Meteorol.* 6: 185-193.
Brandt, D. C. 1980a. Is the mound of *Formica polyctena* Foerst. in origin a simulation of a rock? *Oecologia (Berl.)* 44: 281-282.
Brandt, D. C. 1980b. The thermal diffusivity of the organic material of a mound of *Formica polyctena* Foerst. in relation to the thermo-regulation of the brood. *Neth. J. Zool.* 30: 326-344.
Brasseur, F. & de Sloover, J. R. 1976. L'extinction du rayonnement dans les gammes spectrales bleu, rouge et rouge lointain. Comparaison de deux peuplements forestiers de Haute Ardenne. *Bull. Soc. Roy. Bot. Belg.* 109: 319-334.
Brown, L. 1976. *Birds of prey.* Hamlyn, London.
Brunt, D. 1932. Notes on radiation in the atmosphere. *Quart. J. Roy. Meteorol. Soc.* 58: 389-418.
Buchanan, J. W. 1940. Developmental rate and alternating temperatures. *J. Exp. Zool.* 83: 235-248.
*Budyko, M. J. 1974. *Climate and life.* Academic Press, New York, London.
Bullock, T. H. & Cowles, R. B. 1952. Physiology of an infra-red receptor. The facial pit of pit vipers. *Science* 115: 541-543.
Businger, J. A. 1975. Aerodynamics of vegetated surfaces. In: de Vries, D. A. & Afgan, N. H. (eds.) *Heat and mass transfer in the biosphere*, pp. 139-166. Scripta Book Co., Washington DC.
Büttner, R. 1971. Untersuchungen zur Ökologie und Physiologie des Gasstoffwechsels bei einigen Strauchflechten. *Flora* 160: 72-99.
Büttner, K. J. & Kern, C. D. 1965. The determination of infrared emissivities of terrestrial surfaces. *J. Geophys. Res.* 70: 1329-1337.
Byman, D., Hay, D. B. & Bakken, G. S. 1988. Energetic costs of the winter arboreal microclimate: the gray squirrel in a tree. *Int. J. Biometeorol.* 32: 112-122.
Byrne, G. F. & Rose, C. W. 1972. On the determination of vertical fluxes in field crop studies. *Agric. Meteorol.* 10: 13-17.
*Campbell, G. S. 1977. *An Introduction to Environmental Biophysics.* Springer-Verlag, New York.
Cannell, M. G. R. & Smith, R. I. 1983. Thermal time, chill days and prediction of bud burst in *Picea sitchensis*. *J. Appl. Ecol.* 20: 951-963.
Capon, B. & Asdall, W. V. 1967. Heat pretreatment as a means of increasing germination of desert annual seeds. *Ecology* 48: 305-306.
Caprio, J. M. 1974. The solar thermal unit concept in problems related to plant development and potential evapotranspiration. In: Lieth, H. (ed.) *Phenology and seasonality modeling*, pp. 353-366. Springer, Berlin.
Cernusca, A. 1976. Bestandesstruktur, Bioklima und Energiehaushalt von alpinen Zwergsträuchern. *Oecol. Plant.* 11: 71-102.
Chappell, M. A. 1981. Standard operative temperatures and costs of thermoregulation in the arctic ground squirrel *Spermophilus undulatus*. *Oecologia (Berl.)* 49: 397-403.
Chappell, M. A. & Bartholomew, G. A. 1981. Standard operative temperatures and thermal energetics of the antelope ground squirrel *Ammospermophilus leucurus*. *Physiol. Zool.* 54: 81-93.
Chappell, M. A., Morgan, K. R. & Bucher, T. L. 1990. Weather, microclimate and energy costs of thermoregulation for breeding Adélie penguins. *Oecologia (Berl.)* 83: 420-426.
Chazdon, R. L. & Fetcher, N. 1984. Light environments of tropical forests. In: Medina, E., Mooney, H. A. & Vazquez-Yánes, C. (eds.) *Physiological ecology of plants of the wet tropics*, pp. 27-36. Junk, The Hague.
Chiariello, N. 1984. Leaf energy balance in the wet lowland tropics. In: Medína, E., Mooney, H. A. & Vazquez-Yánes, C. (eds.) *Physiological ecology of plants of the wet tropics*, pp. 85-98. Junk, The Hague.
Chrenko, F. A. & Pugh, L. G. C. E. 1961. The contribution of solar radiation to the thermal environment of man in Antarctica. *Proc. Roy. Soc. B.* 155: 243-265.
Clark, J. A. & Wigley, G. 1975. Heat and mass transfer from real and model leaves. In: de Vries, D. A. & Afgan, N. H. (eds.) *Heat and mass transfer in the biosphere*, pp. 413-426. Scripta Book Co., Washington DC.
Clark, L. 1987. Thermal constraints on foraging in adult European starlings. *Oecologia (Berl.)* 71: 233-238.
Cloudsley-Thompson, J. L. & Chadwick, M. J. 1964. *Life in Deserts.* Foulis, London.
Cone, C. D. 1962. The soaring flight of birds. *Sci. Am.* 206: 130-140.
Corbet, S. A., Willmer, P. G., Beament, J. W. L., Unwin, D. M. & Prys-Jones, O. E. 1979. Postsecretory determinants of nectar sugar concentration. *Plant Cell Environ.* 2: 293-308.

Coulianos, C.-C. 1962. Djur och mikroklimat. *Zool. Revy* 24: 58-70.
Coulson, K. L. 1975. *Solar and terrestrial radiation.* Academic Press, New York.
Creveld, M. C. 1981. *Epilithic lichen communities in the alpine zone of Southern Norway.* Diss., University of Utrecht, Utrecht.
Dahl, E. & Mork, E. 1959. On the relationships between temperature, respiration and growth in Norway Spruce (*Picea abies* (L.) Karst.). *Meddel. Norsk. Skogsforsøksv.* 53.
Dawson, W. R. 1975. On the physiological significance of the preferred body temperature of reptiles. In: Gates, D. M. & Schmeil, R. B. (eds.) *Perspectives of biophysical ecology*, pp. 443-473. Springer, Berlin.
de Felice, P. 1968. Etude des échanges de chaleur entre l'air et le sol sur deux sols de nature différente. *Arch. Meteorol. Geophys. Bioklimatol.* B 16: 70-80.
de Fluiter, H. J., van de Pol, P. H. & Woudenberg, J. P. M. 1963. Fenologisch en faunistisch onderzoek over boomgaardinsekten. *Versl. Landbouwk. Onderz.* 69: 14. Wageningen.
de Vries, D. A. 1963. Thermal properties of soils. In: van Wijk, W. R. (ed.) *Physics of plant environment*, pp. 210-235. North-Holland Publishing Company, Amsterdam.
de Vries, D. A. & Venema, H. J. 1954. Some considerations on the behaviour of the Piche evaporimeter. *Vegetatio* 5-6: 225-234.
den Boer, P. J. 1961. The ecological significance of activity patterns in the woodlouse *Porcellio scaber* Latr. (*Isopoda*). *Arch. Neerl. Zool.* 14: 283-409.
den Boer, P. J. 1967. Zoologisch onderzoek op het Biologisch Station Wijster 1959-1967. *Med. Bot. Tuin. Belmonte Arbor. Wageningen* 9: 161-181.
Dierschke, H. 1974. Saumgesellschaften im Vegetations- und Standortsgefälle an Waldrändern. *Scripta Geobot.* 6: 1-234.
Dierschke, H. 1977. Bibliografie der Arbeiten über das Mikroklima in europäischen Pflanzengesellschaften. *Excerpta Bot.* B 16: 179-234.
Dirmhirn, I. 1964. *Das Strahlungsfeld im Lebensraum.* Akademische Verlagsgesellschaft, Frankfurt am Main.
Dobkin, D. S. 1985. Heterogeneity of tropical flower microclimates and the response of humming bird flower mites. *Ecology* 66: 536-543.
Dobkin, D. S., Olivieri, J. & Ehrlich, P. R. 1987. Rainfall and the interaction of microclimate with larval resources in the population dynamics of checkerspot butterflies (*Euphydryas editha*) inhabiting serpentine grassland. *Oecologia (Berl.)* 71: 161-166.
Dreisig, H. 1980. Daily activity, thermoregulation and water loss in the tiger beetle, *Cicindela hybrida. Oecologia (Berl.)* 44: 376-389.
*Eckert, E. 1959. *Einführung in den Wärme- und Stoffaustausch.* Springer, Berlin.
Edney, E. B. 1957. *The Water Relations of Terrestrial Arthropods.* Cambridge University Press, Cambridge.
Ehleringer, J. R. & Björkman, O. 1978. Pubescence and leaf spectral characteristics in a desert shrub, *Encelia farinosa. Oecologia (Berl.)* 36: 151-162.
Elkins, N. 1983. *Weather and bird behaviour.* Poyser, Calton.
Ellenberg, H. 1974. Zeigerwerte der Gefäßpflanzen Mitteleuropas. *Scripta Geobot.* 9.
Ellenberg, H. 1988. *Vegetation ecology of central Europe.* Translation from the German: Vegetation Mitteleuropas mit den Alpen. Cambridge University Press, Cambridge.
Elton, C. E. 1966. *The pattern of animal communities.* Methuen, London.
*Etherington, J. R. 1982. *Environment and Plant Ecology.* 2nd. ed. Wiley & Sons, Chichester.
Evans, K. E. & Moen, A. N. 1975. Thermal exchange between Sharp-Tailed Grouse (*Pedioecetes phasianellus*) and their winter environment. *Condor* 77: 160-168.
Evans, W. G. 1966. Perception of infrared radiation from forest fires by *Melanophila acuminata* de Geer (*Buprestidae, Coleoptera*). *Ecology* 47: 1061-1065.
Faliński, J. B. 1978. Uprooted trees: their distribution and influence in the primeval forest biotope. *Vegetatio* 38: 175-184.
Faliński, J. B. 1986. *Vegetation dynamics in temperate lowland primeval forests.* Junk, Dordrecht.
Fanger, P. O. 1972. *Thermal Comfort.* McGraw Hill, New York.
Fetcher, N., Oberbauer, S. F. & Strain, B. R. 1985. Vegetation effects on microclimate in lowland tropical forest in Costa Rica. *Int. J. Biometeorol.* 29: 145-155.
Firbas, F. 1931. Untersuchungen über den Wasserhaushalt der Moorpflanzen. *Jahrb. Wissensch. Bot.* 74: 455-496.
Forseth, I. N. & Ehrelinger, J. R. 1982. Ecophysiology of two solar tracking desert winter annuals. *Oecologia (Berl.)* 54: 41-49.
Fridén, L. 1965. *Stipa pennata* and its Companions in the Flora of Västergötland. *Acta Phytogeogr. Suec.* 50: 161-166.
Friedel, H. 1961. Schneedeckendauer und Vegetationsverteilung im Gelände. *Mitt. Forstl. Bundesversuchsanst. Mariabrunn* 59: 317-369.
Fritschen, L. J. & Gay, L. W. 1979. *Environmental instrumentation.* Springer, New York.
Gärdefors, D. 1966. Temperature-humidity 'organ' experiments with three species of grasshopper belonging to the family of *Acrididae. Entomol. Exp. Appl.* 9: 395-401.
*Gates, D. M. 1980. *Biophysical ecology.* Springer, New York.
Gates, D. M. & Papian, L. E. 1971. *Atlas of energy budgets of plant leaves.* Academic Press, London, New York.

Gates, D. M. & Tantraporn, W. 1952. The reflectivity of deciduous trees and herbaceous plants in the infrared to 25 microns. *Science* 115: 613-616.
Gauslaa, Y. 1984. Heat resistance and energy budget in different Scandinavian plants. *Holarct. Ecol.* 7: 1-78.
Geiger, R. 1941. Das Standortklima in Altholznähe. *Mitt. Akad. Dt. Forstwiss.* 1: 148-172.
*Geiger, R. 1961. *Das Klima der bodennahen Luftschicht.* Vieweg, Braunschweig.
*Geiger, R. 1965. *The climate near the ground.* Harvard University Press, Harvard MA. (Translation of Geiger 1961.)
Gessner, F. 1956. Der Wasserhaushalt der Hydrophyten und Helophyten. In: *Handbuch der Pflanzenphysiologie* III, pp. 853-901. Springer, New York.
Gimingham, C. H. 1975. *Heathland Ecology.* Oliver & Boyd, Edinburgh.
Gimingham, C. H. 1988. A reappraisal of cyclical processes in *Calluna* heath. *Vegetatio* 77: 61-64.
Givnish, T. L. & Vermey, G. J. 1976. Sizes and shapes of lianas leaves. *Am. Nat.* 110: 743-778.
Gjessing, Y. & Øvstedal, D. O. 1989. Microclimates and water budget of algae, lichens and a moss on some nunataks in Queen Maud Land. *Int. J. Biometeorol.* 33: 272-281.
Gonschorrek, J. 1977. Aushagerungserscheinungen im *Luzulo-Fagetum* des Wesergebirges. In: Dierschke, H. (ed.) *Vegetation und Klima*, pp. 117-125. Cramer, Vaduz.
Gonzalez, R. R., Nishi, Y. & Gagge, A. P. 1974. Experimental evaluation of standard effective temperature: A new biometeorological index of man's thermal discomfort. *Int. J. Biometeorol.* 18: 1-15.
Goudriaan, J. 1977. *Crop micrometeorology: a simulation study.* Pudoc, Wageningen.
*Grace J. 1977. *Plant response to wind.* Academic Press, London.
Grace, J. 1981. Plants and wind. In: Grace, J., Ford, E. D. & Jarvis, P. G. (eds.) *Plants and their atmospheric environment.* Blackwell, Oxford.
Greenstone, M. H. 1990. Meteorological determinants of spider ballooning: the roles of thermals vs. the vertical windspeed gradient in becoming airborne. *Oecologia* 84: 164-168.
Grime, J. P. 1979. *Plant strategies and vegetation processes.* John Wiley & Sons, Chichester.
Grubb, T. C. 1975. Weather-dependent foraging behaviour of some birds wintering in a deciduous woodland. *Condor* 77: 175-182.
Grubb, T. C. 1977. Weather-dependent foraging behaviour of some birds wintering in a deciduous woodland: horizontal adjustments. *Condor* 79: 271-274.
Guyot, G. & Sequin, B. 1975. Modification of land roughness and resulting microclimatic effects: a field study in Brittany. In: de Vries, D. A. & Afgan, N. H. (eds.) *Heat and mass transfer in the biosphere,* pp. 467-478. Scripta Book Comp., Washington DC.
Gyllin, R., Källander, H. & Sylvén, M. 1977. The microclimate explanation of town centre roosts of Jackdaws *Corvus monedula. Ibis* 119: 358-361.
Hallander, H. 1970. Environments of the wolfspiders *Pardosa chelata* (O. F. Müller) and *Pardosa pullata* (Clerk). *Ekol. Pol.* 18: 41-72.
Heckert, L. 1959. Die klimatischen Verhältnisse in Laubwäldern. *Z. Meteorol.* 13: 211-223.
Heller, H. C. 1972. Measurements of convective and radiative heat transfer in small mammals. *J. Mammol.* 53: 289-295.
Henwood, K. 1975. Field tested thermoregulation model for two diurnal Namib desert tenebrionid beetles. *Ecology* 56: 1329-1342.
Herter, K. 1962. *Der Temperatursinn der Tiere.* A. Ziemsen Verlag, Wittenberg.
Hertz, P. E. & Nevo, E. 1981. Thermal biology of four Israeli agamid lizards in early summer. *Israel J. Zool.* 30: 190-210.
Higgins, L. G. & Riley, N. D. 1970. *Field guide to the butterflies of Britain and Europe.* Collins, London.
Hill, R. W., Beaver, D. L. & Veghte, J. H. 1980. Body surface temperatures and thermoregulation in the Black-capped chickadee (*Parus atricapillus*). *Physiol. Zool.* 53: 305-321.
Holmes, S. (ed.) 1985. *Henderson's Dictionary of Biological Terms.* 9th ed. paperback. Longman, London.
Holzer, K. 1959. Winterliche Schäden an Zirben nahe der alpinen Baumgrenze. *Centralbl. Ges. Forstwesen* 76: 232-244.
Horn, H. S. 1971. *The adaptive geometry of trees.* Princeton University Press, Princeton NJ.
Hundertmark, A. 1939. Uber das Luftfeuchtigkeitsunterscheidungsvermögen und die Lebensdauer der drei in Deutschland vorkommenden Rassen von *Anopheles maculipennis* (*atroparvus, messeae, typicus*) bei verschiedenen Luftfeuchtigkeitsgraden. *Z. Angew. Entomol.* 25: 125-141.
Jackson, H. C. 1978. Low May sunshine as a possible factor in the decline of the Sand Lizard in North-West England. *Biol. Conserv.*13: 1-12.
Jaeggli, M. 1943. Muschi arboricoli del cantone Ticino. *Rev. Bryol. Lich.* 6: 23-67.
Jakovlev, V. 1959. Mikroklimatische Untersuchungen in einigen Acrididenbiotopen. *Z. Morphol. Ökol. Tiere* 48: 89-101.
Jarvis, P. G., James, G. B. & Landsberg, J. J. 1976. Coniferous forest. In: Monteith, J. L. (ed.) *Vegetation and the Atmosphere* II. Academic Press, London.
Johnson, C. G. 1969. *Migration and dispersal of insects by flight.* Methuen & Co, London.
Joosse, E. N. G., Brugman, F. A. & Veld, C. J. 1973. The effects of constant and fluctuating temperatures on the production of spermatophores and eggs in populations of *Orchesella cincta* (Linné) *Collembola, Entomobryidae.*

Neth. J. Zool. 23: 488-502.
Kainmuller, C. 1975. Temperaturresistenz von Hochgebirgspflanzen. *Anz. Math.-Naturw. Kl. Österr. Akad. Wiss.* 1975: 67-75.
Kappen, L. & Zeidler, A. 1977. Seasonal changes between one- and two-phasic response of plant leaves to heat stress. *Oecologia (Berl.)* 31: 45-53.
Kaufmann, J. S. & Bennett, A. F. 1989. The effect of temperature and thermal acclimatization on locomotor performance in *Xantusia vigilis*, the desert night lizard. *Physiol. Zool.* 62: 1047-1058.
Kendeigh, S. C., Dolnik, V. R. & Gavrilov, V. M. 1977. Avian Energetics. In: Pinowski, J. & Kendeigh, S. C. (eds.) *Granivorous birds in ecosystems*, pp. 127-202. Cambridge University Press, Cambridge.
Keppens, H., Wouters, D. S., Impens, I. & Verheyen, R. 1980. A comparitive study of the components of the radiation balance over three types of heathland vegetation. *Oecol. Plant.* 15: 293-298.
Kerr, J. P. & Thurtell, G. W. 1967. An integrating pyranometer for climatological observer stations and mesoscale networks. *J. Appl. Meteorol.* 6: 688-694.
Kessler, A. 1974. Infrarotstrahlungsmessungen in West-Afrika und in der Sahara. *Arch. Meteorol. Geophys. Bioklimatol.* B 22: 135-147.
Kevan, P. G. 1975. Suntracking solar furnaces in high arctic flowers: Significance for pollination and insects. *Science* 189: 723-726.
Kevan, P. G., Jensen, T. S. & Shorthouse, J. D. 1982. Body temperatures and behavioral thermoregulation of high arctic woolly-bear caterpillars and pupae (*Gynaephora rossii*, *Lymantriidae*, *Lepidoptera*) and the importance of sunshine. *Arct. Alp. Res.* 14: 125-136.
Kiese, O. 1971. The measurement of climatic elements which determine production in various plant stands. In: Ellenberg, H. (ed.) *Integrated experimental ecology*. Springer, Berlin.
Kingsolver, J. G. & Watt, W. B. 1983. Thermoregulatory strategies in *Colias* butterflies: thermal stress and the limits to adaptation in temporally varying environments. *Am. Nat.* 121: 32-55.
Knapp, A. K., Smith, W. K. & Young, D. R. 1989. Importance of intermittent shade to the ecophysiology of subalpine herbs. *Funct. Ecol.* 3: 753-758.
Kneitz, G. 1964. *Untersuchungen zum Aufbau und zur Erhaltung des Nestwärmehaushaltes bei Formica polyctena Foerst.* Universität Würzburg, Würzburg.
Kneitz, G. 1970. Saisonale Veränderungen des Nestwärmehaushaltes bei Waldameisen in Abhängigkeit von der Konstitution und dem Verhalten der Arbeiterinnen als Beispiel vorteilhafter Anpassung eines Insektenstaates an das Jahreszeitenklima. *Zool. Anz. Suppl.* 33: 318-322.
Korhonen, K. 1980. Microclimate in the snow burrows of willow grouse (*Lagopus lagopus*). *Ann. Zool. Fenn.* 17: 5-9.
Korhonen, K. 1981. Temperature in the nocturnal shelters of the redpoll (*Acanthis flammea* L.) and the Siberian tit (*Parus cinctus* Budd.) in winter. *Ann. Zool. Fenn.* 18: 165-168.
Kowalski, G. J. & Mitchell, J.W. 1976. Heat transfer from spheres in the naturally turbulent, outdoor environment. *J. Heat Transf.* 95: 649-653.
Kramer, C., Post, J. J. & Woudenberg, J. P. M. 1954. *Nauwkeurigheid en betrouwbaarheid van temperatuur- en vochtigheidsbepalingen in buitenlucht met behulp van kwikthermometers.* K.N.M.I. Meded. Verh. 60, de Bilt.
Kraus, G. 1911. *Boden und Klima auf kleinstem Raum.* Fischer, Jena.
Krečmer, V. 1966, 1967, 1968. Das Mikroklima der Kiefernlochkahlschläge. *Wetter Leben* 18: 133-141, 186-198; 19: 107-115, 203-214; 20: 61-72.
Krog, J. 1955. Note on temperature measurements indicative of special organisation in arctic and subarctic plants for utilisation of radiated heat from the sun. *Physiol. Plant.* 8: 836-839.
Kullenberg, B. 1962. Some points of view on microclimatology as an integrative part of terrestrial ecology. *Zool. Bidr. Uppsala* 35: 456-479.
Lagrew, D. C. & Baskin, J. M. 1975. Evaluation of the anthracene-benzene chemical light meter for ecological research. *Oecologia (Berl.)* 21: 73-84.
*Lamb, H. H. 1977. *Climate, present, past and future.* Methuen, London.
Landsberg, J. J. 1984. Physical aspects of the water regime of wet tropical vegetation. In: Medína, E., Mooney, H. A. & Vazquez-Yánes, C. (eds.) *Physiological ecology of plants of the wet tropics*, pp. 13-27. Junk, The Hague.
Lange, O. L. 1953. Hitze- und Trockenresistenz der Flechten in Beziehung zu ihrer Verbreitung. *Flora* 140: 39-97.
Lange, O. L. 1954. Einige Messungen zum Wärmehaushalt poikilohydrer Flechten und Moose. *Arch. Meteorol. Geophys. Bioklimatol.* B 5: 182-190.
Lange, O. L. 1955. Untersuchungen über die Hitzeresistenz der Moose in Beziehung zu ihrer Verbreitung. I. Die Resistenz stark ausgetrockneter Moose. *Flora* 142: 381-399.
Lange, O. L. 1959. Untersuchungen über Wärmehaushalt und Hitzeresistenz mauretanischer Wüsten- und Savannenpflanzen. *Flora* 147: 595-651.
Lange, O. L. 1961. Die Hitzeresistenz einheimischer immer- und wintergrüner Pflanzen im Jahresverlauf. *Planta* 56: 666-683.
Lange, O. L. 1963. Die Photosynthese der Flechten bei tiefen Temperaturen und nach Frostperioden. *Ber. Dt. Bot. Ges.* 75: 351-352.
Lange, O. L. 1969. Die functionellen Anpassungen der Flechten an die ökologischen Bedingungen arider Gebiete.

Ber. Dt. Bot. Ges.. 82 (1/2): 3-22.
Lange, O. L. 1972. Flechten-Pionierpflanzen in Kältewüsten. *Umschau* 72 (20): 650-654.
Lange, O. L. & Kappen, L. 1972. Photosynthesis of lichens from Antarctica. *Antarctica Res. Ser.* 20: 83-95.
Lange, O. L. & Lange, R. 1963. Untersuchungen über Blattemperaturen, Transpiration und Hitzeresistenz an Pflanzen mediterraner Standorte (Costa Brava, Spanien). *Flora* 153: 387-425.
Larcher, W. 1973. *Ökologie der Pflanzen*. Ulmer, Stuttgart.
Larcher, W. 1977. Produktivität und Uberlebensstrategien von Pflanzen und Pflanzenbeständen im Hochgebirge. Sitzungsber. *Österr. Akad. Wiss. Mat.- Naturwiss. Kl.* Abt. I 186: 373-386.
Larcher, W. 1980. Klimastress im Gebirge. *Rhein.-Westf. Akad. Wiss.Vortr.* N 291: 49-80.
Laurie, I. C. (ed.) 1979. *Nature in Cities*. Wiley, Chichester.
Lauscher, A., Lauscher, F. & Printz, H. 1955. Die Phänologie Norwegens. *Skr. Norsk.Vidensk.-Akad. Oslo, I. Mat.-Naturvid. Kl.* 1955-1.
Lauscher, F. 1953. Die Rolle mikroklimatischer Faktoren beim Massenauftreten von Waldschädlingen. *Wetter Leben* 5: 195-200.
Lee, R. 1978. *Forest microclimatology*. Columbia University Press, New York.
Lee, D. W. & Downing, K. R. 1991. The spectral distribution of biologically active solar radiation at Miami, Florida, USA. *Int. J. Biometeorol.* 35: 48-54.
Lensink, B. M. 1963. Distributional ecology of some *Acrididae* (*Orthoptera*) in the dunes of Voorne, Netherlands. *Tijdsch. Entomol.* 106: 357-443.
Levitt, J. 1972. *Responses of plants to environmental stresses*. Academic Press, New York.
Lewis, M. C. & Callaghan, T. V. 1976. Tundra. In: Monteith, J. (ed.) *Vegetation and the atmosphere*, Vol. 2, pp. 399-433. Academic Press, London.
Lewis, T. & Dibley, G. C. 1970. Air movement near windbreaks and a hypothesis of the accumulation of airborne insects. *Ann. Appl. Biol.* 66: 477-484.
Linacre, E. 1976. Swamps. In: Monteith, J. L. (ed.) *Vegetation and the atmosphere*, Vol. 2, pp. 329-347. Academic Press, London.
List, R. J. 1968. *Smithsonian meteorological tables*. Smithsonian Institution Press, Washington.
Litzow, M. & Pellett, H. 1983. Materials for potential use in sunscald prevention. *J. Arboricult.* 9 (2): 35-38.
Lubin, Y. D & Henschel, J. R.1990. Foraging at the thermal limit: burrowing spiders (*Seothyra, Eresidae*) in the Namib desert dunes. *Oecologia* 84: 461-467.
Luff, M. L. 1965. Morphology and Microclimate of *Dactylis glomerata* tussocks. *J. Ecol.* 53: 771-783.
Luff, M. L. 1966. Cold hardiness of some beetles living in grass tussocks. *Entomol. Exp. Appl.* 9: 192-199.
Lynch, J. M. & Poole, N. J. 1979. *Microbial Ecology*. Blackwell, Oxford.
Machin, J. 1964a. The evaporation of water from *Helix aspersa*. II. Measurement of airflow and the diffusion of water vapour. *J. Exp. Biol.* 41: 771-781.
Machin, J. 1964b. The evaporation of water from *Helix aspersa*. III. The application of evaporation formulae. *J. Exp. Biol.* 41: 783-792.
Mattson, J. O. 1979. *Introduktion till mikro- och lokalklimatologin*. Liber, Malmö.
Melcher, J. C., Armitage, K. B. & Porter, W. P. 1990. Thermal influences on the activity and energetics of yellow-bellied marmots (*Marmota flaviventris*). *Physiol. Zool.* 63: 803-820.
Merriam, G., Wegner, J. & Caldwell, D. 1983. Invertebrate activity under snow in deciduous woods. *Holarct. Ecol.* 6: 89-94.
Mertens, J. A. L. 1980. The energy requirements for incubating in Great Tits and other bird species. *Ardea* 68: 185-192.
Miess, M. 1979. The climate of cities. In: Laurie, I. C. (ed.) *Nature in Cities*, pp. 91-114. Wiley, Chichester.
Minnaert, M. G. J. 1972. *De natuurkunde van het vrije veld*. 3 Vols., 3rd. ed. W.J. Thieme & Cie., Zutphen.
Mitchell, J. W., Beckman, W. A. , Bailey, R. T. & Porter, W. P. 1975. Microclimatic modeling of the desert. In: de Vries, D.A. & Afgan, N. H. (eds.) *Heat and mass transfer in the biosphere*, pp. 275-286. Scripta Book Co., Washington DC.
Mitscherlich, G. 1971. Wald, Wachstum und Umwelt. 2. Band. *Waldklima und Wasserhaushalt*. J. D. Sauerländer's Verlag, Frankfurt/Main.
Monteith, J. L. 1954. Dew. *Quart. J. R. Meteorol. Soc.* 80: 322-341.
*Monteith, J. L. 1973. *Principles of Environmental Physics*. Edward Arnold, London.
Monteith, J. L. 1981. Coupling of plants to the atmosphere. In: Grace, J., Ford, E. D. & Jarvis, P. G. (eds.) *Plants and their atmospheric environment*, pp. 1-30. Blackwell, Oxford.
Monteith, J. L. & Szeics, G. 1961. The radiation balance of bare soil and vegetation. *Quart. J. R. Meteorol. Soc.* 87: 159-170.
Moreno, J. M., Pineda, F. D. & Rivas-Martínez, S. 1990. Climate and vegetation at the Eurosiberian-Mediterranean boundary in the Iberian Peninsula. *J. Veg. Sci.* 1: 233-244.
Morgan, D. C. & Smith, H. 1978. Simulated sunflecks have large rapid effects on plants stem extension. *Nature* 273: 534-536.
Morton, F. 1963. Dolinenklima und Pflanzenwelt. *Wetter Leben* 13: 155-158.
Mulroy, T. W. 1979. Spectral properties of heavily glaucous and non-glaucous leaves of a succulent rosette-plant. *Oecologia (Berl.)* 38: 349-357.

Neilson, R. P. & Wullstein, L. H. 1983. Biogeography of two southwest American oaks in relation to atmospheric dynamics. *J. Biogeogr.* 10: 275-297.
Neubauer, H. F. 1938. Zur Ökologie von in Buchenkronen epiphytisch lebenden Flechten. *Beitr. Biol. Pfl.* 25: 273-289.
Nicolai, V. 1986. The bark of trees: thermal properties, microclimate and fauna. *Oecologia (Berl.)* 69: 148-160.
Nordhagen, R. 1927. Die Vegetation und Flora des Sylenegebietes. *Skr. Norske Vidensk. Akad. I. Mat. Naturv. kl.* 1927(1): 1-162.
Nørgaard, E. 1951. On the ecology of two lycosid spiders (*Pirata piraticus* and *Lycosa pullata*) from a Danish *Sphagnum* bog. *Oikos* 3: 1-21.
Nørgaard, E. 1956. Environment and behaviour of *Theridion saxatile. Oikos* 57: 159-192.
Nyberg, A. 1938. Temperature measurements in an air layer very close to a snow surface. *Geogr. Ann.* 20: 234-275.
*Oke, T. R. 1987. *Boundary layer climates.* Methuen, London, New York.
Olszewski, J. L. 1974. Wind velocity in a deciduous forest stand and in an unwooded area. *Ekol. Pol.* 22: 223-235.
Oomes, M. J. M. 1990. Changes in dry matter and nutrient yields during the restoration of species-rich grasslands. *J. Veg. Sci.* 1: 333-338.
Parkhurst, D. F. & Loucks, O. L. 1972. Optimal leaf size in relation to environment. *J. Ecol.* 60: 505-537.
Peacock, J. M. 1976. Temperature and leaf growth in four grass species. *J. App. Ecol.* 13: 225-232.
Penman, H. L. 1948. Natural evaporation from open water, bare soil, and grass. *Proc. Roy. Soc. Lond.* A. 193: 120-145.
Pettersson, B. 1965. Gotland and Öland. Two limestone islands compared. *Acta Phytogeogr. Suec.* 50: 130-140.
Pieters, G. A. 1972. Measurements of leaf temperature by thermocouples or infrared thermometry in connection with exchange phenomena and temperature distribution. *Mededel. Landbouwhogesch. Wageningen* 72-34: 1-20.
Piggin, J. & Schwerdtfeger, P. 1973. Variations in the albedo of wheat and barley crops. *Arch. Meteorol. Geophys. Bioklimatol.* B 21: 365-391.
Pigott, D. J. & Huntley, J. P. 1978a. Factors controlling the distribution of *Tilia cordata* at the northern limits of its geographical range. I. Distribution in north-west England. *New Phytol.* 81: 429-441.
Pigott, D. J. & Huntley, J. P. 1978b. Factors controlling the distribution of *Tilia cordata* at the northern limits of its geographical range. II. History in north-west England. *New Phytol.* 84: 145-164.
Pinker, R. 1980. The microclimate of a dry tropical forest. *Agric. Meteorol.* 22: 249-265.
Pollard, D. F. W. 1970. The effect of rapidly changing light on the rate of photosynthesis in large-tooth aspen (*Populus grandidentata*). *Can. J. Bot.* 48: 823-829.
Pons, T. L. 1983. *An ecophysiological study in the field layer of ash coppice.* Diss. University of Utrecht, Utrecht.
Porter, K. 1982. Basking behaviour in larvae of the butterfly *Euphydryas aurinia. Oikos* 38: 308-312.
Porter, W. P. & Gates, D. M. 1969. Thermodynamic equilibria of animals with environment. *Ecol. Monogr.* 39: 245-270.
Precht, H., Christophersen, J., Heusel, H. & Larcher, W. 1973. *Temperature and life.* Springer, Berlin.
Pümpel, B. 1977. *Bestandsstruktur, Phytomassevorrat und Produktion verschiedener Pflanzengesellschaften im Glocknergebiet.* Veröff. Österr. M.A.B. Hochgebirgsprogramm Hohe Tauern 1: 88-101.
Quinlivan, B. J. 1966. The relationship between temperature fluctuations and the softening of hard seeds of some legume species. *Austr. J. Agr. Res.* 17: 621-625.
Raman, P. K. 1936. The measurement of the transmission of heat by convection from insolated ground to the atmosphere. *Proc. Ind. Acad. Sci.* 3: 98-106.
Raschke, K. 1956. Uber die physikalischen Beziehungen zwischen Wärmeübergangszahl, Strahlungsaustausch, Temperatur und Transpiration eines Blattes. *Planta* 48: 299-338.
Rauner, J. L. 1976. Deciduous forest. In: Monteith, J. L. (ed.) *Vegetation and the atmosphere* II, pp. 241-264. Academic Press, London
Reichelt, G. 1954. Uber Spätfrostschäden im Grünland in Abhängigkeit vom Relief. *Wetter Leben* 6: 1-9.
Rejmánek, M. 1971. Ecological meaning of the thermal behaviour of rocks. *Flora* 160: 527-561.
Remmert, H. 1985. Crickets in sunshine. *Oecologia (Berl.)* 68: 29-33.
Richards, A. G. 1957. Cumulative effects of optimum and suboptimum temperatures on insect development. In: Johnson, F. H. (ed.) *Influence of temperature on biological systems*, pp. 145-162. Ronald Press, New York.
Robertson, A. 1986. Estimating mean wind flow in hilly terrain from Tamarack (*Larix laricina* (DuRoi) K.Koch) deformation. *Int. J. Biometeorol.* 30: 333-349.
Robinson, N. 1966. *Solar radiation.* Elsevier, Amsterdam.
Root, T. 1988. Energy constraints on avian distributions and abundancies. *Ecology* 69: 330-339.
Rosén, E. 1985. Succession and fluctuations in species composition in the limestone grasslands of South Öland. *Münst. Geogr. Arb.* 20: 25-33.
*Rosenberg, N. J., Blad, B. L. & Verma, S. B. 1983. *Microclimate, the biological environment.* 2nd ed. Wiley Interscience, New York.
Ross, J. 1975. Radiative transfer in plant communities. In: Monteith, J. L. (ed.) *Vegetation and the atmosphere* 1, pp. 13-55. Academic Press, London.
Roth, R. 1961. Konstruktive und thermodynamische Eigenschaften des Piche-Evaporimeters. *Arch. Meteorol. Geophys. Bioklimatol.* 11: 108-125.

Ruinen, J. 1961. The phyllosphere, an ecologically neglected milieu. *Plant Soil* 15: 81-109.
Ryrholm, N. 1988. An extralimital population in a warm climatic outpost: the case of the moth *Idaea dilutaria* in Scandinavia. *Int. J. Biometeorol.* 32: 205-216.
Ryrholm, N. 1989. *The influence of the climatic energy balance on living conditions and distribution patterns of Idaea ssp. (Lepidoptera: Geometridae): an expansion of the species energy theory.* Doctoral Diss. Uppsala University, Uppsala.
Saint Girons, H. 1975. Observations préliminaires sur la thermorégulation des Vipères d'Europe. *Vie Milieu* C 25: 137-168.
Sauberer, F. & Dirmhirn, I. 1953. Uber die Entstehung der extremen Temperaturminima in der Doline Gestettner Alm. *Arch. Meteorol.* (B) 5: 307-326.
Sauberer, F. & Dirmhirn, I. 1958. Ein kleiner Interferenz-Monochromator für Spektralmessungen mit Fernablesung. *Wetter Leben* 10: 159.
Schmeidl, H. 1965. Oberflächentemperaturen in Hochmooren. *Wetter Leben* 17: 87-97.
Schmidbauer, W. 1985. *Die Angst vor Nähe.* Rowohlt, Reinbek.
Schmidt Nielsen, K. 1972. *How animals work.* Cambridge University Press, Cambridge.
Scholander, P. F., Walters, V., Hock, R. & Irving, L. 1950. Body insulation of some arctic and tropical mammals and birds. *Biol. Bull.* 99: 225-271.
Schuepp, P. H. 1977. Turbulent transfer at the ground on verification of a simple predictive model. *Boundary-Layer Meteorol.* 12: 171-186.
Schulze, R. 1970. *Strahlenklima der Erde.* Steinkopff, Darmstadt.
Schurer, K. 1981. Het meten van luchtvochtigheid. *Koeltechniek* 1981: 1-8.
Schwerdtfeger, F. 1963. *Ökologie der Tiere.* Band I. Autökologie. Ed. Parey, Hamburg/Berlin.
Scorer, R. S. 1954. The nature of convection as revealed by soaring birds and dragonflies. *Quart. J. R. Meteorol. Soc.* 80: 68-77.
Sellers, W. D. 1965. *Physical climatology.* The University of Chicago Press, Chicago.
Seybold, A. 1936. Uber den Lichtfaktor photophysiologischer Prozesse. *Jahrb. Wiss. Bot.* 82: 741-795.
Shilov, J. A. 1965. *Heat regulation in birds.* Moscow University Press; also Amerind Publishing Co., New Delhi (1973).
Shreeve, T. G. 1984. Habitat selection, mate location and microclimate constraints on the activity of the speckled wood butterfly *Pararge aegeria. Oikos* 42: 371-377.
Sidorowicz, J. 1959. Observations on directions of air currents over the ground of the wood. *Ekol. Pol.* B5: 345-350.
Simms, E. 1971. *Woodland birds.* Collins, London.
Sinclair, R. 1970. Convective heat transfer from narrow leaves. *Austr. J. Biol. Sci.* 23: 309-321.
Sjögren, E. (ed.) 1988. Plant cover on the limestone Alvar of Öland. Ecology - Sociology -Taxonomy. *Acta Phytogeogr. Suec.* 76: 1-160.
Sjörs, H. 1965. Northern mires: Regional ecology of mire sites and vegetation. *Acta Phytogeogr. Suec.* 50: 180-188.
Slob, W. H. 1982. *Climatological values of solar irradiation on the horizontal and several inclined surfaces at De Bilt.* K.N.M.I. Wetensch. Rapp. 58-7, de Bilt.
Smid, P. 1975. Evaporation from a reedswamp. *J. Ecol.* 63: 299-309.
Söderström, L. & Jonsson, B. G. 1989. Spatial pattern and dispersal in the leafy hepatic *Ptilidium pulcherrimum. J. Bryol.* 15: 793-802.
Solbreck, C. 1976. Flight patterns of *Lygaeus equestris* (*Heteroptera*) in spring and autumn with special reference to the influence of weather. *Oikos* 27: 134-143.
Solomon, M. E. 1951. Control of humidity with potassium hydroxide, sulphuric acid or other solutions. *Bull. Entomol. Res.* 42: 543-554.
Solomon, M. E. 1966. Moisture gains, losses and equilibria of flour mites *Acarus siro* L. in comparison with larger Arthropods. *Entomol. Exp. App.* 9: 25-41.
Sonesson, M. & Lundberg, B. 1974. Late quaternary forest development of the Torneträsk area, North Sweden. 1. Structure of modern forest ecosystems. *Oikos* 25: 121-133.
Sonntag, D. 1966-1968. *Hygrometrie.* Akademie Verlag, Berlin.
Spurr, S. H. 1957. Local climate in the Harvard Forest. *Ecology* 38: 37-46.
Stahel, C. D., Nicol, S. C. & Walker, G. J. 1987. Heat production and thermal resistance in the little penguin *Eudyptula minor* in relation to wind speed. *Physiol. Zool.* 60: 413-423.
Steubing, L. 1955. Studien über den Taufall als Vegetationsfaktor. *Ber. Dt. Bot. Ges.* 68: 55-70.
Stigter, C. J. & Welgraven, A. D. 1976. An improved radiation protected differential thermocouple psychrometer for crop environment. *Arch. Meteorol. Geophys. Bioklimatol.* B 24: 177-187.
Stoutjesdijk, Ph. 1959. Heaths and Inland Dunes of the Veluwe. *Wentia* 2: 1-96.
Stoutjesdijk, Ph. 1961. Micrometeorological measurements in vegetations of various structure. *Proc. Kon. Ned. Akad. Wetensch.* C 64: 171-207.
Stoutjesdijk, Ph. 1966. On the measurement of the radiant temperature of vegetation surfaces and leaves. *Wentia* 15: 191-202.
Stoutjesdijk, Ph. 1970a. Some measurements of leaf temperatures of tropical and temperate plants and their

interpretation. *Acta Bot. Neerl.* 19: 373-384.
Stoutjesdijk, Ph. 1970b. A note on vegetation temperatures above the timber line in southern Norway. *Acta Bot. Neerl.* 19: 918-925.
Stoutjesdijk, Ph. 1972a. Spectral transmission curves of some types of leaf canopies with a note on seed germination. *Acta Bot. Neerl.* 21: 185-191.
Stoutjesdijk, Ph. 1972b. A note on the spectral transmission of light by tropical rainforest. *Acta Bot. Neerl.* 21: 346-350.
Stoutjesdijk, Ph. 1972c. Optical properties of leaves. *Verh. Kon. Ned. Akad. Wet. Afd. Natuurk.* Tweede R. 61: 93-94.
Stoutjesdijk, Ph. 1974a. The microclimate of a reed vegetation. *Verh. Kon. Ned. Akad. Wet. Afd. Natuurk.* Tweede R. 63: 91-93.
Stoutjesdijk, Ph.1974b. The open shade, an interesting microclimate. *Acta Bot. Neerl.* 23: 125-130.
Stoutjesdijk, Ph. 1974c. An improved simple radiation thermometer. *Acta Bot. Neerl.* 23: 131-136.
Stoutjesdijk, Ph. 1975. High surface temperatures. *Verh. Kon. Ned. Akad. Wet. Afd. Natuurk.* Tweede R. 66: 93-97.
Stoutjesdijk, Ph. 1977a. On the range of micrometeorological differentiation in the vegetation. In: Dierschke, H. (ed.) *Vegetation und Klima*, pp. 21-34. Cramer, Vaduz.
Stoutjesdijk, Ph. 1977b. High surface temperatures in the winter and their biological significance. *Int. J. Biometeorol.* 21: 325-331.
Stoutjesdijk, Ph. 1978. Determination of the heat transfer coefficient at the surface of the earth. *Verh. Kon. Ned. Akad. Wet. Afd. Natuurk.* Tweede R. 73: 26-28.
Stoutjesdijk, Ph. 1980. A small thermo-electric psychrometer and its performance. *Verh. Kon. Ned. Akad. Wet. Afd. Natuurk.* Tweede R. 75: 23-24.
Stoutjesdijk, Ph. 1981. *Plantago lanceolata* and its habitat: Micrometeorological aspects. *Verh. Kon. Ned. Akad. Wet. Afd. Natuurk.* Tweede R. 76: 18-20.
Stoutjesdijk, Ph. 1983. Effects of drought on the microclimate in grassland. *Inst. R. Neth. Acad. Art. Sci. Inst. Ecol. Res. Progr. Rep.* 1982: 19-21.
Sutton, O. G. 1953. *Micrometeorology*. MacGraw Hill, London.
Swinbank, W. C. 1963. Long wave radiation from clear skies. *Quart. J. Roy. Meteorol. Soc.* 89: 339-348.
Sykes, M. T. & Wilson, J. B. 1988. An experimental investigation into the response of some New Zealand sand dune species to salt spray. *Ann. Bot.* 62: 159-166.
Szeicz, G. 1965. A miniature tube solarimeter. *J. Appl. Ecol.* 2: 145-147.
Szymkiewicz, D. 1923. Sur l'importance du déficit hygrométrique pour la phytogéographie écologique. *Acta Soc. Bot. Pol.* 1: 8-18.
Szymkiewicz, D.1923-1930. Etudes climatologiques II, IV, V, VIII, X, XI, XV, XVI, XVIII. *Acta Soc. Bot. Pol.* 1 (1923): 244-246, 246-253; 2 (1924): 130-151, 239-253; 4 (1926): 57-58, 60-63, 125-129; 6 (1930): 96-95, 95-100, 102-104.
Taylor, S. E. & Sexton, O. J. 1972. Some implications of leaf tearing in *Musaceae*. *Ecology* 53: 143-149.
Tenow, O. 1975. Topographical dependence of an outbreak of *Oporinia autumnata* Bkh. (Lep., *Geometridae*) in a mountain birch forest in northern Sweden. *Zoon* 3: 85-100.
Thiele, H. U. 1977. *Carabid Beetles in their environments*. Springer-Verlag, Berlin.
Thom, A. S. 1975. Momentum, mass and heat exchange of plant communities. In: Monteith, J. L. (ed.) *Vegetation and the atmosphere* I, pp. 57-110. Academic Press, London.
Thompson, K., Grime, J. P. & Mason, G. 1977. Seed germination in response to diurnal fluctuations of temperature. *Nature* 67:147-149.
Tracy, C. R. 1975. Water and energy relations of terrestrial amphibians: insights from mechanistic modelling. In: Gates, D. M. & Schmier, R. B. (eds.) *Perspectives of biophysical ecology*, pp. 325-346. Springer-Verlag, Berlin.
Tranquilini, W. 1960. Das Lichtklima wichtiger Pflanzengesellschaften. In: Ruhland, W. (ed.) *Handbuch der Pflanzenphysiologie*, Vol. 2, pp. 304-338. Springer-Verlag, Berlin.
Trites, A. W. 1990. Thermal budgets and climate spaces: the impact of weather on the survival of Galapagos (*Arctocephalus galapagoensis* Heller) and northern fur seal pups (*Callorhinus ursinus* L.). *Funct. Ecol.* 4: 753-768.
Turin, H., Haeck, J. & Hengeveld, R. 1977. Atlas of the carabid beetles of the Netherlands. *Verh. Kon. Ned. Akad. Wet. Afd. Natuurk.* Tweede R. 65: 1-288.
Türk, A. & Arnold, W. 1988. Thermoregulation as a limit to habitat use in alpine marmots (*Marmota marmota*). *Oecologia (Berl.)* 76: 544-548.
Turner, H. 1958. Maximaltemperaturen oberflächennaher Bodenschichten an der alpinen Waldgrenze. *Wetter Leben* 10: 1-11.
Turner, H. 1961. Jahresgang und biologische Wirkungen der Sonnen- und Himmelsstrahlung an der Waldgrenze der Ötztaler Alpen. *Wetter Leben* 13: 93-113.
Turner, H. & Tranquilini, W. 1961. Die Strahlungsverhältnisse und ihr Einfluss auf die Photosynthese der Pflanzen. *Mitt. Forstl. Bundesversuchsanst. Mariabrunn* 59: 69-104.
Turner, J. R. G., Gatehouse, C. M. & Corey, C. A. 1987. Does solar energy control organic diversity. *Oikos* 48: 195-205.

Tüxen, R. 1977. Beobachtungen über Schnee-Verteilung im Buchenwald, ihre Ursachen und Wirkungen. In: Dierschke, H. (ed.) *Vegetation und Klima*. pp. 127-162. Cramer, Vaduz.
Unkasevic, M. 1989. Some improvements in calculating the plant stand surface albedo and its influence on ground surface temperature. *Int. J. Biometeorol.* 33: 184-196.
Unwin, D. M. 1980. Microclimate measurements for ecologists. Academic Press, London.
Vaartaja, O. 1949. High surface soil temperatures. *Oikos* 1: 6-29.
van der Hage, J. C. H. 1984. A small optical line sensor for radiation measurements in vegetation. *J. Exp. Bot.* 35: 762-766.
van der Held, E. F. M. 1937. De psychrometerconstante. *Warmtetechniek* 8.
van der Maarel, E. 1965. Beziehungen zwischen Mollusken und Pflanzengesellschaften. In: Tüxen, R. (ed.) *Biosoziologie*, pp. 184-198. Junk, The Hague.
van der Poel, A. J. & Stoutjesdijk, Ph. 1959. Some microclimatological differences between an oak wood and a *Calluna* heath. *Meded. Landbouwhogesch. Wageningen* 59 (2): 1-8.
van der Toorn, J. 1972. *Variability of Phragmites australis (Cav.) Trin. ex Steudel in relation to the environment.* Van Zee tot Land, 48. Staatsuitgeverij, The Hague.
van Dieren, W. 1934. *Organogene Dünenbildung*. Nijhoff, The Hague.
van Eimern, J. 1986. Vergleich der Strahlungsverhältnisse über einem Buchenwald, zwei Kiefernwäldern und einer Grasfläche. *Int. J. Biometeorol.* 31: 217-235.
van Heerdt, P. F., Isings, J. & Nijenhuis, E. D. 1956. Temperature and humidity preferences of various *Coleoptera* from the duneland area of Terschelling. *Proc. Kon. Ned. Akad. Wet.* C 59: 668-676.
van Os, L. J.1981. *Fenologisch onderzoek aan graslandplanten.* R.I.N. Rapport, Bilthoven.
van Schaik Zillesen, P.G. & Brunsting, A. M. H. 1983. Capacity for flight and egg production in *Lochmaea suturalis* (Co., Chrysomelidae). *Neth. J. Zool.* 33: 266-275.
van Wijk, W. R. & Derksen, W. J. 1963. Sinusoidal temperature variation in a layered soil. In: van Wijk, W. R. (ed.) *Physics of plant environment*, pp. 171-209. North-Holland Publishing Company, Amsterdam.
van Wijk. W. R. & de Vries, D. A. 1963. Periodic temperature variations in a homogeneous soil. In: van Wijk, W. R. (ed.) *Physics of plant environment*, pp. 102-143. North-Holland Publishing Company, Amsterdam.
van Wingerden, W. K. R. E. & van Kreveld, R. 1989. Vegetation structure and distribution patterns of grasshoppers (*Orthoptera Acrididae*). *Econieuws* 2(4): 5.
van Zanten, B. O. 1978. Experimental studies on trans-oceanic long-range dispersal of moss spores in the southern Hemisphere. *J. Hattori Bot. Lab.* 44: 455-482.
van Zanten, B. O. & Gradstein, S. R. 1988. Experimental Dispersal Geography of Neotropical Liverworts. *Nova Hedwigia Beih.* 90: 41-94.
Verwijst, Th. 1988. Environmental correlates of multiple-stem formation in *Betula pubescens* ssp. *tortuosa. Vegetatio* 76: 29-36.
Viitanen, P. 1967. Hibernation and seasonal movements of the viper, *Vipera berus berus* (L) in southern Finland. *Ann. Zool. Fenn.* 4: 472-546.
Volk, O. H. 1934. Ein neuer für botanische Zwecke geeigneter Lichtmesser. *Ber. Dt. Bot. Ges.* 52: 195-202.
Vugts, H. F. & van Wingerden, W. K. R. E. 1976. Meteorological aspects of aeronautic behaviour of spiders. *Oikos* 27: 433-444.
Vulto, J. C. & van der Aart, P. J. M. 1983. Salt spray and its influence on the vegetation of the coastal dunes of Voorne and Goeree (the Netherlands) in relation to man-made changes in coastal morphology. *Verh. Kon. Ned. Akad. Wet. Afd. Natuurk.* Tweede R. 81: 65-73.
Wachter, H. 1976. Die Unruhe der Lufttemperatur als biometeorologischer Faktor. *Arch. Meteorol. Geophys. Bioklimatol.* B 24: 41-55.
Wallace, A. R. 1878. *Tropical nature, and other essays.* London.
Wallén, C. C. 1969. *Definitions and scales in climatology as applied to agriculture.* Proc. Regional Training Seminar on Agrometeorology, pp. 207-212. Agricultural Univ. Wageningen, Wageningen.
Wallgren, H. 1954. Energy metabolism of two species of the genus *Emberiza* as correlated with distribution and migration. *Acta Zool. Fenn.* 84: 5-110.
Walter, H. 1960. Grundlagen der Pflanzenverbreitung I. Standortslehre. Einführung in die Phytologie III (1) 2. Aufl. Ulmer Verlag, Stuttgart.
Walter, H. 1977. *Vegetationszonen und Klima.* 3. Aufl. Ulmer Verlag, Stuttgart.
Walter, H. 1985. *Vegetation of the Earth and Ecological Systems of the Geo-biosphere.* 3rd. ed. Springer-Verlag, Berlin.
Walter, H. & Breckle, S.-W. 1985. *Ecological Systems of the Geobiosphere.* Springer-Verlag, Berlin.
Walter, H. & Straka, H. 1970. *Arealkunde.* 2. Aufl. Ulmer, Stuttgart.
Warenberg, K. 1982. Reindeer forage plants in the early grazing season. *Acta Phytogeogr. Suec.* 70: 1-71.
Warren Wilson, J. 1957. Observations on temperature of arctic plants and their environment. *J. Ecol.* 45: 499-531.
Wartena, L. 1968. *The evaporation of lakes.* In: Proc. Regional training Seminar on Agrometeorology, pp. 243-253. Agricultural Univ. Wageningen, Wageningen.
Wartena, L., Palland, C. L. & van der Vossen, G. H. 1973. Checking of some formulae for the calculation of long wave radiation from clear skies. *Arch. Meteorol. Geophys. Bioklimatol.* B 21: 335-348.
Watson, I. D. & Johnson, G. T. 1988. Estimating person view-factors from fish-eye lens photographs. *Int. J.*

Biometeorol. 32: 123-128.
Webb, D. R. 1980. Environmental harshness, heat stress and *Marmota flaviventris*. *Oecologia* 44: 390-395.
Webster, M. D. & Weathers, W. W. 1990. Heat produced as a by-product of foraging activity contributes to thermoregulation by Verdins, *Auriparus flaviceps*. *Physiol. Zool.* 63: 777-794.
Weiss, S. B., Murphy, D. D. & White, R. R. 1988. Sun, slope, and butterflies: topographic determinants of habitat quality for *Euphydryas editha*. *Ecology* 69: 1486-1496.
Went, F. W. 1944. Plant growth under controlled conditions. III. Correlation between various physiological processes and growth in the tomato plant. *Am. J. Bot.* 31: 597-618.
Went, F. W. 1957. The experimental control of plant growth. *Chron. Bot.* 17: 1-343.
Westhoff, V. & den Held, A. J. 1969. *Plantengemeenschappen in Nederland.* Thieme, Zutphen.
White, G. 1789. *The natural history of Selborne.* 1911 ed. Cassel & Cy., London.
Wigglesworth, V. B. 1953. *The principles of insect physiology.* 5th. ed. Methuen, London.
Williams, E. H. 1981. Thermal influences on oviposition in the Montane Butterfly *Euphydryas gillettii*. *Oecologia (Berl.)* 50: 342-346.
Willis, A. J. 1964. Investigations on the Physiological Ecology of *Tortula ruraliformis*. *Trans. Br. Bryol. Soc.* 4: 558-682.
Williams, W. A. & Elliott, J. R. 1960. Ecological significance of seed coat impermeability to moisture in crimson, subterranean and rose clovers in a Mediterranean type climate. *Ecology* 41: 733-742.
Willmer, P. G. & Unwin, D. M. 1981. Field analysis of insect heat budgets: Reflectance, size and heating rates. *Oecologia (Berl.)* 50: 250-255.
Wilmers, F. 1968. Kleinklimatische Untersuchungen von Laubwaldrändern bei Hannover. *Ber. Inst. Meteorol. Klimatol. Tech. Univ. Hannover* 1: 1-162.
Wittig, R. & Durwen, K.-J. 1982. Ecological indicator-value spectra of spontaneous urban floras. In: Bornkamm, R., Lee, J. A. & Seward, M. R. D. (eds.) *Urban Ecology*, pp. 23-31. Blackwell, Oxford.
Woodell, S. R. J. 1974. Anthill vegetation in a Norfolk saltmarsh, *Oecologia (Berl.)* 16: 221-225.
*Woodward, F. I. 1987. *Climate and plant distribution.* Cambridge University Press, Cambridge.
Woodward, F. I. 1988. Temperature and the distribution of plant species. In: Long, S. P. & Woodward, F. I. (eds.) *Plants and Temperature*, pp. 59-75. The Company of Biologists, Cambridge.
Woudenberg, J. P. M., 1969. *Nachtvorst in Nederland.* K.N.M.I. Wetensch. Rapp. 68-1, de Bilt.
Wright, D. H. 1983. Species-energy theory: an extension of species-area theory. *Oikos* 41: 496-506.
Wuenscher, J. E. 1970. The effect of leaf hairs of *Verbascum thapsus* on leaf energy exchange. *New Phytol.* 69: 65-73.
Wygoda, M. L. 1984. Low cutaneous evaporative water loss in arboreal frogs. *Physiol. Zool.* 57: 329-337.
Yom-Tov, Y., Imber, A. & Otterman, J. 1977. The microclimate of winter roosts of the Starling *Sturnus vulgaris*. *Ibis* 119: 366-368.
*Yoshino, M. M. 1975. *Climate in a small area.* University of Tokyo Press, Tokyo.
Zahn, M. 1958. Temperatursinn, Wärmehaushalt und Bauweise der Roten Waldameise (*Formica rufa* L.). *Zool. Beitr.* N. F. 3: 127-194.
Zangerl, A. R. 1978. Energy exchange phemomena physiological rates and leaf size variation. *Oecologia* 34: 107-112.
Zanstra, P. E. 1976. Welding uniform sized thermocouple junctions from thin wires. *J. Phys. E. Scient. Instr.* 9: 526-528.
Zanstra, P. E. & Hagenzieker, F. 1977. Comments on the psychrometric determination of leaf water potentials in situ. *Plant Soil* 48: 347-367.
Zavitkovski, J. 1983. Characterisation of light climate under canopies of intensively cultured hybrid poplar plantations. *Agric. Meteorol.* 25: 245-255.

Additional References

Allander, K & R. Dufva (1997). Parasiter och sjukdomar hos fåglar. In: S. Ekman& A.Lundberg, *Fåglarnas ekologi, Vår Fågelvärld* suppl.26, Sveriges Ornitologiska Förening, Stockholm.
Beintema, A., O. Moedt & D. Ellinger (1995). *Ecologische Atlas van de Nederlandse weidevogels.* Schuyt & Co, Haarlem.
Carey, C.(1996). Avian Energetics and nutritional Ecology. Chapman & Hall v-xiv, 1-543
Carascal, L.M., J.A. Diaz, D.L. Huertas & I. Mozetich, (2001): Behavioural thermoregulation by treecreepers; trade-off between saving energy and reducing crypsis. *Ecology* 82: 1642-1654.
Dubska, L., I. Literak, E. Kociahava, V. Taragelova, V. Sverakova, O. Sychra & Hromadko (2011). Synanthropic birds influence the distribution of Borrelia species: analysis of Ixodes ricinus ticks feeding on passerine birds. *Appl. Environm. Microbiol.* 77(3): 1115-1117. Published online 2010 December 10.doi 10.1128/AEM 02278-10.
Gavrilov, V.M. (2013). Origin and development of homoiothermy: a case study of avian energetics. *Advances in bioscience and biotechology* 4: 1-17.
M. Haftorn, S. (1986). *Fuglekongen, vår minste fugl.* NKS - Forlaget.
Kendeigh, S.C. , V.R. Dolnik & V.M. Gavrilov (1977): Avian energetics: 127-202, in *Granivorous birds in ecosystems* (, J. & S.C. Kendeigh, eds.). Cambridge University Press, Cambridge.
Landsborough Thomson, A., ed. (1965): *A new dictionary of birds.* Nelson, London and Edinburgh, 2nd impression, the British Ornithologists' union.
Marchand, P.J. (1991). *Life in the cold, 2nd ed. An introduction to winter ecology.* Univ.Press New England, Hanover and London.
McNab, B.K. (2009) Ecological factors affect the level and scaling of avian BMR. *Comparative Biochemistry and Physiology*, Part A: 22-45. Downloaded from www.elsevier.com/locate/cbpa
Münch, H. (2003). *Die Kreuzschnäbel.* Neue Brehm-Bücherei Bd.634. Westarp Wissenschaften, Hohenwarsleben.
Schmidt-Nielsen, K. (1997): *Animal Physiology*. Cambridge University Press. .
Stoutjesdijk, Ph. (1974). The open shade, an interesting microclimate. *Acta Botanica* 23: 125-130.
Stoutjesdijk, Ph. (1977). High surface temperatures in the winter and their biological significance. *Int. J. Biometeorol.* 21 : 325-331.
Stoutjesdijk, Ph. (2002). The ugly duckling. A thermal viewpoint. *J. Therm. Biol.* 27: 413-422.
Stoutjesdijk, F. (2003). Birds in the cold: The effects of plumage structure and environment on operative temperature, shown by spherical models. *Ornis Svecica* 13: 123-136.
Stoutjesdijk, F. (2009). Vogels in de winter: hoe dik is hun jas en andere bespiegelingen. *Het Vogeljaar* 57(1): 16-24.
De Vries, J. & M.R. van Eerden (1995). Thermal conductance in aquatic birds in relation to the degree of water contact, body mass and body fat: energetic implications of living in a strong cooling environment. *Physiol. Zool.* 68 (1995): 1143-1163.
Waldbauer, G. (1998). *The Birder's bug book.* Harvard University Press: i-xi, 1- 290. Wiersma, P. & T. Piersma (1994). Effects of microhabitat, flocking, climate and migratory goal on energy, expenditure in the annual cycle of red knots. *The Condor* 96: 257-279.
Whittow, G.G. 2000. *Sturkie's Avian Physiology*, 5th edition. Academic Press San Diego. vii-x, 1-479.

List of symbols

Symbol	Description	Unit (or range)
A	view factor	(0-1)
A^*	coefficient of turbulent transport	m^2/s
a	thermal diffusivity	m^2/s
BMR	basal metabolic rate	W
C	specific heat	$JK^{-1}g^{-1}$
C_p	specific heat of air with constant pressure	$JK^{-1}g^{-1}$
ρC	specific heat by volume	$Jm^{-3}K^{-1}$
ρC_p	specific heat by volume of air	$Jm^{-3}K^{-1}$
c	concentration of water vapour	g/m^3
D	diffusion coefficient of water vapour in air	m^2/s
d	depth	m
d_D	damping depth	m
E	evaporation	$gm^{-2}s^{-1}$
e	vapour pressure	mbar
e_{air}	vapour pressure in air	mbar
e_{leaf}	e_{max} at leaf temperature	mbar
e_{max}	saturation vapour pressure at a certain temperature	mbar
e_s	e_{max} at surface temperature	mbar
H	energy flux	W/m^2
H_{con}	energy flux by conduction	W/m^2
H_{ev}	energy flux by evaporation (or condensation)	W/m^2
$H_{ev/veg}$	energy flux by evapotranspiration of closed vegetation	W/m^2
$H_{ev/wet}$	energy flux by evapotranspiration of a wet surface	W/m^2
H_{soil}	energy flux in soil	W/m^2
h	height (of trees)	m
k	thermal conductivity	$Wm^{-1}K^{-1}$
LCT	lower critical temperature for BMR, basal metabolic rate	°C, K
l	length	m
n_r	resistance against water loss as compared with a free water surface under the same conditions	(>1)
P	water potential	bar
PAR	photosynthetic active radiation	W/m^2, mol (Einstein)
p	atmospheric pressure	bar
R	radiation flux	W/m^2
R_r	reflected radiation	W/m^2
R_{em}	emitted long-wave radiation	W/m^2
R_{atm}	long-wave radiation from the atmosphere	W/m^2
R_{net}	net radiation	W/m^2
R_{neta}	net radiation for a surface at air temperature	W/m^2

R_{netl}	net radiation for a transpiring leaf	W/m^2
R_{sun}	solar radiation	W/m^2
RH	relative humidity	%
r	resistance (heat and mass transport)	s/m
	(r can have the dimension Km^2W^{-1} or mbar m^2W^{-1}; see pp. 34, 37.)	
r^*	radius	m
SD	saturation deficit	mbar
s	slope of saturation vapour pressure versus temperature curve	mbar/°C
T	absolute temperature	K
T^*	turbidity	(>1)
t	temperature	°C, K
t_{air}	temperature, air	°C, K
t_b	temperature, body	°C, K
t_{dew}	temperature, dewpoint	°C, K
t_e	operative temperature	°C, K
t_s	temperature, surface	°C, K
t_w	temperature, wet bulb	°C, K
u	wind velocity	m/s
V	evaporation heat of water	J/g
VPD	vapour pressure deficit = saturation deficit	mbar
z	vertical distance	m
α	coefficient of convective heat transfer	$Wm^{-2}K^{-1}$
α_{con}	conductance (coefficient of heat transfer by conduction)	$Wm^{-2}K^{-1}$
α_{rad}	coefficient of radiative heat transfer (increment of σT^4 per K)	$Wm^{-2}K^{-1}$
γ	psychrometer 'constant'	mbar/°C
Δ	difference of temperature, etc. (Δt, Δe)	
δ	amplitude of temperature wave	°C, K
ε	emissivity for long-wave radiation	(0-1)
λ	wave-length	nm
$\lambda_{T,\max}$	wave-length at which the emitted long-wave radiation is maximal for surface temperature T	nm
σ	Stefan Boltzmann constant	$Wm^{-2}K^{-4}$

Added symbols Chapter 6

C	the thermal conductivity in Watt per degree centigrade/Celsius (see also specific heat)
H	heat loss in Watt
M	mass (weight)
Top	operative temperature

General index

Index of Latin names